DATE DUE

Advances in

ECOLOGICAL RESEARCH

VOLUME 27

Advances in

ECOLOGICAL
RESEARCH

Edited by

M. BEGON

*Department of Environmental and Evolutionary Biology,
University of Liverpool, UK*

A.H. FITTER

Department of Biology, University of York, UK

VOLUME 27

ACADEMIC PRESS

Harcourt Brace and Company, Publishers

San Diego   London
Boston   New York
Sydney   Tokyo   Toronto

Academic Press, Inc.
525 B Street, Suite 1900, San Diego, California 92101–4495, USA
http://www.apnet.com

Academic Press Limited
24–28 Oval Road, London NW1 7DX, UK
http://www.hbuk.co.uk/ap/

ISBN 0–12–013927–8

A catalogue record for this book is available from the British Library

Typeset by Saxon Graphics Ltd, Derby
Printed in Great Britain by Hartnolls Limited, Bodmin, Cornwall

97 98 99 00 01 02 EB 9 8 7 6 5 4 3 2 1

Contributors to Volume 27

D. BINKLEY, *Department of Forest Sciences, Colorado State University, Fort Collins, CO 805523, USA.*

P.J.A. BURT, *Natural Resources Institute, University of Greenwich, Chatham Maritime, Kent ME4 4TB, UK.*

D.M. EISSENSTAT, *Department of Horticulture, 103 Tyson Building, The Pennsylvania State University, University Park, PA 16802–4200, USA.*

J.H. FOWNES, *Department of Agronomy and Soil Science, University of Hawaii at Manoa, Honolulu, HI 96822, USA.*

L.L. HANDLEY, *Scottish Crop Research Institute, Invergowrie, Dundee DD2 5DA, Scotland, UK.*

P. LAVELLE, *Laboratoire d'Ecologie des Sols Tropicaux Centre ORSTOM, 93143–Bondy Cedex, France.*

D.E. PEDGLEY, *formerly of Natural Resources Institute, Chatham Maritime, Kent ME4 4TB, UK.*

M.G. RYAN, *USDA Forest Service, Rocky Mountain Experiment Station, 240 West Prospect Road, Fort Collins, CO 80526–2098, USA.*

C.M. SCRIMGEOUR, *Scottish Crop Research Institute, Invergowrie, Dundee DD2 5DA, Scotland, UK.*

R.D. YANAI, *College of Environmental Science and Forestry, The State University of New York, Syracuse, NY 13210, USA.*

Preface

This volume contains the traditional variety of material, ranging from physiological and evolutionary ecology to ecosystem ecology. Eissenstat and Yanai address the remarkable uncertainties surrounding the longevity of roots. Ecologists have long been accustomed to considering the lifespan of leaves in cost–benefit models of plant performance, but, although roots are at least as labile as leaves, they have rarely featured in such models. There is still a need to improve the basic data, but they show that there is enough information for a cost–benefit model of root behaviour that can be used to highlight key areas for new research: studies on interactions with pathogens and herbivores are likely to be especially important.

Burt and Pedgley also consider ecology at the level of the individual organism, by showing the importance of wind dispersal to the population ecology of nocturnal insects. They show that distances moved by insects can be much greater than would be inferred from unsophisticated use of weather data, because of the complexity of air movement patterns. One result is increased clumping of originally dispersed populations, which has profound implications for management of pest species. Even so, there remains ignorance about atmospheric patterns and the consequences for insect behaviour.

Lavelle examines the ways in which members of the soil fauna perform distinct functions, as predatory regulators in foodwebs (microfauna), litter transformers (mesofauna) and 'ecosystem engineers' (macrofauna). He shows that the operation of these functional groups at distinct temporal and spatial scales can explain many properties of the soil system and suggests that this approach can be used to answer questions about the ability of soils to maintain ecosystem functions under anthropogenic pressures that lead to depauperization of species. The analysis of such functional properties requires sophisticated technologies, and one that is increasingly used is measurements of natural abundance of stable isotopes such as ^{13}C and ^{15}N.

Handley and Scrimgeour analyse both existing data and a new set of data for ^{15}N values in an old field ecosystem; in doing so, they undertake a critical analysis of the technique and its interpretation, concluding that the use of natural abundance measurements as a tracer of N movement is rarely feasible and offer an important critique of the widely used two-source model for determining pathways of N movement in natural ecosystems. They do show, however, the nature of the insights that can be obtained from carefully designed and well characterized studies.

Ryan's article is also concerned with an ecosystem-level phenomenon, which has implications both in commercial forestry and in the current debate over the role of forests as sinks for carbon. Productivity typically peaks in the middle age of a forest, and this phenomenon has long puzzled foresters, who have generally explained this as an increase in respiratory load. Ryan shows that reductions in leaf area and photosynthetic capacity are more likely to be responsible, but the causes of these changes remain to be established.

While this volume was being prepared, Mike Begon, my co-editor for the last nine volumes, decided to retire from the editorship. I would like to thank him for his careful and perceptive editing which has helped maintain *Advances in Ecological Research* as one of the leading ecological journals.

Alastair Fitter

Contents

The Ecology of Root Lifespan

D.M. EISSENSTAT AND R.D. YANAI

Nocturnal Insect Migration: Effects of Local Winds

P.J.A. BURT AND D.E PEDGLEY

Faunal Activities and Soil Processes: Adaptive Strategies That Determine Ecosystem Function

P. LAVELLE

Terrestrial Plant Ecology and ^{15}N Natural Abundance: the Present Limits to Interpretation for Uncultivated Systems with Original Data from a Scottish Old Field

L.L. HANDLEY AND C.M. SCRIMGEOUR

Age-Related Decline in Forest Productivity: Pattern and Process

M.G. RYAN, D. BINKLEY AND J.H FOWNES

The Ecology of Root Lifespan

D.M. EISSENSTAT AND R.D. YANAI

ADVANCES IN ECOLOGICAL RESEARCH VOL. 27
ISBN 0–12–013927–8

I. SUMMARY

Theories of plant evolution and adaptation have posited relations between tissue deployment and environmental conditions which have yet to be substantiated below ground. Root production is quantitatively important, exceeding above ground productivity in a range of ecosystems. In addition, carbon expended for root maintenance often exceeds that used for root production. Root production and turnover have consequences for carbon and nutrient cycling, water and nutrient acquisition, competition between plants and the survival and reproduction of species under changing environmental conditions. Despite its importance, relatively few studies have examined factors controlling root lifespan.

This review identifies some competing theories of root lifespan and reviews the evidence available to support them. New methods of root observation and analysis produce data appropriate to testing these theories, but the results to date are few and often conflicting. Tentative generalizations include a suggestion that small diameter roots with low tissue density tend to have short lifespans. Root lifespan appears to be longest in cold environments, but data are lacking for tropical species. There is a strong seasonal variation in lifespan, with roots produced in the fall surviving longest, at least in temperate climates. Species differences are difficult to quantify because of seasonal and interannual variation, but root lifespans of deciduous fruit crops seem to be shorter than those of temperate deciduous forest trees or citrus, a broadleaf evergreen.

Until more data are available on root lifespan, simulation modeling offers an approach to developing testable hypotheses and identifying research needs. We applied a cost–benefit analysis to determine the lifespan that would maximize root efficiency, defined as the amount of nutrient acquired per unit of carbon expended. The analysis suggests that roots should have long lifespans if they have low maintenance respiration or are located in favorable patches of nutrient-rich soil. Shedding roots in dry soil may not be necessary for root efficiency if reductions in maintenance respiration can match reductions in uptake. Accurate predictions of optimal root lifespan are presently limited by insufficient information on the changing carbon costs and nutrient uptake capacity of roots with age. The costs and benefits of root hairs, root exudates and mycorrhizal fungi also need to be included.

The current model of root efficiency omits some important factors that may exert control over root lifespan. Fine roots have other functions in addition to absorption, including transport of water and nutrients. Seasonality of climate and the need for carbon and nutrient storage could constrain the root lifespan that optimizes plant fitness to differ from that which maximizes root efficiency. Roots compete for carbon and nutrients with other plant organs; roots may be shed when demands on roots are low relative to reproductive demands or to the demands of new leaf production. Death of roots may not entail the loss of all the carbon and nutrients they contain, if material is resorbed from senescing tissues or if nutrient cycling is tight and the lost material eventually benefits the plant.

Finally, root herbivory and parasitism are probably much greater than generally appreciated. Young roots generally lack structural defenses and are at high risk of attack by soil organisms. The lifespan of stressed roots with low carbohydrate reserves is probably affected by weak pathogens and primary saprophytes that reside in the rhizosphere. Plant–animal (including plant–fungus) interactions are likely dominant factors influencing the lifespan of roots.

More long-term studies are needed, both to detect existing patterns of root lifespan with resource availability and plant strategy, and to progress beyond short-term responses to treatments, which may be misleading. Factors that should be included in studies of root lifespan in addition to resource availability include root age, plant carbon status, and pathogen pressure. Because root lifespan differs from leaf lifespan in many species, better knowledge of root lifespan may result in revision of current theories regarding plant adaptation and growth strategies.

II. INTRODUCTION

Root lifespan has important consequences for plant growth and productivity, plant competition, and carbon and nutrient cycling. Roots, like other plant organs, have a life history: they are born, age and die (Harper, 1977). The root system can be thought of as a population of foraging units of different ages responsible for acquiring water and nutrients. Growth of a root system, the ability of a root system to relocate in favorable patches, and the eventual architecture of a root system are largely determined by the birth and death of its parts. Root construction and maintenance also influence carbon (C) and mineral nutrient consumption, while root death influences the partial return of these resources to the soil.

The mass and energy involved in the birth and death of roots may be at least as great as that involved in the birth and death of leaves. Net primary production (NPP) is greater below ground than above ground in a range of ecosystems (Caldwell, 1987). Even in forests, which can have enormous above-ground biomass, below-ground NPP was consistently higher than above-ground NPP, especially early in stand development (Gower *et al.*, 1994). This greater allocation may reflect differences in the intensity of competition below and above ground. In shady as well as sunny environments, below-ground competition commonly has a greater influence on plant species growth and survival than above-ground competition (Donald, 1958; Wilson, 1988; Dillenburg *et al.*, 1993). Compared with the population dynamics of above-ground parts, however, very little is known about below ground systems. Recent technological advances, especially video-imaging in mini-rhizotrons and video-processing with personal computers, have made it more possible to observe and quantify the demography of roots. In this review, we draw together current information on root lifespan, illustrate a cost–benefit modelling approach to predict root lifespan, reveal areas of ignorance, and provide direction for future demographic studies of roots.

III. ESTIMATING ROOT LIFESPAN

Accurately quantifying root turnover has been one of the most intractable problems in studies of terrestrial plant ecosystems. The most common method has been to collect soil cores periodically (usually monthly) and determine average mass of live and dead roots at each interval (Vogt *et al.*, 1986). The change in mass of live and dead roots through the season can then be used in various formulae to estimate root production and turnover. Average lifespan can be estimated by dividing production by the mean standing biomass. Estimates of root lifespan by soil coring have been reviewed by Schoettle and Fahey (1994) and Bloomfield *et al.* (1996). Other ways to estimate below-ground NPP include a nitrogen-balance approach (Nadelhoffer *et al.*, 1985; Nadelhoffer and Raich, 1992), C-isotope approaches (Milchunas *et al.*, 1985; Milchunas and Lauenroth, 1992; Nepstad *et al.*, 1994), root ingrowth techniques (Fabião *et al.*, 1985) and direct observation (Cheng *et al.*, 1990). There are several papers that discuss approaches for estimating annual root turnover (Milchunas *et al.*, 1985; Caldwell and Eissenstat, 1987; Hendrick and Pregitzer, 1992; Milchunas and Lauenroth, 1992; Hendricks *et al.*, 1993; Fahey and Hughes, 1994), which reveal no consensus on the best approach for estimating root turnover. This review focuses primarily on results of observational techniques in which roots are tracked individually, allowing specific demographic information on different cohorts of roots and direct estimates of root lifespan. In recent years, root tracking has been greatly facilitated by the use of miniature video cameras or borescopes inserted into transparent tubes (minirhizotrons) buried in the soil. Personal computers assist in processing the video images, allowing the fate of thousands of individual roots to be followed (Hendrick and Pregitzer, 1992, 1993). Direct observation avoids the assumptions used in ingrowth, budgeting, and isotope approaches. Direct observation also avoids important problems in the interpretation of root data collected by sequential coring, such as simultaneous birth and death of roots during a sampling interval and spatial and sampling variation being confounded with temporal variation. In addition, it is now realized that roots can disappear rapidly (in less than a few weeks), either by rapid decomposition or by herbivory (Hendrick and Pregitzer, 1992; Fahey and Hughes, 1994; Kosola *et al.*, 1995). Such rapid disappearance can cause serious overestimates of root longevity by sequential coring approaches. The most serious problem associated with minirhizotrons and other direct observation approaches is the artificial environment at the surface of the observation windows, which may alter root behavior (see Harper et al., 1991). For example, root length densities observed with minirhizotrons can be considerably lower than estimates based on soil coring (McMichael and Taylor, 1987). Care in the installation of the observation tubes or windows and subsequent protection from light and thermal gradients can help minimize this problem.

A difficulty in assessing root lifespan by any method is the definition of death. A fine root may exhibit signs of necrosis in portions of its length while the remainder of its length is still healthy. Moreover, in some species, roots continue to absorb water and nutrients after death of epidermal and cortical cells. Even portions of roots whose entire epidermis and cortex has sloughed may still provide important transport functions in the stele and be capable of new lateral formation from a viable pericycle (Spaeth and Cortes, 1995). Investigators usually use a combination of indicators in assessing death, such as color changes, loss of cortex and disappearance. Investigators who use nondestructive visual techniques such as minirhizotrons occasionally find that a root thought to be dead initiates new laterals upon continued observation. Vital staining, such as the use of tetrazolium dye, can be used to compare other methods of assessing death, such as ultraviolet fluorescence, color (dark brown, very dark brown or black), partial decay, and the presence of nonmycorrhizal fungal mycelia. Root death assessed by u.v. or visual approaches underestimated the number of dead roots by 12–15% (range 8–30%) compared to the tetrazolium method (Wang et al., 1995). Root lifespan may therefore be overestimated with minirhizotrons.

IV. VARIATION IN ROOT LIFESPAN

Estimates of root lifespan using direct observation techniques are now quite numerous (Table 1). This table lists the median lifespan of different cohorts of roots of 14 plant species (seven of which are rootstocks of citrus) and five forest communities. Included in the table are the season the cohort was produced, the location of the study and the method of estimating lifespan.

Root lifespan of a given species may vary considerably among cohorts produced in different seasons. In temperate climates, lifespan is often shortest for roots produced in late spring. In England, median root lifespan was as little as 14 days in apple roots produced in June whereas roots produced in January lived an average of 84 days (Head, 1966). In this case, lifespan refers to the duration that roots remained white. Apparently, in apple and strawberry roots, the cortex is sloughed soon after the root turns brown (Atkinson, 1985). It is likely, however, that assessment of root death based on color change alone probably underestimates root lifespan compared with definitions of death based on root decay or disappearance. In Wisconsin, hybrid poplar roots

Table 1

Median lifespans (days to 50% mortality) of cohorts of roots determined by direct observation in a range of plants and communities

Plant species/community (treatment)	Cohort	Lifespan (days)	Location	Comments and References (M = minirhizotron; R = rhizotron)
Annual crops				
Grain sorghum (*Sorghum bicolor*)	Spring Summer	42–47 24–26	Georgia Piedmont	M, Data from Cheng *et al.* (1990) till and no-till combined. Spring: early vegetative phase. Summer: beginning seed fill
Groundnut (*Arachis hypogaea*)	All roots	24–31	Growth room study	Five cultivars. Grown in transparent tubes, 95 cm long, 5 cm diameter (Krauss and Deacon, 1994)
Perennial herbaceous crops				
Alfalfa (*Medicago sativa*)	22 June–20 July 3–17 Aug	58–131 47–92	St Paul, Minnesota	M, Four alfalfa germplasms (G.D. Goins, M.P. Russell, 1996) Lifespan increased with depth
Strawberry (*Fragaria × ananassa*)	All roots	17	Kent, England	R, East Malling Research Station. Dead roots defined as roots that turned 'brown' (from Figure 9 in Atkinson, 1985)
Woody fruit crops				
Apple (*Malus domestica*) M.VII rootstock	June–Sept	14–21	Kent, England	R, Dead roots defined in accordance with Atkinson (1985) as roots that turned 'brown' (East Malling Res. Station, Head, 1966)
Kiwifruit (*Actinidia deliciosa*)	Oct–Nov All roots	28–84 28	Gisborne, New Zealand	R, No difference among cohorts produced in spring, summer, fall or winter (Reid *et al.*, 1993)
Citrus Trifoliate orange (*Poncirus trifoliata*)	Apr–Dec, 1992	90	Avon Park, Florida	M, Unhedged trees, Valencia sweet orange (*C. sinensis*) scion on range of rootstocks. Approximately monthly sampling (D.M. Eissenstat, unpublished data)
Sour orange (*Citrus aurantium*)		90		
Cleopatra mandarin (*C. reshni*)		116		
Volkamer lemon (*C. volkameriana*)		152		
Carrizo citrange (*P. trifoliata × C. sinensis*)		124		
Swingle citrumelo (*P. trifoliata × C. paradisi*)		99		

Table 1 cont.

Plant species/community (Treatment)	Cohort	Lifespan (days)	Location	Comments and References (M = minirhizotron; R = rhizotron)
Volkamer lemon (*C. volkameriana*)	May, 1993 Aug, 1993 Sept, 1993	51 33 16	Avon Park, Florida	M (0–17 cm depth, deeper roots lived about the same in May but 10–20 days longer in August and September) Differences in survival among cohorts during the season and between rootstocks related to infection by *Phytophthora nicotianae* (Kosola et al., 1995)
Rough lemon (*C. jambhiri*)	May, 1993 Aug, 1993	30 20		
Forest trees/communities				
Hybrid poplar (*Populus generosa inter-americana*) var. Beupre	All		Growth chamber	M, (20 cm diameter, 4.3 l pots) (Hooker et al., 1995)
Arbuscular mycorrhizal		19 49		
Nonmycorrhizal				
Hybrid poplar (*Populus tristis* × *P. balsamifera*) cv. Tristis no. 1	May June July	62 118 130	Rhinelander, Wisconsin	M, M. Coleman, unpublished data
Pin cherry (*Prunus pensylvanica*) forest	21–31 July	59–67	Pellston, Michigan	R, Observations made every 2 days (Pregitzer et al., 1993). Median life-span extrapolated after 82 days based on Figure 2
20–40 days pulse of water		82–90		
20–40 days pulse of water + N		42		
No addition of water or N (control)				
Sugar maple (*Acer saccharum*) forest	June–September 1989	200	N site, Michigan.	M, Sites 80 km apart, N site (Lat 44° 23' Long 85° 50'), S site (Lat 43° 40' Long. 86° 09') otherwise very similar (Hendrick and Pregitzer, 1993; Figure 2).
	June–September 1989	125	S site, Michigan	
Sugar maple (*Acer saccharum*) forest	27 April 11 June, 1989	340 250	Michigan (N site)	M, (Hendrick and Pregitzer, 1992)
0–30 cm depth				
30–110 cm depth				
Sugar maple-Beech-Yellow birch forest (*Acer saccharum–Fagus grandifolia–Betula alleghaniensis*)	Late spring	180	Hubbard Brook, New Hampshire	Mesh-screen technique for tracking roots at the interface of the mineral soil surface (Fahey and Hughes, 1994)
Sugar maple–Beech forest	Late spring	195	South-central New York	Assumed mortality rate was constant with root age and extrapolated from 59% survivorship after 150 days (Fahey and Hughes, 1994)

produced in May typically only lived until July (62 days), whereas 50% or more of the roots produced in July lived past November (130 days) (Table 1). Similarly, in two sugar maple forests in northern Michigan, roots produced in the fall lived longer than those produced in the summer (Hendrick and Pregitzer, 1993), but in these forests, root lifespan was also long for roots produced in April through early June (Hendrick and Pregitzer, 1992).

In contrast to temperate species, in which roots produced in the fall seem to live the longest, citrus roots produced in September in subtropical Florida may have shorter lifespans than those produced in the spring (Kosola et al., 1995). This result was linked to an increase in propagules of the parasitic root-rot fungus, Phytophthora. The variability in median root lifespans of different cohorts is further illustrated by our work on citrus. One rootstock, Carrizo citrange, in early May had a median lifespan of 141 days whereas a late July flush of Carrizo roots had a median lifespan of 348 days. Median lifespan of all the roots produced over the two-year period ranged from 90 days in the trifoliate orange rootstock to 152 days for Volkamer lemon (Table 1), a rootstock that had been specifically selected for resistance to the root-rot fungus, Phytophthora nicotianae (Kosola et al., 1995).

Because of the variability in lifespan among different root cohorts, differences in the average lifespan of all the roots produced in a year can be difficult to detect between species. Nonetheless, it appears that other fruit crops have considerable shorter root lifespans than does citrus. In apple and in kiwifruit overall median lifespans of fine roots were only about 28 days (Atkinson, 1985; Reid et al., 1993). Strawberry roots had average lifespans of only 17 days (Atkinson, 1985). Median lifespan of groundnut (Arachis hypogaea L.) is also only about 24–31 days (Krauss and Deacon, 1994).

Roots of temperate deciduous trees in mature forests have lifespans similar to citrus, a broadleaf evergreen, but longer lifespans than deciduous fruit trees. In a Michigan forest dominated by the late successional species Acer saccharum, median lifespans of the roots produced from June to September varied from about 125 to 200 days, depending on site (Hendrick and Pregitzer, 1993). However, if only the early spring flush of roots (roots before June) was considered, lifespans were much longer (about 250–340 days, depending on the depth of roots in the soil) (Hendrick and Pregitzer 1992). In forests dominated by the early successional species, Prunus pensylvanica, median lifespans were only 40–80 days (Pregitzer et al., 1993). Roots in patches enriched with water or with water and nitrogen had significantly greater lifespans than those in unamended patches (Table 1).

In two northern hardwood forests, Fahey and Hughes (1994) estimated fine root survivorship by tracking surface roots that had grown into mesh screens just below the litter layer. In the Hubbard Brook forest in New Hampshire, which was dominated by Acer saccharum, Fagus grandifolia and Betula alleghaniensis, median lifespan of a late-spring flush of roots was about 180 days. Similar values were measured in a forest dominated by *A. saccharum* and *F. grandifolia* in south-central New York (59% of the roots had survived about 150 days). Both sites exhibited shorter lifespans than the spring cohort of roots (250–320 days) of Hendrick and Pregitzer (1993) but longer lifespans than their season-long average (80–120 days). There is some evidence that roots near the soil surface live longer than those deeper in the soil (Hendrick and Pregitzer, 1992; Schoettle and Fahey, 1994), which may also contribute to the differences found by the root screen method of Fahey and Hughes compared with the observation tubes used by Pregitzer and coworkers in Michigan.

In summary, although recent investigations have greatly increased the amount of reliable information on root lifespan, data are still too sparse to make many generalizations. The roots of woody species such as apple and kiwifruit may live no longer than those of annual herbaceous species. Roots in natural ecosystems such as northern hardwood forests seem to live longer that those in agricultural systems (also see Table 4). Within a species, cohorts of roots produced at different times of the year can differ greatly in lifespan.

V. ANALOGIES TO LEAF LIFESPANS

Leaves are more readily observed than roots, and theories concerning leaf lifespan are better developed than theories of root lifespan. Fine roots, whose primary function is the acquisition of water and nutrients with little role in storage and support, are in many ways similar to leaves, whose role is primarily acquisition of C. Like leaves, most fine roots typically exhibit determinant growth, extending only a few centimeters after emerging from woody laterals, and never undergo secondary thickening. Their ephemeral character is illustrated by the fact that fine roots often have shorter lifespans than leaves (Schottle and Fahey, 1994). The comparison of roots and leaves, however, is an imperfect analogy. Lateral roots arise internally from the pericycle, rupturing the cortex as they emerge. Branches and leaves, on the other hand, are born from external meristems and often die following formation of a distinct abscission layer, which has not been described in roots. Fine roots that branch extensively also serve important transport functions, causing their fate to affect the fate of all more distal segments; leaves have no such dependencies. Finally, a part of the function of roots can be consigned to symbionts, such as mycorrhizal fungi or nitrogen-fixing bacteria. The role of the root is then to exchange C for nutrients from the symbiont, rather than to absorb nutrients from the soil. Despite these differences in roots and leaves, analogies to leaves have provided much of the theory of root lifespan.

One organizing theme in the ecology of leaf lifespan is the influence of habitat resource availability on the evolution of a suite of interdependent plant characteristics, including leaf longevity (Grime, 1977; Chabot and Hicks,

1982; Coley *et al.*, 1985; Chapin *et al.*, 1993). For example, in a study of 23 Amazonian tree species, those with shorter leaf lifespan had higher specific leaf area (leaf area/leaf dry mass), leaf diffusive conductance, maximum net C assimilation rate, mass-based leaf nitrogen (N) and phosphorus (P) and lower leaf toughness (Reich *et al.*, 1991). Similarly, leaf lifespan in a lowland tropical rainforest in Panama was positively correlated with shade tolerance and plant defensive compounds such as tannins and lignins and negatively correlated with plant growth rate (Coley *et al.*, 1988). In leaves, short lifespan is thought to be important for rapid growth and morphological plasticity, whereas long lifespan is advantageous to nutrient conservation and nutrient-use efficiency (Monk, 1966; Chapin, 1980; Chabot and Hicks, 1982; Coley, 1988; Reich *et al.*, 1991, 1992; Grime, 1994). Root lifespan may be linked to a similar suite of traits as found in leaves (Eissenstat, 1992, 1997; Fitter, 1994; Grime, 1994). If so, plants of infertile habitats should produce coarse, well-defended, absorptive roots of long lifespan whereas plants from high-resource environments should produce short-lived roots with rapid potential rates of nutrient uptake (high V_{max}), rapid potential growth rates, and little allocation to certain kinds of defensive compounds such as lignins.

An important trait linked to lifespan in leaves and roots may be the mor-phological plasticity of the plant, that is, the ability to position leaves and roots in resource-rich patches. In unproductive habitats with unpredictable and short-lived pulses of resource supply, leaves and roots should have long lifespans and low morphological plasticity (Grime *et al.*, 1986). In more pro-ductive habitats, it is more advantageous to be able to grow rapidly in resource-rich patches (Grime, 1994). These generalizations are based on a comparison of the growth of species adapted to different habitats (Grime, 1994). Less is known about differences in selective mortality of tissues with-in plants or in plants adapted to different environments.

There have been some observations of selective mortality of branches and leaves in relationship to patchy distribution of light. For example, the light-demanding pioneer species, *Rosa canina*, selectively sheds shoots located in deep shade, which reduces maintenance costs and thus maintains a more favorable whole-plant C balance (Küppers, 1994). Selective mortality of shaded branches and leaves low on the bole of the tree is common in shade-intolerant species (Millington and Chaney, 1973). Shade tolerant, late-successional species tend to retain more leaves – an example of low morph-ological plasticity and long lifespan in low-resource environments. The analogous hypothesis below ground is that selective mortality of roots in unfavorable patches of soil is greater in plants from productive than un-productive habitats. Current information is inadequate to test this hypothesis.

The processes by which roots are shed are poorly understood. We know of no study showing that roots form a distinct abscission layer and thus are actively shed in the manner of leaves or stems. For example, B. Huang (personal communication) found no evidence of abscission layer formation prior to death of rain roots in desert succulents (North *et al.*, 1993). Roots may die simply by restriction of the flow of carbohydrates until their reserves are exhausted. Alternatively, root death may be caused primarily by biotic factors extrinsic to the plant, such that root lifespan is a function of herbivore pressure in the rhizosphere and the degree of root defense. These alternative mechanisms of plant control of root lifespan are not easily separated experimentally.

VI. CONTROLS AND CONSTRAINTS ON ROOT LIFESPAN

A. Hypotheses

Many factors presumably shape the lifespan of a root. Below is a list of hypotheses that have received a certain degree of support, including some derived from hypotheses concerning the lifespan of leaves (Chabot and Hicks, 1982). While the hypotheses each describe a different aspect of root biology, they are not mutually exclusive. Some describe proximate causes of root death; others do not specify a mechanism.

(1) Root lifespan is a function of herbivore pressure and the degree to which roots are defended from herbivores and pathogens.
(2) Long lifespan is important in roots with functions other than water and nutrient acquisition, such as storage, transport, and structural support.
(3) Root lifespan is a function of competition for carbon among various plant parts.
(4) Root lifespan corresponds to the length of the favorable growing season.
(5) Long root lifespan is a form of mineral nutrient conservation.
(6) Root lifespan maximizes the efficiency of resource acquisition per unit C expended.

B. Root Herbivory and Parasitism

We suspect that extrinsic biotic factors contribute far more to root mortality than is generally recognized in the ecological literature. Examples of dramatic plant mortality associated with root pathogens are common in agriculture; root herbivory in natural ecosystems can also be substantial. Some data suggest that root-feeding nematodes can reduce net primary productivity in grasslands by 12–28% (Stanton, 1988). In Eucalypt forests in the Brisbane Ranges in Australia, *Phytophthora cinnamomi* has caused resistant species to expand at the expense of susceptible species such as *Eucalyptus* (Weste, 1986). Experimentally reducing root-feeding insects by applying insecticides to soil has been shown to enhance plant species richness and reduce seedling mortality, especially of perennial forbs compared with grasses (Brown and Gange, 1991). Consequently, in studies using sterilized soil, root death might be substantially delayed by the absence of herbivores and pathogens.

Low levels of heterotrophic feeding on roots may be more the rule than the exception. In addition to root-feeding insects and nematodes, parasitic fungi may increase rates of root mortality, even in apparently healthy plants. For example, in Florida, healthy citrus trees had highest mortality of roots during periods when *Phytophthora* activity is greatest, such as late summer (Graham, 1995; Kosola *et al.*, 1995) Fine roots of the *Phytophthora*-susceptible rootstock (*Citrus jamibhiri*) had shorter median lifespans and supported larger populations of *Phytophthora* than the fine roots of the more tolerant rootstock (*Citrus volkameriana*) (Kosola *et al.*, 1995). Weak pathogens or primary saprophytes are also likely to be important in accelerating the death of stressed roots. For example, *Fusarium solani*, a fungus whose inoculum is ubiquitous in root tissues of citrus, is able to develop only when starch reserves in the citrus roots are depleted, for example following canopy loss or during heavy fruit set (Graham *et al.*, 1985).

A cost–benefit analysis of root retention and shedding would predict low mortality in young roots, with peak mortality at the optimal lifespan. But survivorship curves of roots indicate either fairly constant mortality with root age (Kosola *et al.*, 1995) or high mortality of young roots (Hendrick and Pregitzer, 1993; Pregitzer *et al.*, 1993). Young roots are generally more susceptible to parasitism and herbivory than older roots, which have developed thick secondary cell walls in the epidermis or hypodermis (Graham, 1995). Woody roots that have established a cork periderm have additional structural defenses against root-feeding organisms. Extrinsic biotic factors are probably a better explanation for mortality of young roots than active shedding by the plant.

Rates of root losses to herbivory and parasitism are not entirely beyond the control of the plant. Healthy roots typically produce a wide range of chemicals to defend against root-feeding organisms. Many of these organisms would not be considered pathogenic unless the roots were very stressed. The development of new pesticides routinely involves screening root compounds for insecticidal, fungicidal and nematicidal activity. Numerous compounds produced by plant roots have shown pesticidal efficacy, including phytoalexins, polythienyls, alkaloids, acetylenes and terpenoids (Veech, 1982; Chitwood, 1992). That C stress increases a root's susceptibility to biotic attack is well-accepted among plant pathologists (Dodd, 1980). Although roots may not be actively shed, reducing expenditures for root defense could be a mechanism whereby plants control root mortality.

Many factors influence the activity of soil herbivores, such as season, soil moisture, soil temperature and plant C status. Correlations of these factors with root lifespan may be mediated by the activity of soil organisms, which are not usually considered in models of root foraging strategy. Roots of some species may live longer in dry or cold soil not so much because their maintenance expenses are lower, but because the activity of root-feeding organisms is less. The different lifespans of roots produced at different times of year, reviewed above, may likewise reflect the activity of soil organisms. A more indirect effect of plant and environmental factors on root lifespan acts through the strength of root defense. More research is obviously needed on the role of herbivores and pathogens in natural communities and the control plants have on root defense.

C. Roots with Multiple Functions

Fine roots have functions other than nutrient and water absorption, which are sometimes ignored in discussions of the costs and benefits of retaining or shedding roots. They serve as a reservoir of meristems, and they are involved to varying degrees not only in absorption but also in transport of water and nutrients. Roots may be classified in terms of external links which have no laterals and internal links which have one or more laterals (Fitter, 1991). Analyses of optimal root lifespan apply most clearly to external links, but root demographic studies include both internal and external links. The higher the order of branching of the fine roots of a plant, the more the survival of an internal link influences other root elements that depend on it for transport to and from the shoot. An external link should, consequently, have shorter lifespans than internal links, as found in kiwi (Reid et al., 1993).

Certain roots emerging in seedlings eventually build the structural framework of the root system. Structural roots are often genetically distinct from the fine laterals and must have a very different life history. The well-studied seminal roots in cereals fall into this category. In citrus, 'pioneer' roots are produced which are quite different from the fine laterals. Pioneer roots are somewhat thicker, have a prominent root tip, and extend rapidly and indeterminately in the soil with typically little branching of laterals near the tip. These roots generally have little mycorrhizal infection, are not readily infected by fungal pathogens or nematodes and readily extend through dry soil (D.M. Eissenstat and J.H. Graham, personal observation). Thus, plants may have evolved special mechanisms to protect certain roots from early mortality even though they appear similar to the fine laterals at an early age.

D. Competing Sinks for Carbon

Whether it is advantageous to plants to retain roots should depend on the relative pay-off of other possible investments of the C. For example, reproduction is an important competing sink for C. In many annual plants, most root growth occurs prior to flowering and most root death occurs during and after flowering. For example, total root length declines during and after flowering in wheat (Box and Johnson, 1987), sorghum (Cheng et al., 1990), cotton (Klepper et al., 1973) and soybean (Hoogenboom et al., 1987). Cheng et al. (1990) did a minirhizotron study in the Georgia piedmont where they tracked individual roots. Lifespan of sorghum roots produced in the summer at beginning seed fill was only half of those produced in the spring (W. Cheng, unpublished data, Table 1). High root mortality has been associated with very heavy fruit crops in *Prunus* (Chandler, 1923) and *Citrus* (Smith, 1976; Graham et al., 1985). Conversely, there is no evidence that root mortality is linked to seed fill in rice (Beyrouty et al., 1987), groundnut (Krauss and Deacon, 1994), or dry bean (Snapp and Lynch, 1996), based on root length dynamics.

Carbon shortages in plants, such as those caused by defoliation, should affect C allocation to roots. Plants often respond to leaf loss by reducing root respiration (Culvenor et al., 1989), slowing or ceasing root growth (Crider, 1955), or reallocating C to re-establish a functional equilibrium between roots and shoots (Brouwer, 1981). For example, 4 weeks after removal of the top third of the canopy in Valencia orange trees, there was at least a 20% loss of roots at a soil depth of 9–35 cm (Eissenstat and Duncan, 1992). In apple, leaf removal 6 weeks prior to natural leaf fall caused high root mortality within 2 weeks (Head, 1969). A similar response was observed in blackcurrant (Atkinson, 1972). Rapid root shedding following defoliation is less evident in grasses (Crider, 1955; Richards, 1984). The roots of two cold-desert tussock grasses differed in their responses to defoliation (Richards, 1984). In *Pseudoroegneria spicata*, roots of clipped plants continued to grow at the same rate as unclipped plants during the growing season but suffered greater mortality over winter. The expenditure of C for root growth in the clipped plants may have reduced the reserves available for root maintenance over winter. In a more grazing-tolerant species, *Agropyron desertorum*, root growth rates were slower in clipped than unclipped plants during the growing season, suggesting reduced draw on plant C reserves; there was little difference between clipped and unclipped plants in root survivorship over winter.

Carbon storage is another competing sink for plant resources. In strongly seasonal climates, stored C and nutrients are essential to the growth of roots and leaves at the beginning of the growing season. Stored C may be mobilized at other times for defense, repair, or replacement of tissues. Whether it is more advantageous to replenish storage or to increase growth of absorptive organs will depend not only on the costs and benefits of the tissue produced but also on the competing demands for stored C, including future risk of demand (Chapin et al., 1990).

That root longevity is affected by the C status of plants is illustrated by experimental manipulation of C supply. Of the two studies found in the literature, however, the lifespan response reported has been in opposite directions. In a 2-year study of ponderosa pine (*Pinus ponderosa*), roots lived longer at elevated than ambient CO_2 (D.T. Tingey, O.L. Phillips, M.G. Johnson, M.J. Storm and J.T. Ball, unpublished data). In contrast, in a study of hybrid poplar (*Populus* × *euramericna* cv. Eugenei), roots in low-N soil conditions lived longer under ambient than elevated CO_2 (Pregitzer *et al.*, 1995). In high-N soil, root longevity was not affected by increases in CO_2. The availability of C may interact with nutrient availability in controlling root dynamics in complicated ways that are presently poorly understood.

E. Seasonality

In strongly seasonal climates, the length of the growing season often dictates the lifespan of leaves (Harper, 1989); roots, however, often live less long. Although there is typically a strong flush of roots in the spring, often prior to leaf emergence (Lyr and Hoffman, 1967), these roots may live less than a month (Head, 1969). In droughty environments, cacti and many species in the Proteaceae have long-lived leaves that tolerate the dry period, but roots that are more ephemeral. North American desert cacti are noted for producing roots quickly after a rain event that die soon after the soil dries (Huang and Nobel, 1992; North *et al.*, 1993). In seasonally dry parts of western Australia and South Africa, many evergreen woody plants produce cluster roots which proliferate in the surface organic layers during the wet season but are shed by the time of the dry season (Lamont, 1995). Leaf and root longevity may differ so strongly in these environments because of the differences in timing of light and soil resource availability.

Thus, limited data suggest that in temperate climates, there is no strong relationship of root longevity to length of season with favorable temperatures, as commonly observed with leaves. Conversely, in very dry climates or climates of strongly seasonal rainfall, root longevity may be strongly linked to the length of time the soil is wet even though leaf longevity may not be. These conclusions are based on few data; more species comparisons are needed to better understand the relationship between length of growing season and root longevity.

F. Mineral Nutrient Conservation

Under nutrient-limited conditions, root foraging strategy should avoid unnecessary nutrient loss. Clearly, shedding roots is a pathway of nutrient loss, as nutrient resorption from dying roots is probably minimal (Nambiar, 1987; Dubach and Russelle, 1994). Long root lifespan in infertile environments could be a mechanism of nutrient conservation. In later sections, we explore the hypothesis that optimal root deployment should maximize the efficiency of nutrient uptake per unit C expended and review specific studies of root lifespan to soil fertility. In this section, we consider whether long lifespan could be maximizing nutrient uptake relative to nutrient expenditure rather than C expenditure. Carbon may not be the best measure of 'currency' in strongly nutrient-limited environments where plant density is low and light availability is high. Under these circumstances, C may not be limiting or may be only weakly limiting, as indicated by a lack of growth responses or increased nutrient uptake when plants are exposed to elevated CO_2 (Arnone and Körner, 1995).

Observations above ground suggest that long tissue lifespan contributes to nutrient conservation in infertile environments. For example, in temperate deciduous forest regions, evergreen-dominated vegetation types such as pine forests and bog vegetation often occur in low-nutrient soils. In Mediterranean climates, very infertile regions support exclusively evergreen shrubs whereas more fertile regions have a mix of evergreen and drought-deciduous vegetation (Lamont, 1995). Across a wide range of ecosystems, annual nutrient losses in litterfall are smaller in species with lower leaf turnover rates and lower leaf nutrient concentrations (Vitousek, 1982). In addition, evergreen leaves are more resistant to decomposition than deciduous leaves, causing nutrients to be released more slowly to the soil solution. Polyphenols in long-lived leaves on infertile soils may inhibit mineralization of organic nitrogen (Northup et al., 1995). The ectomycorrhizal and ericoid mycorrhizal plants common to these infertile soils may be able to break down this organic N and absorb the resulting amino acids, bypassing competition for mineral N with roots and other microbes (Chapin, 1995).

These patterns of nutrient retention and tight nutrient cycling in above-ground tissues may also occur in roots. The challenges are to understand the amounts and fate of nutrient losses from the root and whether nutrients are reabsorbed by the plant in a preferential manner. The 'tightness' of nutrient cycling back to the plant after the root dies may be higher in low-nutrient environments, which would diminish the advantages of long root lifespan in nutrient conservation. Recent literature suggests diverse means by which nutrients may be tightly cycled within a plant–soil system (Newman, 1988; Northrup et al., 1995). In low-nutrient environments, slow mineralization rates and a lack of mixing can cause a thick organic layer to develop at the soil surface, which becomes the main source of N and P. Plants adapted to low-nutrient environments, such as ericoid and ectomycorrhizal plants, or plants that produce cluster roots, are particularly adapted to the proliferation of roots or hyphae in this organic surface horizon, excretion of extracellular enzymes, and uptake of N and P (Read 1993; Lamont, 1995). Because roots and mycorrhizal hyphae are concentrated in this layer, nutrients are more likely to be recaptured by other roots of the same plant when a surface root dies. This may occur directly by neighboring roots or by roots linked by shared mycorrhizal hyphae (Newman, 1988). In the latter case, nutrients are preferentially retranslocated from the dying root to other living roots via the hyphal strands that link the roots, thus bypassing competition with soil microbes. The ability of ectomycorrhizal and ericoid fungi to take up organic N from soil, also bypassing microbes, further enhances nutrient return to the plant.

VII. TRADE-OFFS BETWEEN ROOT MAINTENANCE AND ROOT CONSTRUCTION

As in leaves, there are tradeoffs between maintaining old and growing new roots; C can be conserved or expended for lesser or greater gain in water or nutrients. One important factor in calculating such trade-offs is the relative cost of maintaining existing roots versus shedding roots and rebuilding them at a more favorable time or location.

In the previous section, we considered mineral nutrients as the currency for cost–benefit analyses of root lifespan. In the following sections, we examine root costs in terms of C. Carbon is the preferred currency in cost–benefit analyses and plant allocation models, perhaps because it describes the energy status of the plant. There are, however, some drawbacks to using C units for cost. First, estimating C losses from the roots by respiration, exudation, cell sloughing and mycorrhizal fungi can be difficult. Second, plants in infertile soils may not be very C-limited, according to results of CO_2 enrichment studies (Arnone and Körner, 1995); conserving C may not be important to these plants. Third, when the benefit is a resource other than C, such as a nutrient or water, the exchange value of that resource to C should be known; otherwise, it is difficult to compare plants in different circumstances (Bloom et al., 1985). For example, a nutrient-limited plant can presumably afford to expend more C per unit of nutrient acquired, because C is less valuable to it than to a C-limited plant.

The costs of maintaining roots, which may include exuding organic compounds and constructing and maintaining mycorrhizas, is often as great or greater over the lifetime of the root than the cost of building the roots in the first place. One way to compare these costs is to calculate the age at which cumulative maintenance respiration equals construction cost. In well-watered citrus seedlings grown at high N supply, the time required for maintenance respiration to equal the cost of root construction ranges from 20 to 26 days, depending on P supply and mycorrhizal status (Peng et al., 1993; Table 2). In a slash-pine ecosystem in Florida (Cropper and Gholz, 1991), root respiration rates measured at the soil surface would equal construction costs after 45 days, assuming that the roots were 45% C and the growth efficiency was 0.77. For desert succulents, which have quite slow maintenance respiration rates, it would take about 90 days for maintenance respiration of roots at 20°C to equal root construction costs (Nobel et al., 1992). Because root respiration increases exponentially with temperature, the C expended in respiration exceeds that used in construction sooner in warmer soil. At 30°C, a common

D.M. EISSENSTAT AND R.D. YANAI

Table 2

Summary of respiratory costs of Volkamer lemon colonized by *Glomus intraradices* (M) or uninoculated (NM). Plants were grown at either high-P supply (5P = 5 mM KH_2PO_4) or low-P supply (1P = 1 mM) (calculations made from Peng *et al.*, 1993. Copyright held by the American Society of Plant Physiologists.)

| | Daily cost (μmol CO_2 d^{-1}) | | | |
| | 1P | | 5P | |
	M	NM	M	NM
Construction cost, $CONST_t$	468	174	671	563
Total respiration, $R_{T(t)}$	555	234	820	600
Growth respiration, $R_{G(t)}$	114	40	153	117
Ion-uptake respiration, $R_{I(t)}$	111	56	109	111
Maintenance respiration, $R_{M(t)}$	330	138	559	372
$R_{M(t)}/CONST_t$	0.71	0.79	0.83	0.66
	Cost per unit root dry wt. mmol CO_2 (g new root)$^{-1}$			
Construction cost, $CONST_w$	48.7	44.7	45.3	42.0
	mmol CO_2(g whole-root system)$^{-1}$ d^{-1}			
Maintenance respiration, $R_{M(w)}$	2.46	1.77	2.05	1.62
	Days			
Number of days for maintenance respiration of 1 g of root ()$R_{M(w)}$) to equal the cost of constructing 1 g of root	19.8	25.3	22.1	25.9

condition in the desert, it would take only 48 days for maintenance respiration to equal construction costs (Q_{10} = 1.9; Palta and Nobel, 1989). Using root maintenance respiration of a range of plant species (Amthor, 1984) and an average root construction cost of 45 mmol C (g dry wt)$^{-1}$ gives estimates of 13–32 days for maintenance respiration to equal the construction costs of roots.

Root lifespan commonly exceeds the point at which respiratory costs equal construction costs. In Volkamer lemon seedlings, the maintenance costs over the lifetime of the root are seven-fold greater than the construction cost for roots that live 152 days (Table 1). In slash pine, roots appear to live for an average of about 1.5 years (Schoettle and Fahey, 1994); thus, maintenance costs over the lifetime of the root are about 12-fold greater than the cost of root construction in this species. In desert succulents, maintenance respiration is 2.1- to 2.5-fold greater than root construction costs after the first year for nodal roots of *Agave deserti* and established roots of *Ferocactus acanthodes* and *Opuntia ficus-indica*; these roots can apparently live at least 2 years (Nobel *et al.*, 1992). Clearly, the C allocated for maintenance over the lifetime of a root often far exceeds the C allocated for root construction. Considering that annual root biomass production often exceeds 50% of total

NPP, the annual C costs associated with root maintenance are impressive. From the perspective of optimal lifespan, the high C requirements of maintenance respiration indicate the potential disadvantage of maintaining a root that is inefficiently acquiring water or nutrients.

VIII. MODELING OPTIMAL ROOT LIFESPAN

A successful theory of root deployment would explain the observed variation in lifespan with reference to environmental conditions, such as temperature, moisture and soil fertility, and plant factors such as life form, life stage, C status and symbioses. Such a theory is in its infancy, at best. The previous section illustrated some of the trade-offs, in terms of C costs, between ending the life of a root to construct a new one and maintaining existing roots for longer lifespans. A more comprehensive cost–benefit analysis would attempt to predict optimal root lifespan based on both C expended and resources gained.

Simulation models of root C expenditure and nutrient uptake provide a means to test specific hypotheses of optimal root strategies (Yanai et al., 1995). Such models are not restricted to describing optimal behavior. Carbon costs and nutrient uptake can be simulated over time; optimal lifespan may be indicated by the maximal lifetime ratio between nutrient uptake and C costs. Departures of observed root longevity from theoretical predictions of maximal efficiency may provide insights into the trade-offs and constraints on both root and whole-plant function. There are many reasons, as discussed above, why root lifespan might not optimize the efficiency of nutrient acquisition.

In this section we introduce a model of root efficiency and describe how this model can be used to explore environmental conditions that might influence optimal lifespan. Previous models (McKay and Coutts, 1989; Fahey, 1992; Yanai et al., 1995) have not included some factors important to root longevity, such as the variation in C expenditures and ion uptake rates with root age.

A. Root Efficiency

We define root efficiency, E, as the rate of resource acquisition divided by the rate of C expenditure for root growth and maintenance. Analysis of C cost and nutrient benefit should indicate the optimal lifespan for a root in a given soil environment; specifically, the lifespan that maximizes E.

$$E = (UPTAKE)(COST)^{-1} \qquad (1)$$

where E = the efficiency of nutrient acquisition by roots [(mol nutrient)(mol C expended)$^{-1}$], $UPTAKE$ = nutrient gain [(mol nutrient) (g fine root)$^{-1}$ day^{-1}], and $COST$ = carbon cost [(mol C)(g fine root)$^{-1}$ day^{-1}].

The calculation of $COST$ includes the C contained in the root and C expended in growth respiration and maintenance respiration. Carbon costs are averaged over the lifespan of the root to obtain the cost per unit root per day.

$$(COST) = (C_{root} + R_{G(w)} + LR_{M(w)}) \; L^{-1} \qquad (2)$$

where C_{root} = root C content (mol C)(g root)$^{-1}$, $R_{G(w)}$ = growth respiration [(mol C)(g root)$^{-1}$], L = root longevity (days), and $R_{M(w)}$ = maintenance respiration [(mol C)(g root)$^{-1}$(day)$^{-1}$].

This equation does not separate mycorrhizal fungal construction and maintenance from that of the root proper; these costs are included in C_{root}, $R_{G(w)}$ and $R_{M(w)}$.

Likewise, root exudates that are rapidly metabolized are included in the maintenance component, according to the way $LR_{M(w)}$ is usually measured.

Root efficiency can be calculated using lifetime average values for root and soil properties or using shorter intervals, with cumulative costs and benefits determining lifetime efficiency. Previous efforts at exploring optimal lifespan of roots used only the lifetime averages of root and soil properties, producing a single average efficiency for each root considered (Yanai et al., 1995). This approximation failed to simulate the effects of changes in root and soil properties over the lifetime of the root. In this paper, C costs and nutrient uptake are calculated on a daily timestep; efficiency is reported on a daily basis, and cost and uptake are accumulated to show changes in lifetime efficiency as the root ages. In addition to allowing parameters to be varied as the root ages, implementing the model on a daily timestep allows soil conditions to change over the life of the root.

In the following simulations, we varied the following parameters over time, depending on the simulation: inter-root distance, uptake kinetics, root respiration, radial water velocity, and soil moisture (which affects the effective diffusion coefficient). Importantly, we did not vary soil solution concentrations to simulate depletion of soil nutrients over time. We assumed that root radius and specific length were constant over the life of the root.

B. Carbon Costs

1. Construction Cost

Root construction cost is the sum of the C content of the root (C_{root}) and growth respiration ($R_{G(W)}$). In the model, root construction exacts a one-time cost, assessed for each day's growth. It is, of course, a simplification to assume that root construction ceases after root elongation. As roots age, the secondary walls become more lignified and suberized. The values of C_{root} and $R_{G(W)}$ depend on the age at which the roots were measured. Likewise, root construction costs may reflect mycorrhizal colonization and plant nutrient status. The effects of mycorrhizas on C costs are discussed under section XI, Further Considerations. The effect of plant nutrient status is illustrated by citrus seedlings (Table 2), in which low P conditions promote thinner roots, with a lower starch concentration and a higher percentage of the more lignified epidermal, hypodermal and stelar cells and fewer and smaller cortical cells (Peng et al., 1993; D.M. Eissenstat, unpublished data).

High tissue construction cost is commonly assumed to indicate long life-span, but this assumption may be unjustified. Short-lived leaves may have fewer defensive compounds than long-lived leaves but more proteins for rapid C assimilation. As a result, construction costs of tissues of different lifespans tend to be quite similar (Chapin, 1989; Poorter, 1994). Likewise, in roots, short-lived, highly absorptive roots with high V_{max} and high hydraulic con-ductivity may use C for energetically expensive proteins and enzymes asso-ciated with rapid ion uptake and assimilation, while well-defended roots may allocate similar amounts of C to defense. A refinement on the COST term of the efficiency equation would partition root construction costs into those associated with defense of the tissue (such as formation of a woody periderm, high C:N ratio, and high concentrations of phenolic compounds such as lignins, tannins and suberins), those associated with water and ion uptake (such as ATP, reductant, and carrier proteins) and those associated with stor-age (such as starch and fructans). Such a scheme has been proposed for leaves (Lerdau, 1992).

2. Ion-Uptake Respiration

Respiration associated with the uptake of nutrient ions is excluded from our estimates of root C costs. Respiration associated with ion uptake may repre-sent a substantial portion of total root respiration (Veen, 1981; de Visser, 1985; van der Werf et al., 1988; Johnson, 1990; Bloom et al., 1992). For example, in barley mutants, about 14% and 23% of total root respiration was associated with ammonium and nitrate uptake, respectively (Bloom et al., 1992). These values are similar to estimates derived by more indirect meth-ods (references in Bloom et al., 1992). Nitrogen uptake represents approxi-mately 90% of the total ion uptake respiration (Veen, 1981). Little of the res-piration for ion uptake is expended to acquire P. Under P-limited conditions, the respiration associated with N uptake should be treated as a whole-plant cost, and should not enter into the cost–benefit analysis of a particular root. Consequently, we exclude the costs of ion-uptake respiration in applying the model to optimize P uptake by citrus. In a system where N is the limiting resource, costs of absorption, transport and assimilation would need to be included and would depend on the form of N taken up. Nitrate assimilation can represent a substantial carbon cost. In some species, however, under high light conditions, nitrate is reduced in the leaves by photosynthetic electron transport, with minimal cost to plant carbon stores. The costs of N uptake are therefore not easily represented in a model of roots alone.

3. Maintenance Respiration

Maintenance respiration can be operationally defined as the respiration not attributable to growth or ion uptake. It is often determined from a regression of total root respiration as a function of root growth rate, where maintenance respiration is the Y-intercept, or the respiration at zero growth (Szaniawski and Kielkiewicz, 1982). This approach assumes that maintenance respiration is constant and unaffected by root age, ion gradients, plant nutrient and water status or numerous other factors. An approach more amenable to roots in soil is to determine total root respiration and then subtract from the total the growth respiration (using estimates of root construction costs) and ion-uptake respiration (from measured rates of nutrient uptake and their theoretical costs) (Poorter et al., 1991; Peng et al., 1993). In both approaches, $R_{M(w)}$ represents residual respiration and, thus, includes root exudates that are metabolized by microbes and growth and maintenance respiration of the extramatrical hyphae.

4. Sensitivity of E to Variation in Cost Parameters

The sensitivity of E to values of the cost parameters is straightforward to analyze because the equation for cost is so simple (Eqn 2). Where construction costs ($C_{root} + R_{G(w)}$) are small relative to maintenance costs ($LR_{M(w)}$), changes or errors in these parameters will have little effect on the magnitude of E. Where L is very low, E will be more sensitive to construction costs. For the lemon seedling parameters used in these modeling illustrations (Table 3), a 50% change in $C_{root} + R_{G(w)}$ causes a 1% change in E if L is 10 days and a 0.1% change in E if L is 100 days. Maintenance respiration, $R_{M(w)}$, is much more influential, with E nearly inversely proportional to $R_{M(w)}$. Increasing $R_{M(w)}$ by 50% causes a 33% decrease in E at a longevity of 10 or 100 days. Alterations in L have little effect on E when L is large relative to construction costs, because the average daily cost will be nearly equal to $R_{M(w)}$ (L appears in both the numerator and the denominator of Eqn 2). Increasing L from 10 to 100 days increases E by only 2%. In the simulations we present, longevity is not a parameter estimated from measurements; we calculate E for any value of L. The cost–benefit analysis of root longevity is essentially an exploration of the sensitivity of E to L; we seek the L that maximizes E under various combinations of values of the other parameters. E varies strongly with L in our simulations due to changes in other parameters over the life of the root, such as V_{max}, soil moisture, or $R_{M(w)}$, not due to the structure of the cost part of the efficiency equation.

C. Uptake of Nutrients

In the model of cost-effectiveness of roots, nutrient uptake represents the benefit term in the equation. We estimate nutrient uptake using a steady-state model of solute uptake (Nye and Tinker, 1977; Yanai, 1994), which includes active uptake at the root surface and transport through the soil by diffusion and solution flow. The advantage of assuming a steady-state condition, in

Table 3

Parameter values used in simulating Volkamer lemon seedlings. Details of the uptake model can be found in the Appendix. Mycorrhizal (M) and nonmycorrhizal (NM) plants were grown in pots at two levels of P fertility for 92 days (Peng *et al.*, 1993). Soil was a Chandler series hyperthermic, uncoated Typic Quartzipsamment. Specific root length (λ) was calculated from measured dry weight and length of fine roots at 58 days for 1P plants and 65 days for 5P plants (Peng *et al.*, 1993) (Table 2). Root radius (r_0) was calculated from specific root length using a tissue density of 0.167 g cm^{-3} (Eissenstat, 1991). Maintenance respiration ($R_{M(w)}$) was measured at 25°C and excludes ion-uptake respiration; construction cost ($CONST_w$) is the sum of growth respiration ($R_{G(w)}$) and the carbon in the root (C_{root}) (Peng *et al.*, 1993). Construction cost was assessed on the first day. Uptake kinetics (V_{max} and K_m) were measured on excised roots using ^{32}P (K.R. Kosola and D.M. Eissenstat, unpublished data). Water uptake rate (v_0) was calculated from whole-plant transpiration and root surface area.

Parameters relating to root length and density were varied over time. Root mass was estimated from regression analysis of sequential harvests. Root length was calculated from root mass and λ. Inter-root distance, r_x, was calculated from root length and a pot volume of 130 cm^3. In the later simulations, r_x was 0.4 cm.

Soil parameters were held constant for the duration of the simulation. These include buffering capacity (b), diffusion coefficient of P in water (D_l), effective diffusion coefficient (D), solid-liquid partitioning coefficient (K_d), the impedance factor (describes tortuosity, f), and soil bulk density, ρ. In the pot study, P was added to soil weekly at concentrations of 1 mM (= 1P) and 5 mM P (= 5P); these concentrations were used for average solution concentration (C_{av}).

	NM 1 P	NM 5 P	M 1 P	M 5 P	Source
λ (m g^{-1})	37.7	23.2	32.8	21.6	Peng *et al.* (1993)
r_0 (cm)	0.02249	0.02866	0.02411	0.02971	Eissenstat (1991)
$R_{M(w)}$ (mmol g^{-1} d^{-1})	1.77	1.62	2.46	2.05	Peng *et al.* (1993)
$CONST_W$ (mmol g^{-1})	44.7	42.0	48.7	45.3	Peng *et al.* (1993)
V_{max} (pmol cm^{-2} s^{-1})	2.3				K.R. Kosola & D.M. Eissenstat, unpublished data
K_m (mM)	36				K.R. Kosola & D.M. Eissenstat, unpublished data
v_0 (cm s^{-1})	3.70×10^{-6}	3.90×10^{-6}	6.70×10^{-6}	3.70×10^{-6}	Peng *et al.* (1993)
b (unitless)	4.61				
D_l (cm^2 s^{-1})	0.89×10^{-6}				Barber (1984)
D (cm^2 s^{-1})	3.40×10^{-6}				
K_d	3				Ballard and Fiskell (1974)
θ_v (ml cm^{-3})	0.17				Unpublished data
f	$3.13* \theta^{1.92}$				van Rees *et al.* (1990)
C_{av} (mM)	1	5	1	5	Peng *et al.* (1993)
ρ (g cm^{-3})	1.48				Unpublished data

which the rate of solute uptake equals the rate of transport to the root, is that a single rate of nutrient uptake can be estimated for a root in a given soil environment, without a dynamic simulation of the development of the concentration profile over time. Parameters in the model can be changed as the root ages to explore factors affecting optimal root lifespan.

1. Model Assumptions

When the model is applied to a whole root system, roots are assumed to be homogeneous and uniformly distributed; parameter values describe the average root. Alternatively, the model can be used to describe a single root. The soil is also assumed to be homogeneous, except for the variation radial to the root, which is the steady-state profile of solute accumulation or depletion around each root. Roots are characterized by the specific root length, λ, the root density (half-distance to the next root r_x), the radius of the root, r_0, the rate of water influx into the root, v_0, and the root uptake kinetics (the maximal rate, V_{max}, and the half-saturation constant, K_m). The soil is characterized by the average soil solution concentration, C_{av}, the buffer capacity, b, and the effective diffusion coefficient, D, which is a function of soil water content, soil bulk density, and the tortuosity of the path a nutrient ion must take to reach the root. The equations used to calculate uptake from these parameters are given in the Appendix. The parameter values used in our simulations are given in Table 3.

2. Model Sensitivity

The sensitivity of E to values of the uptake parameters is complex because the uptake equations are complex (Eqns 3, 4, and 5 in the Appendix). The importance of any parameter in determining uptake depends on the values of the other parameters. As a result, any exploration of model behavior must use appropriate values for all parameters. Williams and Yanai (1996) conducted a systematic exploration of model sensitivity across all parameters using statistical analysis of variance. They found that the most important variables in explaining variation across the parameter space defined by the ranges of parameter values reported in the literature were C_{av} and V_{max}. There were important interactions of two, three, and four parameters, which illustrates the need to conduct sensitivity analyses with the relevant values for other parameters. To show the nature of the main effects and the interactions, Williams and Yanai (1996) used a graphical approach, displaying five dimensions of parameter space by using rows and columns of graphs on multiple pages.

The results of this sensitivity analysis can be interpreted to indicate optimal foraging strategies for roots in different environmental conditions. Investment in increased uptake capacity (by increasing V_{max} or decreasing K_m) will have little effect on uptake if the rate of delivery to the root (limited by C_{av}, Db, and v_0) is limiting to uptake. In this case, adding to root length is a better strategy for increasing nutrient uptake. Conversely, when the rate of uptake at the root surface is limiting, then increased uptake capacity will be advantageous. In these circumstances, increased r_0 will also improve nutrient uptake per unit length of root, whereas in the transport-limited case, increased r_0 has less effect. In all cases, however, the same biomass invested in longer, finer roots rather than increased root thickness results in greater efficiency.

For the simulations that follow, representing lemon seedlings in fertilized sandy soil, uptake of P is most sensitive to D and b. Increasing Db results in decreased uptake, because the steady-state solution to solute transport and influx under these conditions results in an accumulation of P at the root surface, rather than a depletion zone. Diffusion is away from the root, and more rapid diffusion allows a lower C_0. Increases in each of the other parameters (r_0, v_0, C_{av}, V_{max}, K_m, λ, and r_x) cause at least proportional increases in uptake, with the greatest effects due to r_0 and v_0. In this parameter set, despite the accumulation of P at the root surface, V_{max} is not saturated, because K_m is high relative to C_0, in keeping with the linear uptake kinetics we observed in this concentration range (K. Kosola, unpublished data).

D. Other Factors

Root hairs, root exudates and mycorrhizal fungi influence nutrient uptake and C costs but have not been explicitly included in this model. The importance of these factors in influencing the costs and benefits of root deployment will be described later.

For illustration, we restrict our simulations in this paper to the acquisition of P, the limiting nutrient in our model system of citrus plants growing in sandy soil. Other resources, such as water, ammonium, or nitrate, could also be modeled as the primary benefit accruing from root deployment.

IX. MODEL APPLICATION

To illustrate the application of a cost–benefit analysis to optimal root lifespan, we parameterized the efficiency model with data based on our research in citrus. We tested the ability of the model to simulate P uptake by lemon seedlings over a two-month period as measured by Peng *et al.* (1993). Then we applied the model to roots over time from birth to six months, with nutrient uptake varying as a function of root age. Roots must be simulated from the time they are constructed to allow an analysis of optimal lifespan based on maximal efficiency. We use this simulation to introduce calculations of cumulative root uptake and root cost and lifetime average root efficiency. In the following sections, we further examine how morphological (root diameter) and environmental (fertility, soil temperature and drought) factors affect root uptake, cost, efficiency and optimal lifespan.

A. Validation of Simulated P Uptake

The experiment by Peng *et al.* (1993) using Volkamer lemon seedlings involved four treatments, with mycorrhizal and nonmycorrhizal plants and two levels of P fertilization. We simulated uptake from day 33 to day 92 of the experiment, the period for which data on root length were available. The values of parameters used in the model are shown in Table 2. Average solution concentrations (C_{av}) were assumed to be constant at the concentrations added weekly to the pots. Uptake capacity of the roots was assumed to be constant with root age at the rate measured on seedling root systems (described by V_{max} and K_m). Phosphorus uptake per gram of root varied with treatment, because of differences in solution P concentration, specific root length, root radius, and water uptake rate, and decreased slightly over time because of root competition in the pot (average inter-root distance decreased). The simulated uptake summed over the 59 days agreed very well (within 5%) with the change in measured P content of harvested seedlings at days 33 and 92 for three of the treatments (Figure 1). The exception was the mycorrhizal plants at low P. The model underpredicted P uptake in this treatment by 23%, suggesting that the effect of mycorrhizal colonization, which was not included in the model, was important to P uptake. We did not calculate cumulative efficiency for these simulations because the first 33 days were not simulated (root length data were not available). Daily efficiency, given constant maintenance costs, is easily calculated from uptake.

B. Simulation with Root Age

To simulate lifetime average root efficiency, and thereby explore optimal root lifespan, requires simulating roots from the time they are constructed. The

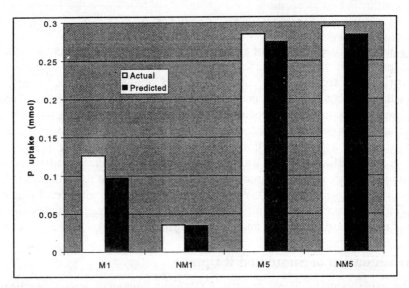

Fig. 1. Actual and predicted uptake of mycorrhizal (M) and nonmycorrhizal (NM) Volkamer lemon seedlings grown with weekly additions of 1 (M1, NM1) and 5 mmol P (M5, NM5) (from Peng *et al.*, 1993).

cost of constructing roots is one reason for plants to retain roots rather than build new ones: the average lifetime cost of a root decreases over time, if maintenance costs are constant, because the initial investment in constructing them is amortized over a longer period. In fact, if uptake were constant with root age, the optimal lifespan would be infinite. The decrease in nutrient uptake as roots age is presumably one of the primary reasons for root turnover. Declining nutrient uptake could be due to changes in the root or to changes in the supply of nutrient as the soil is depleted by uptake.

Estimates of changes in nutrient uptake capacity with root age come mainly from ion-depletion or ion-uptake techniques in solution culture. For a range of tree species, suberized woody roots typically have rates of nutrient absorption 30 to 80% of white roots (see references in van Rees and Comerford, 1990). When the cortex disintegrates, as is commonly found in older cereal roots in the field, phosphate uptake may be only 5% of that of roots where the cortex is still intact (Clarkson et al., 1968). In 2-week old barley plants, P uptake 1 cm from the root tip was similar to that in basal portions of the root, suggesting that uptake capacity does not decline in the first two weeks (Clarkson et al., 1968).

Although there are good indications that uptake rates are lower in older roots, uptake kinetics as a function of root age are not well known for citrus or for any other species. In our simulations, we assume a curve for V_{max} as a function of root age that starts at 60% of maximum at day 1, achieves maximum at 2 weeks, and declines asymptotically to 20%, passing 50% at about 3 months. For citrus seedlings, maximum V_{max} (3.8 pmol cm^{-2} s^{-1}) was calculated by determining the V_{max} (2.3 pmol cm^{-2} s^{-1}) of a population of roots (whose average age was determined by video analysis of sequential images taken with minirhizotrons). Other parameters were estimated as indicated from Table 2.

Simulated daily uptake per gram of root (Figure 2) is highest in the high-P treatments; other differences between treatments are due to differences in specific root length and water uptake rates. The changes in uptake with root age are driven by the relationship we assumed between V_{max} and root age. Cumulative uptake reflects the changes in daily uptake.

Daily costs are constant because maintenance respiration was assumed to be constant and the respiration associated with ion uptake is excluded from the model. Respiration is higher in mycorrhizal than nonmycorrhizal plants, and it is higher in the low-P treatment than in the high-P treatment. Cumulative cost, the sum of construction costs plus daily maintenance costs from zero to the age of the root, increases linearly because of the constancy of daily cost. Differences in construction costs are small compared with lifetime maintenance costs.

Daily efficiency shows a pattern with root age that is driven by the pattern of uptake with root age. Plants at high P are more efficient than plants at low P; the least efficient plant is the nonmycorrhizal low-P plant. Lifetime efficiency is the cumulative cost divided by the cumulative efficiency; this shows the average efficiency of the root were it to die at the given age. The high-P treatments show maximal lifetime efficiency at a lifespan of 65–70 days. Simulated lifetime efficiency declines less steeply in the nonmycorrhizal treatments because the costs are lower. The optimal lifespan is sensitive to the assumed pattern of uptake with root age: if P uptake were not assumed to be so low in older roots, efficiency would not drop off as quickly and the optimal lifespan would be longer. The pattern of nutrient uptake with root age is an important uncertainty which deserves further experimental determination.

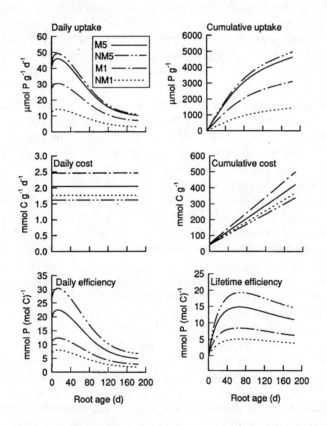

Fig. 2. Model simulations of daily and cumulative uptake, cost and efficiency of mycorrhizal (M1, M5) and nonmycorrhizal (NM1, NM5) Volkamer lemon seedlings at low (M1, NM1) and high (M5, NM5) phosphorus supply. Model parameterization is based on Table 3.

X. MORPHOLOGICAL AND ENVIRONMENTAL FACTORS AFFECTING ROOT COSTS AND BENEFITS

A. Specific Root Length, Root Diameter and Tissue Density

Specific root length (λ) is the ratio of root length to root mass. It has been used as a simple index of root benefit to root cost (Fitter, 1991), assuming that resource acquisition is proportional to length and root cost (construction and maintenance) is proportional to mass. Many studies equate λ with a measure of root fineness, assuming root density is constant. Here we discuss separately root diameter and tissue density and their relationship to root lifespan.

1. Root Diameter

Previous attempts using an earlier version of a cost–benefit efficiency model (Yanai *et al.*, 1995) failed to establish an optimal diameter of roots for maximum nutrient uptake efficiency. Thinner roots were always more efficient than coarser roots, because of the importance of root length and root surface area in nutrient uptake. The simulation assumed that finer roots had the same C cost as coarse roots per gram of root. This assumption may not be justified. Low-P nonmycorrhizal citrus roots are thinner than high-P mycorrhizal roots, and have about 6% higher construction cost and 9% higher maintenance respiration (Table 2). More importantly, thinner roots may place other constraints on root lifespan not represented in the present efficiency model, including increasing risk of herbivory and constrained axial water transport because of smaller xylem vessels. When C costs of fine roots were assumed to be higher than coarse roots, Yanai *et al.* (1995) found that optimal root diameter was no longer infinitely small. More accurate simulations of the effects of root diameter under different environmental conditions await more quantitative information on the relative costs of thin and thick roots.

Experimental evidence that root diameter is linked to root lifespan is limited. Obviously, roots that undergo secondary growth are normally longer-lived than absorptive roots that never undergo secondary thickening. The question is whether root longevity is related to root diameter among absorptive roots. Root diameter varies continuously within a plant's root system and can exhibit considerable morphological plasticity (Fitter, 1985). Absorptive roots typically decrease in diameter from internal to external links, with the terminal roots having the smallest diameter and shortest lifespan (Reid *et al.*, 1993). Root diameter also ranges widely among species and appears to be strongly influenced by plant phylogeny (Fitter, 1991; Eissenstat, 1992). The diameter of the finest elements of the root system can be less than 100 μm in many graminoid species found in the Juncaceae, Cyperaceae and Poaceae, in ericoid mycorrhizal species in the Ericaceae and Epacridaceae and in many annual dicots such as the well-studied species *Arabidopsis thaliana* (Harley and Smith, 1983; Fitter, 1991; Eissenstat, 1992; D.M. Eissenstat, unpublished data). At the other extreme, the fine root elements of many woody species in the Magnoliaceae, Rutaceae and Pinaceae and herbaceous species in the Aliaceae are at least 500–1000 μm diameter. There is some evidence that species with thin roots have a shorter lifespan than those with coarse roots, as indicated by limited comparisons of cold-desert shrubs (Caldwell and Camp,

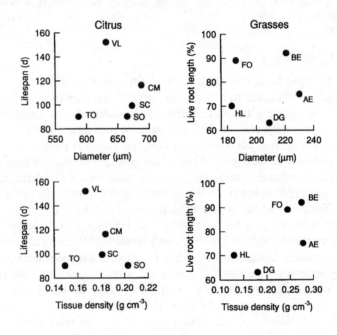

Fig. 3. The relationship of root diameter and tissue density to root longevity in citrus (Eissenstat, 1991) and perennial grasses (Ryser, 1996). Citrus roots were sampled in a Valencia orange rootstock trial. The perennial grasses were *Arrhenatherum elatius* (AE), *Dactylis glomerata* (DG), and *Holcus lanatus* (HL), species common to nutrient-rich grasslands, and *Festuca ovina* (FO) and *Bromus erectus* (BE), species characteristic of nutrient-poor grasslands. Citrus root diameters and tissue density are averages of roots about 70–90 days old. Citrus lifespan data and rootstock abbreviations are from Table 1. Grass root diameters and tissue density are from roots harvested at the end of the first growing season. Per cent live root was estimated from roots harvested at the end of the second growing season.

1974; Fernandez and Caldwell, 1975), chaparral shrubs (Kummerow *et al.*, 1978) and tundra species (Shaver and Billings, 1975). Among citrus rootstocks and perennial grasses, however, there was no relationship between mean root diameter and median root lifespan (Figure 3). Comparisons are needed of species with a broader range in root diameters before the relationship of root lifespan to genetic differences in root diameter can be fully assessed. Phenotypic variation in root diameter and its relation to root lifespan also deserves further study.

2. Tissue Density

Root tissue density is not constant and its variation may be ecologically important. Among citrus rootstocks, tissue density varied about 30%, tended to be lowest in rootstocks with the fastest root extension (Eissenstat, 1991), but exhibited no relationship to root lifespan (Figure 3). In perennial grasses, tissue density was again lowest in species with the fastest growth rates (Ryser and Lambers, 1995; Schläpfer and Ryser, 1996). Root density among these species differed by more than 100% and was positively correlated with the percentage of roots still alive after two growing seasons (Ryser, 1996; Figure 3). Root survivorship was determined by vital staining (triphenyl-tetrazolium chloride). These results are interesting but should be confirmed by methods that clearly separate root birth from death and that account for roots that are eaten or decomposed during the study period. Similar relationships between site productivity, growth rate, tissue density and lifespan have also been observed in leaves (Garnier and Laurent, 1994; Ryser and Lamber, 1995; Ryser, 1996; Schläpfer and Ryser, 1996). In leaves, tissue density has been related to high amounts of lignins and tanins and other secondary wall materials (Garnier and Laurent, 1994), which may affect tissue palatability and increase tissue toughness against adverse environmental conditions such as frost heaving.

B. Soil Fertility

Roots deployed in nutrient-rich soil will clearly be more efficient at nutrient uptake than roots in less fertile soil, if they acquire more nutrient for the same C expended (Bloom *et al.*, 1985). It is less clear how root lifespan should vary with soil fertility. Should roots be shed more rapidly in fertile or infertile soils? The effect of soil fertility on optimal root lifespan may depend on whether all the roots of the plant are exposed to uniformly fertile soil or whether a small portion of the roots are exposed to a fertile patch. This distinction is often overlooked when interpreting root responses to fertilization. If all the roots are supplied uniformly with nutrients, then competition for carbohydrates occurs principally between shoots and roots; for localized supply, carbohydrate competition will occur between roots in the nutrient-rich patches and roots in the less-fertile bulk soil. Thus, optimal root lifespan is not only affected by nutrient supply but by the heterogeneity of supply among the root axes.

1. Uniform Nutrient Supply

Experimental investigations on the effect of fertility on root lifespan have mainly addressed community-level rates of root turnover in forested ecosystems. The evidence that habitat fertility is inversely related to root lifespan is conflicting (see review by Hendricks et al., 1993). Much of the controversy may result from the different methods used to estimate root turnover. Aber et al. (1985) found decreasing root lifespan with increasing nitrogen availability when they estimated turnover using nitrogen-budgeting approaches (Table 4), but they found no relationship using a 'max-min' approach based on sequential harvesting of roots. This method assumes asynchronous birth and death and therefore probably underestimated root turnover on sites with low seasonal fluctuations. The nitrogen-budget approach, however, assumes that net mineralization is adequately estimated in buried bags, that all mineralized N is taken up, that N allocation above ground is accurately measured, and that the difference between uptake and above-ground allocation is used by fine roots and mycorrhizas. Although indirect, the method seemed to give reasonable estimates of whole stand rates of root turnover. For example, lifespan of sugar maple in Wisconsin ranged from 181 to 389 days by the N-budgeting approach (Table 4), which agrees well with lifespan of surface roots of sugar maple estimated from direct observation in Michigan, New York and New Hampshire (180 to 340 days, Table 1). Stands with high rates of nitrification had the highest N availability and shortest root lifespan (Aber et al., 1985) (Table 4). These hardwoods were typically mixed, containing both ecto- (oaks, birch, hickory) and arbuscular mycorrhizal (maple, cherry) species. On Blackhawk Island, Wisconsin, ectomycorrhizal hardwoods had considerably longer root lifespan than those trees in the forest type in which arbuscular-mycorrhizal sugar maple was a major component. At the University of Wisconsin Arboretum, the main difference in root lifespan was between hardwoods and conifers, not between ectomycorrhizal and arbuscular-mycorrhizal species. Conifers tended to be on sites with the lowest N availability.

Few studies of the effect of soil fertility on lifespan have used direct observation techniques. In a study of hybrid poplar (*Populus* × *euramericna* cv. Eugenei), lifespan of a spring cohort of roots was significantly diminished in trees fertilized with nitrogen (Pregitzer et al., 1995). Similar results were found in ponderosa pine (*Pinus ponderosa*). With no nitrogen addition 64% of the roots lived longer than 2 months; at 100 kg N ha^{-1} 51% lived past 2 months and at 200 kg N ha^{-1} 48% of the roots lived longer than 2 months (D.T. Tingey, D.L. Phillips, M.G. Johnson, M.J. Storm and J.T. Ball, unpublished data). Thus, direct

Table 4

The relationship of root lifespan to nitrification rate and nitrogen availability (after Aber et al., 1985, and Nadelhoffer et al., 1985). Lifespan = Standing fine root biomass/annual fine root production. Annual fine root production calculated by the nitrogen-budgeting approach

Forest Community	Nitrification (% of mineralization)	Nitrogen Availability (kg N ha⁻¹ yr⁻¹)	Average Lifespan (days)	Location	Comments and soil type
Black oak–White oak–Black cherry–Shagbark hickory–Red maple forest (Quercus velutina–Q. alba–Prunus serotina–Carya ovata–A. rubrum)	100	143	167	South-central Wisconsin University of Wisconsin Arboretum	Roots collected from 0–20 cm. Alfisol
Red oak (Q. rubrum)–White oak–Black cherry–Shagbark hickory–Sugar maple forest	100	133	188		Alfisol
White and black oak forest	100	107	301		Alfisol
Paper birch (Betula papyrifera) forest	—	92	358		Alfisol
Sugar maple (Acer saccharum) forest	100	102	389		Alfisol
Red (P. resinosa) and white pine forest	—	69	507		Alfisol
White pine (Pinus strobus) forest	71	79	528		Alfisol
White Spruce (Picea glauca) forest	—	66	760		Alfisol
Red and jack pine (P. banksiana) forest	50	47	812		Alfisol
Sugar maple and red oak forest	100	133	181	Blackhawk Island, Wisconsin	Alfisol
White and red oak forest	30	86	553		Alfisol
Red and white oak forest	4	92	568		Alfisol
White pine, red oak, white oak forest	46	60	753		Spodosol
Red and white pine forest	82	36	1223		Entisol

observation lends greater support to the hypothesis that root longevity diminishes with increased soil fertility.

2. Heterogeneous Nutrient Supply

Studies of nutrient-amended patches on root lifespan have given mixed results. In mixed hardwood forests, roots that proliferated in response to additions of water or water and nutrients lived longer than roots in unamended patches of soil (Pregitzer *et al.*, 1993; Fahey and Hughes, 1994). Conversely, localized water and nutrient addition diminished root lifespan in a pot study using four old-field herbaceous species (Gross *et al.*, 1993). Dense populations of young roots resulting from local proliferation may be especially vulnerable to consumption by soil organisms (Graham, 1995). Citrus roots proliferating in zones of nutrient enrichment, for example, were more heavily infected with *Phytophthora* than roots outside the soil patches (K.R. Kosola and D.M. Eissenstat, unpublished data). Short lifespan in these conditions may not be optimal for nutrient uptake.

The duration of the fertile patch should also affect optimal root lifespan (Fitter, 1994). If the patches are very short-lived, nutrient uptake may not be sufficient to repay the costs of rapid root proliferation. In such a case, increasing the uptake capacity of existing roots may be a better foraging strategy (Grime, 1994). For example, roots of *Arrhenatherum elatius*, a plant associated with fertile environments, proliferated only in resource-rich patches if the patch was present for a day or more (Campbell and Grime, 1989). *Festuca ovina*, a low-nutrient-adapted species, was less responsive than *A. elatius* in root proliferation but had higher specific rates of N absorption for the nutrient pulses that were shorter than 1 day. This research led to the hypothesis that plants adapted to infertile soils where pulses tend to be short-lived should produce long-lived roots, whereas plants in fertile environments should produce short-lived roots which can be located in new patches as they become available (Campbell and Grime, 1989).

3. Modeling Simulations of Soil Fertility

The effects of soil fertility on optimal root lifespan were simulated in a previous application of the cost–benefit model of root efficiency (Yanai *et al.*, 1995). That earlier version of the model calculated lifetime efficiency based on lifetime average properties of roots; in contrast, the current version of the model calculates daily values of uptake and cost and sums them to calculate lifetime efficiency. As in the current model, some advantage of young roots over old roots had to be included to give a decrease in efficiency at long lifespan, otherwise, efficiency increased continuously with lifespan owing to the amortization of construction costs. The two presumed advantages of young roots are high water uptake rates and high nutrient uptake rates (Queen, 1967;

Chung and Kramer, 1975; Eissenstat and van Rees, 1994). The effect of soil fertility on optimal lifespan differed depending on which of these mechanisms was implemented. In all simulations, lifetime root efficiency was lower in infertile than in more fertile soils, because simulated uptake was lower. Clearly, in heterogeneous soil, roots should be deployed in fertile patches.

If nutrient uptake capacity was assumed to decline with lifespan (V_{max} decreased linearly with lifespan), maximal efficiency occurred at shorter lifespans in more fertile soil. In fertile soil (simulated by a higher average concentration of solute in the soil solution, (C_{av})), the rate of nutrient uptake was more limited by V_{max}, and the effect of lowered V_{max} with longer lifespan was disadvantageous: optimal lifespan was low. In infertile soils, uptake is more likely to be limited by the supply of nutrient than by V_{max}, and the disadvantage of retaining roots was lessened: optimal lifespan was longer.

In contrast, if nutrient uptake capacity was held constant but water uptake rates were dependent on root lifespan (decreased linearly with lifespan), the effect of soil fertility on optimal lifespan was different. The rate of water movement toward the root surface is more important in infertile than fertile soils in determining the rate of nutrient uptake (Yanai, 1994; Williams and Yanai, 1996). The advantage of short-lived roots, therefore, was greater in infertile soils, and the simulations showed optimal lifespan to be shorter in infertile than in fertile soils.

Current knowledge of water and nutrient uptake rates in roots of different ages and different lifespans is still too limited to distinguish which of the simulated mechanisms is more realistic. Certainly, uptake kinetics and water uptake rates will be more or less limiting to nutrient acquisition in different environments. Differences in the limitations to root efficiency in different circumstances may explain some of the variation observed in patterns of root lifespan with respect to soil fertility. The duration of fertile patches will also affect optimal root lifespan, with efficiency declining as nutrients are depleted.

C. Soil Temperature

Observations of native plants in the field suggest that roots may live longer in cooler environments. Roots tend to exhibit low mortality rates over winter (Head, 1969; Hendrick and Pregitzer, 1993). In a study by Hendrick and Pregitzer (1993), roots of sugar maple lived, on average, 75 days longer at the more northerly of two sites. The effect of soil temperature was also compared between two *Festuca* grasslands, one at an elevation of 845 m and the other at 170 m (Self *et al.*, 1995). There was a 5°C difference in annual temperature between the two sites. Root standing number and root production were greater at the low- than the high-elevation site and root longevity was greater at the high-elevation site. Increasing soil temperature with a heating grid at the high-elevation site reduced root longevity. Except for these two studies, no one has examined root lifespan along either latitudinal or elevational gradients.

To simulate the effect of temperature on root costs and efficiency, we assumed an exponential increase in maintenance respiration with temperature, doubling $R_{M(w)}$ with each increase of 10°C (Ryan, 1991). We did not simulate the effect of increased temperature on uptake, although temperature may affect nutrient diffusivities and uptake kinetics. Above 30°C, uptake kinetics are usually much less affected by increases in temperature than is maintenance respiration (Barber, 1984; Marschner, 1986). We used parameters describing the nonmycorrhizal high-P treatment of the Volkamer lemon study (Table 3) because the effects of mycorrhizas are not included in this model and because the nonmycorrhizal low-P plants were P deficient and grew poorly.

Simulated lifetime efficiency, the ratio of cumulative uptake to cumulative cost, is highest at low temperature (Figure 4) because increased temperature was assumed to increase C cost without a corresponding benefit in increased uptake rates. The age at which lifetime efficiency is maximized is about 80 days at the lowest temperature and only about 50 days at the highest temperature. Thus, a 20-degree increase in soil temperature only resulted in a 30-day decrease in optimal root lifespan.

The results of this simulation suggest that differences in maintenance respiration alone can not account for the 75-day greater median lifespan at a 'northern' than 'southern' site in Michigan (Hendrick and Pregitzer, 1993), because the temperature difference between the two sites at a soil depth of 15 cm was only 2°C (growing season average). Secondary effects, such as greater activity of root pathogens and higher N mineralization at warmer temperatures may have contributed to differences observed in median lifespan.

D. Soil Dryness

In many ecosystems, a large fraction of the total root length is in the surface soil. For example, in pine ecosystems, 30–80% of root length may occur in the top 10 cm (Eissenstat and van Rees, 1994); in northern hardwoods, 50% of root length may occur in the top 7 cm (the forest floor) (Fahey and Hughes, 1994). This portion of the soil also experiences more fluctuations in moisture and more extreme dryness than deeper soil horizons, with consequent effects on root lifespan. In a rhizotron study, cotton root length declined by more than 50% at the 15-cm soil depth following exposure to dry soil for about 21 days (Klepper et al., 1973). Fine roots of soybean also tended to die after about 20 days of drought (Huck et al., 1987). In a tall-grass prairie dominated by big bluestem (Andropogon gerardii), root mortality was high in the top 10 cm in the first couple of weeks of drought (Hayes and Seastedt, 1987). In the following year when no drought occurred, mortality rates were fairly constant. Despite the frequency and importance of dry surface soil, the effects of local-

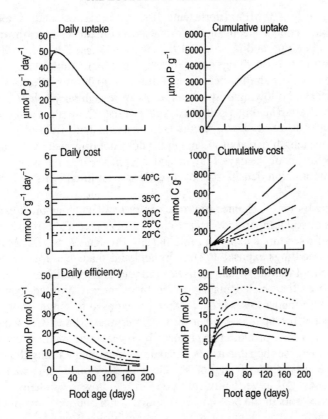

Fig. 4. The effects of increasing maintenance respiration on optimal lifespan. Model simulations of daily and cumulative uptake, cost and efficiency were for nonmycorrhizal (NM5) Volkamer lemon seedlings at soil temperatures that range from 20 to 40°C. For purposes of illustration, uptake was assumed constant at this range of temperatures whereas maintenance respiration was assumed to have a $Q_{10} = 2.0$.

ized drought on root behavior are poorly understood. The cost of retaining roots in dry soil should be compared with the benefits derived, both during and after the drought, to give an indication of optimal lifespan.

If optimal lifespan depends on the likely duration of drought, roots should be shed most readily in species adapted to conditions of prolonged drought. The desert succulent *Agave deserti* grows new roots rapidly after rain and then sheds them when the soil dries again (Huang and Nobel, 1992). Species not adapted to drought exhibit much greater tolerance of dry soil (Molyneux and Davies, 1983; Etherington, 1987; Jupp and Newman, 1987; Meyer *et al.*, 1990). Corn and tomato seedlings exhibited slow but continuous root growth in quite dry soil (–4.0 MPa) (Portas and Taylor, 1976; root death is not mentioned in this report). Death of epidermal and cortical tissues, however, is quite common (Jupp and Newman, 1987; Stasovski and Peterson, 1991). In seminal roots of corn seedlings exposed to drought for 34 days, the extrastelar tissues died back radially, beginning with the epidermis (Stasovski and Peterson, 1991). Following rehydration, roots were generally able to produce new laterals despite having a collapsed cortex, if the whole plant had not been severely water stressed. These results are consistent with field observations that corn root length density (30 cm depth) does not diminish after exposure to dry soil for about 25 days (Taylor and Klepper, 1973). However, observations of population density do not reveal compensating shifts in birth and death rates.

Other factors may influence root death in dry soil, such as plant carbohydrate status or soil temperature. Mean root mortality increased from 8 to 16% in Douglas-fir seedlings exposed to 22 days of drought ($P > 0.05$; Marshall, 1986). Tree seedlings exposed to both drought and shade had four-fold faster rates of fine root mortality than well-watered, unshaded seedlings ($P < 0.05$), suggesting that the shaded plants, which were more C-limited, could less afford to retain their roots. Dry soils that are exposed to direct sunlight can reach temperatures of greater than 60°C (Ehleringer, 1985). Consequently, roots near the soil surface often experience both water and temperature stress, which may increase the advantages of shedding over maintaining the roots.

Plant age may also affect root shedding in dry soil. Citrus roots were grown in sandy soil in vertically split pots: irrigation was withheld from the top pot, while roots in the bottom pot supplied water and nutrients in quantities sufficient to prevent signs of stress. Six-month-old citrus seedlings exhibited less than 3% root mortality after exposure to dry soil for over 80 days (Kosola and Eissenstat, 1994). Roots of mature citrus trees, however, often live less than 80 days (Kosola et al., 1995; Table 1: note that values in this table are for median lifespan, i.e. the time it takes for 50% of a population of roots to die). Consequently, a second study was established to compare death of fine roots in 1-year-old seedlings with those of 6-year-old bearing grapefruit trees (Espeleta, 1995). For the first 35 days of drought, root behavior was quite similar in juveniles and adults with neither exhibiting much mortality (about 2%). After 56 days, however, fine root mortality was 28% in adults but only 6% in juveniles; juvenile roots were also respiring more actively than those of adults, partly because of greater growth respiration. By 105 days, mortality was 33% in adults but only 8% in juveniles.

There are several differences between juvenile and adult citrus that could explain their different responses to dry surface soil. First, roots of adult citrus trees tended to be finer and less branched than the roots of juveniles (Espeleta, 1995). Juvenile fine roots may need to be thicker to serve additional functions, such as the transport of water and nutrients and the formation of the future structural root system of the tree. Second, compared with adults, juvenile trees typically allocate proportionally more biomass to roots (Ledig, 1983); this allocation of C may inhibit shedding of inefficient roots. Finally, the adult trees at the time of rapid root shedding were in a period of high C allocation to the fruit. Whatever the reasons, it is clear that patterns of root lifespan exhibited by juveniles may contrast greatly with those of adults in response to environmental stress.

We simulated the effect of localized drought on optimal root lifespan, using parameter values for the nonmycorrhizal high-P treatment of Volkamer lemon seedlings, as in the illustrations of temperature effects (Table 3). Root costs were based on data from the experiment described above, where fine roots of grapefruit trees on Volkamer lemon rootstocks were exposed to moist and dry soil (Espeleta, 1995). Maintenance respiration was found to drop to 5% of predrought conditions after 30 days; we assumed this reduction was achieved linearly, beginning on the first day of drought. Nutrient-uptake kinetics (V_{max} and K_m) were assumed to be unchanged by drought, as found for sour orange roots exposed to dry soil for up to 43 days (Whaley, 1995). The effect of drought on P uptake was simulated, not through variation in uptake kinetics, but through the effects of drought on nutrient transport through soil. Drought increases soil resistance to diffusion by increasing soil tortuosity. We recalculated the effective diffusion coefficient (D) and the buffer power (b) assuming a volumetric water content of 1% (D changed from 3.42×10^{-8} to 2.0×10^{-8} cm s^{-1}; b changed from 4.61 to 4.54; see Appendix for equations). The rate of water movement towards the root (v_0) also affects nutrient transport; this was set to zero during drought. In addition to simulating the effects of drought on nutrient transport and root uptake, we also simulated the effect of dry soil on P uptake per unit root length directly by diminishing uptake to 5% of that taken up by well-watered roots – the reduction Whaley (1995) observed in a study of sour orange seedlings using ^{32}P and ^{33}P. We assumed no residual effects of drought on respiration or uptake after soil moisture was restored.

We simulated various durations of drought at different stages of root development. Here we present simulations of an 80-day drought and an unremitted drought, both beginning when roots were 15 days old (Figure 5). We compare them to the case of no drought and to the case where uptake was fixed at 5% of normal.

Daily uptake shows the effect of drought on nutrient uptake. Of the three parameters that we modeled as affected by drought, the most important in limiting P uptake was the cessation of water movement toward the root, which is normally driven by the difference between root and rhizosphere soil water potential (in fact, under conditions of localized drought, water would tend to move from the root to the soil; v_0 would be negative). The effect of drought on v_0 alone would have reduced uptake by 68%. The drop in the

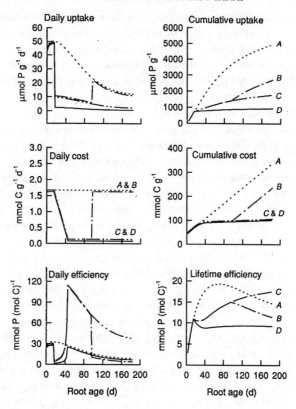

Fig. 5. The effects of drought on daily and cumulative uptake, cost and efficiency of nonmycorrhizal (NM5) Volkamer lemon seedlings. Model simulations were for conditions of no drought (*A*), 80 days of drought (*B*) beginning 15 days after the root was constructed, and continuous drought. For the continuous drought treatment, predicted responses are illustrated for P uptake in sandy soil based on the uptake model (*C*) and based on experimental data (*D*) (Whaley, 1995).

effective diffusion coefficient, *D*, accounted for an additional reduction of 11%; the drop in *b* caused only a 0.3% further reduction. The total 79% reduction in uptake was smaller than that observed using radioisotopes of P, where uptake in dry soil was only 2–5% of that in wet soil (Whaley, 1995); a 95% reduction is shown in simulation D.

The daily cost shows the assumed decline of maintenance respiration during the drought. Respiration recovers to normal rates at the end of the 80-day drought and remains low in unremitted drought.

Daily efficiency shows the combined effects of reduced respiration and reduced uptake. Daily efficiency drops sharply with the onset of drought because nutrient uptake was affected immediately by reduced soil moisture, whereas root maintenance respiration was assumed to respond to drought more slowly. Efficiency recovers rapidly during drought as maintenance respiration diminishes. Indeed, in simulations B and C, in which uptake declines by 79% during drought while respiration declines by 95%, daily efficiency is higher in dry than in moist soil. In simulation D, respiration and uptake both decline by 95%, and the daily efficiency of roots in dry soil is the same as those in moist soil. Clearly, the result of drought on root efficiency depends on the relative reductions of C costs and nutrient benefits.

Lifetime efficiency is generally greatest in roots never exposed to drought (simulation A). An exception is simulation C, where roots experience high daily efficiency, due to savings in respiration, that outweigh losses in uptake; these roots eventually overtake the non-drought roots in lifetime efficiency. The relatively more important savings in costs mean that efficiency declines when the drought ends (simulation B), implying that roots should be shed at the cessation of drought. If the loss in uptake were more severe than the savings in respiration costs, efficiency would decline during the drought and recover afterwards; whether roots should be shed at the onset of drought would depend on the likely duration of the drought. Where uptake and respiration were equally affected by drought (simulation D), lifetime efficiency never recovered; roots should have been shed at the onset of drought.

This simulation illustrates the importance of the relative reductions in costs and uptake to the efficiency of shedding roots in dry soil. Reduced maintenance respiration may be the reason that the roots of some plant species are so tolerant of dry surface soil.

XI. FURTHER CONSIDERATIONS

The lifespan of roots may be influenced by many factors beyond the scope of the efficiency model in its present form, such as seasonality, herbivore and pathogen pressure, competing sinks for carbohydrates in the plant, nutrient resorption and recycling, and other functions of roots such as transport and storage. In addition to these factors, which we discussed previously, there are other factors relating to root hairs, root exudates and mycorrhizal fungi that contribute to the C costs and nutritional benefits of roots. Costs and benefits will be discussed from the perspective of how they change with root age, especially for roots in a soil environment. Much is known about their costs and benefits in qualitative terms, but quantitative information is insufficient to include them explicitly in the current model.

A. Root Hairs

Root hairs are specially modified epidermal cells which develop in the elongation zone of the root, usually within a few centimeters of the root tip. An important benefit of root hairs is an increase in absorptive surface area beyond the depleted zone near the root surface (Bhat and Nye, 1973). Root hairs can be sites of extensive mucilage production (Dawes and Bowler, 1959), which has additional costs and benefits. The production of mucilage by root hairs can enhance the ability of the hair to attach to soil particles and thereby prevent air gaps from developing between the soil and root surface when the soil dries (Greaves and Darbyshire, 1972; Sprent, 1975). The prevention of air gaps and maintenance of a continuous film of water between the soil and root surface can strongly influence rates of water and nutrient uptake in soils of low moisture status (Newman, 1974).

Simulation models have provided some insights into the benefits of root hairs for P uptake (Itoh and Barber, 1983a, b). Root hairs can be included in the single-root model similar to that described above, where each hair is treated like a small root. Itoh and Barber found that root hair length, root hair radius and root hair density all influenced predicted P uptake, with root hair length being particularly significant. Prediction of P uptake without taking into account root hairs resulted in underpredictions of more than 50% in species with long root hairs such as Russian thistle and tomato; including root hairs resulted in a much closer fit to the 1:1 line of predicted with observed uptake (slope = 0.98, r = 0.89).

The costs of root hairs compared with unmodified epidermal cells have not been examined in detail. Despite the putative role of hairs in mucilage production, one study indicates that epidermal cells with hairs do not exude more C than older portions of the root without hairs (McCully and Canny, 1985). Once the cells develop heavily lignified and suberized secondary cell walls, rates of exudation would likely decrease. In *Arabidopsis*, cell volume of a root hair epidermal cell was about 50% greater than that of an unmodified cell (T.R. Bates and J.P. Lynch, unpublished data). The proportion of the cell that was vacuole, however, was not measured. It is not known if epidermal cells with hairs have higher respiration rates than those without hairs.

Longevity of root hairs does not coincide with longevity of the root. Root hair formation typically occurs within a few days of root formation, but root hair death can occur well before the root dies. Estimates of the lifespan of root hairs depends in part on whether the investigator defines death as loss of the viability of the root hair cell or as collapse of the skeletal cell wall structure, which may persist after the cell has died. In maize, Fusseder (1987) found evidence of cytoplasmic disintegration in hairs only 2 to 3 days old using electron microscopy, but by vital staining of nuclei, he concluded that hairs lived 1–3 weeks. Similar lifespans of root hairs have been reported for wheat (Henry and

Deacon, 1981) and barley (Holden, 1975) using the vital staining method. Consequently, the assumption that the nucleus dies at the same time as other cell organelles may lead to an overestimation of root hair lifespan. Investigators who define death by the loss of the entire root hair structure would produce even longer estimates of hair lifespan. Root hairs associated with soil sheath formation can have quite long lifespans (Goodchild and Myers, 1987; McCully and Canny, 1988). Some root hairs develop very thick walls that persist for many months after the cell has died (Head, 1973; Fusseder, 1987). These hairs may continue to function in helping bind soil particles to the root surface even though they are no longer living. Eventually, not only root hairs but the entire epidermis can be sloughed or abraded. In the primary roots of maize, epidermal senescence, which includes loss of root hairs, usually becomes extensive in the bare root portion beyond the soil sheath (McCully and Canny, 1988). In citrus roots, it is common to find sloughed epidermis in older fibrous roots that otherwise are quite healthy. For example, in Florida about 90% of the length of roots 2 to 3 months old had less than 30% of the epidermis still intact (D.M. Eissenstat and D. Achor, unpublished data).

Although there have been advances in quantifying the benefits of root hairs in short-term experiments, to model root lifespan we must understand the benefits of root hairs over the entire lifetime of the root. For many roots, root hairs may only exist for a small fraction of the root's lifetime. The costs of root hairs are poorly understood, even in a qualitative way.

B. Root Exudates

The costs of root exudation are better known than the benefits. Root exudates include water-soluble C compounds such as sugars, amino acids and organic acids and water-insoluble C compounds such as root cap cells, mucilage and limited cell wall debris (Lambers, 1987). Another term, rhizodeposition, has been used to express the loss of C from roots, but its meaning differs among investigators. Lynch and Whipps (1991) include dead roots and root respiration in rhizodeposition, whereas Newman (1985) did not. For purposes of C budgets, it is important neither to miss the C exuded nor to count it twice. Where cortical and epidermal walls are sloughed, these costs may already have been estimated in root construction costs. Mucilage and root cap cells, however, normally would not be included in construction costs. When maintenance respiration is calculated using total soil respiration and subtracting respiration attributable to growth or ion uptake, the remaining or 'residual' respiration includes not only root maintenance respiration but also respiration by soil microbes and mycorrhizal fungi. Some of the C thus respired can be assumed to have been exuded by roots.

Excluding root respiration and death, Newman (1985), in a review of the literature, estimated that from 10–100 mg of soluble exudates and

100–250 mg insoluble organic material, of which 60 mg was root cap cells and mucilage, was lost from young roots about 3 weeks old. This cost can be compared with estimates of 'residual' respiration (Table 2). If soluble C were metabolized at the rate at which it was exuded, it would represent about 1–10% of the 'residual' maintenance respiration of high-P nonmycorrhizal citrus plants. If the insoluble C were also metabolized, exudates could account for 12–38% of root 'residual' maintenance respiration or about 7–23% of total root respiration.

Some investigators have estimated higher rates of soluble C exudation. Two barley cultivars grown in sterile solution culture for 25 days had soluble C exudation rates of 390–465 μmol C (g root dry wt)$^{-1}$ day^{-1} (Xu and Juma, 1994) (this rate would comprise 14–17% of total root respiration in citrus; total respiration of barley seedlings was not measured). Cheng et al. (1993) estimated that microbial respiration of exudates represented nearly 60% of total root and soil respiration in 3-week-old barley plants. They used an isotopic dilution technique, in which unlabeled glucose fed to microbes allowed microbial respiration to be partitioned from root respiration; plant carbohydrates were uniformly labeled with ^{14}C. In a subsequent study with 3-week-old plants, Cheng et al. (1994) attributed 45% of the total root-soil respiration to microbial metabolism of root exudates in tall fescue and 31% in buffalo gourd. Although there are still several assumptions that need to be examined, this technique holds promise for estimating root exudation in situ. This work suggests that root exudation may be a substantial component of root C costs. Better estimates are needed of exudation by roots of different ages to define these costs more clearly.

Root exudation is not without benefit: it may enhance the ability of the plant to acquire P and other nutrients of low solubility such as iron. Some plants have specialized mechanisms to enhance P solubility in the soil solution. Certain arctic species produce extracellular phosphatases (Kroehler and Linkins, 1988). Species adapted to very low-P soils may exude chelating compounds such as citrate and piscidic acid (Gardner et al., 1983; Ae et al., 1990). Cluster roots in particular produce abundant exudates that promote P uptake in very infertile soils (Gardner et al., 1983). Because of the extremely high value of P relative to C in these soils, this represents an efficient use of C to maximize the growth or fitness of the plant. Chelating compounds such as siderophores can also be produced by rhizospheric microorganisms that feed on root exudates. In alkaline soils, many plants can enhance P solubility by exuding protons and organic acids, which reduces rhizosphere pH (Nye, 1992). Lastly, colonization by mycorrhizal fungi is generally higher in roots with faster rates of exudation (Graham et al., 1981; Schwab et al., 1991).

C. Mycorrhizas

The roots of most plant species can form a symbiotic relationship with mycorrhizal fungi (Newman and Reddell, 1987). Although mycorrhizas are widespread and known to have important effects on nutrient uptake and C costs of roots, they are rarely explicit in calculations of either nutrient or C fluxes. Incorporating mycorrhizal fungi in an analysis of root longevity is made difficult by our limited understanding of the processes by which the mycorrhizal symbiosis enhances nutrient uptake, the biomass, construction costs and turnover rates of mycorrhizal fungi, the age at which roots become colonized by mycorrhizas and the respiration associated with the colonization, and the changes in nutrient uptake and respiratory costs of mycorrhizal roots with root age. In addition, mycorrhizal fungi may have direct effects on root lifespan; there is some evidence that mycorrhizal roots live longer than uncolonized roots on the same plant (Harley and Smith, 1983; Espeleta, 1995; but see Hooker et al., 1995). This observed effect on root longevity might be consistent with increased efficiency of nutrient acquisition, as nutrient uptake can be enhanced by mycorrhizal fungi. Alternatively, the mycorrhizal fungus may increase the root sink for C, extending root longevity beyond that optimal for the plant. Finally, mycorrhizal roots may be better defended against root pathogens than non-mycorrhizal roots (Gange et al., 1994; Newsham et al., 1995).

1. Nutritional Benefits

The nutritional benefits of mycorrhizas are well known, especially where the limiting nutrient is P. Mycorrhizal uptake of nutrients involves three distinct processes: uptake from the soil by extraradical fungal hyphae, translocation through the hyphae to the fungal structures within the root, and transfer from the fungus to the plant across the interface between them (Smith et al., 1994). Any of these processes may limit nutrient uptake. Although the high absorptive surface area of the external hyphae certainly contributes to enhanced acquisition of nutrient by the plant, active extramatrical hyphal length alone is not a good predictor of nutrient acquisition. For example, the arbuscular mycorrhizal fungus, Glomus caledonium, transported 50 times more P per metre of hyphae than an ineffective species, Scutellospora calospora, in a study of cucumber plants (Pearson and Jakobsen, 1993a, b). Per unit of hyphal C expended, the more effective species was 38 times more efficient at P transport. The ineffective species contributed only about 7% of the total P taken up by the plant whereas the effective species contributed all of P uptake. For transfer from the fungus to the plant, Smith et al. (1994) found about four-fold differences between two species of arbuscular mycorrhizal fungi in the rate P was transported from the fungal arbuscules and intercellular hyphae to the cortical cells in the roots of the onion host. In addition to the added surface area for nutrient absorption, mycorrhizal fungi can enhance nutrient acquisition by the secretion of enzymes that transform organic N and P into more available forms (Read, 1993).

2. Fungal Biomass

To calculate the effect of mycorrhizas on optimal root lifespan requires information on the fungal biomass per gram of root. This biomass, however, is difficult to determine, both for fungal structures within the root and also for the extraradical hyphae. The arbuscular mycorrhizal fungus, *Glomus fasciculatum*, occupied 4.3% of onion root volume at 13 weeks; approximately 76% of the root length was colonized (Toth *et al.*, 1991). The per cent of the total fungal volume that was arbuscule was 37% in these same roots. Arbuscules degenerate rapidly; typically having a lifespan of about 8–10 days (Harley and Smith, 1983; Toth and Miller, 1984). It is not known whether the C is reabsorbed by the root or fungus after the arbuscules degenerate. The lipid-rich vesicles live much longer, develop after arbuscules, and are commonly abundant in older roots at high levels of colonization (Brundrett *et al.*, 1985). For the highly vesicular fungus, *Glomus intraradices*, construction costs of citrus roots were 8–9% higher in mycorrhizal than nonmycorrhizal roots 3 months after transplanting (Peng *et al.*, 1993); most of this cost difference was due to the presence of the vesicles. Thus, internal structures may represent an additional 5–10% of the cost of root construction.

The biomass of extraradical hyphae can be estimated indirectly from the hyphal length per length of root and some assumptions of tissue density and hyphal radius. Hyphal lengths vary widely. In tallgrass prairie and permanent pasture, for example, which have arbuscular mycorrhizas, extraradical hyphal lengths ranged from 60 to 270 cm (cm root)$^{-1}$ in the top 10 cm of soil (Miller *et al.*, 1995), which is in the range found by other investigators (Smith and Gianinazzi-Pearson, 1988; Sylvia, 1990; Miller and Jastrow, 1991). Miller *et al.* (1995) determined hyphal biomass and length. Standing biomass of extraradical arbuscular hyphae was estimated at 6–7% of root biomass in the top 10 cm of soil (Miller *et al.*, 1995). Deeper in the soil, the amount of hyphae per unit of root would likely diminish (Koide and Mooney, 1987).

Ectomycorrhizas show a similar range of fungal biomass per unit mass of root. For example, Jones *et al.* (1990) found ectomycorrhizal hyphal lengths associated with *Salix* roots to range from 100 to 300 cm (cm root)$^{-1}$, which amounts to 0.5–2% of the root biomass, assuming the density of fungal and root tissue are similar and that their diameters are 3 and 400 μm, respectively (Eissenstat, 1991; Eissenstat and van Rees, 1994). However, ectomycorrhizal fungi are probably longer-lived than arbuscular mycorrhizal fungi and the mass of the mycorrhizal mantle increases with root age. For a range of ectomycorrhizal fungi on *Pinus sylvestris*, Colpaert *et al.* (1992) estimated that between 2 and 6 months, the density of extramatrical mycelia increased two- to eight-fold. Fungal biomass at 6 months ranged from 8% to 44% of total root biomass. These results are consistent with data summarized by Harley and Smith (1983) where about 30–40% of the biomass of an ectomycorrhizal root may consist of fungal sheath.

In summary, in arbuscular mycorrhizal and young ectomycorrhizal plants, extraradical hyphal biomass probably represents less than 10% of root biomass. Importantly, much of this biomass may be dead, representing the tough chitin-impregnated fungal walls, not living tissue. For example, only about 20% and 10% of the external hyphae of *Glomus mosseae* and *Glomus intraradices* was active 6 and 13 weeks after planting *Paspalum notatum* in pots (Sylvia, 1988). The fraction of hyphae that is living has major implications in estimating myco-rrhizal C costs, if the majority of hyphal cost is not due to hyphal construction but to the respiration associated with maintenance and ion uptake.

3. Carbon Expenditures

Mycorrhizas can increase below-ground C costs typically by 10–20% (see references in Peng *et al.*, 1993). Some of these host-plant expenditures on mycorrhizal fungi can be supported by neighboring plants. Approximately 5–10% of mycorrhizal root carbon may come from surrounding plants (Watkins *et al.*, 1996; Simard *et al.*, 1995). Host-plant C expenditure on mycorrhizal fungi depends on fungal biomass, construction costs, and respi-ration. Below-ground respiration can be 1.3–3 times higher in mycorrhizal plants than in nonmycorrhizal plants of equivalent nutritional status (Peng *et al.*, 1993; Rygiewicz and Andersen, 1994). Much of this increase may be due to the high respiration rates of extramatrical hyphae; ^{14}C pulse-chase experi-ments revealed that 50% more C was respired than retained in young, active ectomycorrhizal hyphae in ponderosa pine (Rygiewicz and Andersen, 1994).

The length of time that roots are mycorrhizal and the lifespan of hyphae are additional factors affecting the costs and benefits of mycorrhizas. Roots may take one or more weeks to be colonized by mycorrhizal fungi in dis-turbed soil or in pot cultures where root inoculum is used, but where roots are growing in an established hyphal network, colonization may take only a few days (Brundrett *et al.*, 1985). In the field, mycorrhizal colonization of the roots probably occurs very rapidly. The longevity of hyphae has not been determined, which greatly limits estimation of C costs. In the field, longevity of the hyphae may be diminished by grazing animals, especially fungal-feed-ing insects like *Collembola* and fungal-feeding nematodes (McGonigle and Fitter, 1988; Setälä, 1995).

Mycorrhizas can increase the fraction of whole-plant biomass allocated to roots (Peng *et al.*, 1993). Consequently, where mycorrhizas are not providing a nutritional benefit, these C costs can cause mycorrhizal plants to grow more slowly than nonmycorrhizal plants. Normally, however, the nutritional bene-fits of mycorrhizas outweigh the C costs, with consequent decreases in root: shoot allocation and specific rates of root respiration and increases in leaf C assimilation and overall plant growth (Ingestad and Ågren, 1991; Eissenstat *et al.*, 1993; Peng *et al.*, 1993).

XII. FUTURE DIRECTIONS

The biomass of roots at the ecosystem scale indicates the magnitude of below-ground production, and physiological studies show that roots are expensive to maintain. Competition below ground influences plant survival and reproductive success. Despite these basic truths about their importance, we know very little about how the deployment of roots has evolved in an adaptive fashion.

In this paper we posed a number of hypotheses that may explain variation in root lifespan. Assessing the relative importance of these hypotheses is not yet possible because of important unknowns. Studies of individual plants have mainly focused on crop species. Observations of root productivity at the community or ecosystem level do not clearly indicate how different species have adapted root lifespan to their environment. It is also not known to what degree root death is under the control of the plant. If roots are mainly lost to root herbivory and parasitism, then selective root death may be accomplished more by limiting root defenses than by active shedding.

Recent developments in minirhizotrons, video imaging and computer processing have created new opportunities to address key questions associated with the ecology of root lifespan at the species level. For example, do species from low-nutrient habitats retain roots longer than species from fertile habitats? Is root lifespan correlated with leaf lifespan? If not, general theories on the relationship of tissue longevity to nutrient-use efficiency will require revision. Can plants optimize root foraging in a spatially and temporally heterogeneous environment not only by proliferating roots in favorable soil patches but also by shedding roots in unfavorable patches?

Observation of roots will always be more difficult than observation of leaves. For this reason, much of the theory regarding the ecology of root lifespan has been inspired by analogy to leaf lifespan. Modeling is another source of hypotheses about the adaptive advantages of root foraging strategies. In this review, we used an approach that treats C as cost and P as a benefit. Advances in this direction will require a better understanding of maintenance respiration and nutrient uptake kinetics, especially as they change with root age and environmental conditions. In addition, the costs and benefits of root hairs, mycorrhizal fungi and exudation in the natural environment must be clarified.

Other modeling approaches deserve consideration. A cost–benefit analysis that treats the limiting nutrient, rather than C, as the cost might prove fruitful. This approach would require information on the amount of nutrient recovery from roots after death. Above-ground sinks compete with roots for both carbon and nutrients; the optimal allocation of resources to leaves and roots can be explored with whole-plant models. The optimal root lifespan should do more than maximize nutrient uptake; it should maximize plant success in the environments to which it is adapted. Given the difficulty of observing roots, simulation models will continue to be useful tools for selecting research questions and advancing understanding of the ecology of root lifespan.

ACKNOWLEDGEMENTS

We thank James Graham and Kurt Pregitzer for useful comments on this manuscript; Alastair Fitter and David Robinson were especially helpful in providing direction and constructive review. We also wish to thank W. Cheng, M. Coleman, M. Russelle, M. Miller, K. Pregitzer, P. Ryser and D. Tingey for providing unpublished data or manuscripts. We are grateful for the financial support of our experimental and modeling work provided by grants from the U.S. Dept. of Agriculture (NRICGP-94-37107-1024) and the National Science Foundation (BSR-911824, IBN-9596050).

REFERENCES

Ae, N., Arihara, J., Okada, K., Yoshihara, T. and Johansen, C. (1990). Phosphorus uptake by pigeon pea and its role in cropping systems of the Indian subcontinent. *Science* **248**, 477–480.

Aber, J.D., Melillo, J.M., Nadelhoffer, K.J., McClaugherty, C.A. and Pastor, J. (1985). Fine root turnover in forest ecosystems in relation to quantity and form of nitrogen availability: a comparison of two methods. *Oecologia* **66**, 317–321.

Amthor, J.S. (1984). The role of maintenance respiration in plant growth. *Plant Cell Environ.* **7**, 561–569.

Arnone, III, J.A. and Körner, Ch. (1995). Soil and biomass carbon pools in model communities of tropical plants under elevated CO_2. *Oecologia* **104**, 61–71.

Atkinson, D. (1972). Seasonal periodicity of black currant root growth and the influence of simulated mechanical harvesting. *J. Hort. Sci.* **47**, 165–172.

Atkinson, D. (1985). Spatial and temporal aspects of root distribution as indicated by the use of a root observation laboratory. In: *Ecological Interactions in Soil: Plant, Microbes and Animals* (Ed. by A.H. Fitter, D. Atkinson, D.J. Read and M.B. Usher), pp. 43–65. Special Publ. Ser. British Ecol. Soc. **4**. Blackwell Scientific Publications, Oxford.

Baldwin, J.P., Nye, P.H. and Tinker, P.B. (1973). Uptake of solutes by multiple root systems from soil. III. A model for calculating the solute uptake by a randomly dispersed root system developing in a finite volume of soil. *Plant Soil* **38**, 621–635.

Ballard, R. and Fiskell, J.G.A. (1974). Phosphorus retention in the Coastal Plain. *Proc. Soil Sci. Soc. Amer.* **38**, 250.

Barber, S.A. (1984). *Soil Nutrient Bioavailability: a Mechanistic Approach.* John Wiley and Sons, Inc., New York.

Beyrouty, C.A., Wells, B.R., Norman, R.J., Marvel, J.N. and Pillow, Jr., J.R. (1987). In: *Minirhizotron Observation Tubes: Methods and Applications for Measuring Rhizosphere Dynamics* (Ed. by H.M. Taylor), pp. 99–108. ASA Special Publ. **50**, American Society of Agronomy, Madison.

Bhat, K.K.S. and Nye, P.H. (1973). Diffusion of phosphate to plant roots in soil. I. Quantitative autoradiography of the depletion zone. *Plant Soil* **38**, 161–175.

Bloom, A.J., Chapin, III, F.S., and Mooney, H.A. (1985). Resource limitation in plants: An economic analogy. *Annu. Rev. Ecol. System.* **16**, 363–392.

Bloom, A.J, Sukrapanna, S.S, Warner, R.L. (1992). Root respiration associated with ammonium and nitrate absorption and assimilation by barley. *Plant Physiol.* **99**, 1294–1301.

Bloomfield, J., Vogt, K. and Wargo, P.M. (1996). Tree root turnover and senescence. In: *Plant Roots: The Hidden Half* (Ed. by Y. Waisel, A. Eshel and U. Kafkafi), 2nd edn., pp. 363–382. Marcel Dekker, Inc., New York.

Box, J.E., Jr, and Johnson, J.W. (1987). Minirhizotron rooting comparisons of three wheat cultivars. In: *Minirhizotron Observation Tubes: Methods and Applications for Measuring Rhizosphere Dynamics* (Ed. by H.M. Talor), pp. 123–130. ASA Special Publ. **50**, American Society of Agronomy, Madison.

Brouwer, R. (1981). Co-ordination of growth phenomena within a root system of intact maize plants. *Plant and Soil* **63**, 65–72.

Brown, V.K. and Gange, A.C. (1991). Effects of root herbivory on vegetation dynamics. In: *Plant Root Growth: An Ecological Perspective* (Ed. by D. Atlkinson), pp. 453–470. Blackwell Scientific, Boston.

Brundrett, M.C., Piché, Y. and Peterson, R.L. (1985). A developmental study of the early stages in vesicular-arbuscular mycorrhiza formation. *Can. J. Bot.* **63**, 184–194.

Caldwell, M.M. (1987). Competition between roots in natural communities. In: *Root Development and Function.* (Ed. by P.J. Gregory, J.V. Lake and D.A. Rose), pp. 167–185. Cambridge University Press, New York.

Caldwell, M.M. and Camp, L.B. (1974). Belowground productivity of two cool desert communities. *Oecologia* **17**, 123–130.

Caldwell, M.M. and Eissenstat, D.M. (1987). Coping with variability: examples of tracer use in root function studies. In: *Plant Response to Stress* (Ed. by J.D. Tenhunen, F.M. Catarino, O.L. Lange and W.C. Oechel), pp. 95–106. NATO ASI Series 15G, Springer-Verlag, Berlin.

Campbell, B.D. and Grime, J.P. (1989). A comparative study of plant responsiveness to the duration of episodes of mineral nutrient enrichment. *New Phytologist* **112**, 261–267.

Chabot, B.F. and Hicks, D.J. (1982). The ecology of leaf life spans. *Annual Rev. Ecol. System.* **13**, 229–259.

Chandler, W.H. (1923). *Results of Some Experiments in Pruning Fruit Trees.* N.Y. Agricultural Exp. Sta. Bull. 415, Ithaca.

Chapin, III, F.S. (1980). The mineral nutrition of wild plants. *Annu. Rev. Ecol. Systemat.* **11**, 233–260.

Chapin, III, F.S. (1989). The costs of tundra plant structures: Evaluation of concepts and currencies. *American Naturalist* **133**, 1–19

Chapin, III, F.S. (1995). New cog in the nitrogen cycle. *Nature* **377**, 199–200.

Chapin, III, F.S., Schulze, E-D and Mooney, H.A. (1990). The ecology and economics of storage in plants. *Ann. Rev. Ecol. System.* **21**, 423–447.

Chapin, III, F.S., Autumn, K., and Pugnaire, F. (1993). Evolution of suites of traits in response to environmental stress. *American Naturalist* **142**, S78–S92.

Cheng, W., Coleman, D.C. and Box, J.E. Jr. (1990). Root dynamics, production and distribution in agroecosystems on the Georgia Piedmont using minirhizotrons. *J. Appl. Ecol.* **27**, 592–604.

Cheng, W., Coleman, D.C., Carroll, C.R. and Hoffman, C.A. (1993). *In situ* measurement of root respiration and soluble C concentrations in the rhizosphere. *Soil Biol. Biochem.* **25**, 1189–1196.

Cheng, W., Coleman, D.C., Carroll, C.R., and Hoffman, C.A. (1994). Investigating short-term carbon flows in the rhizospheres of different plant species, using isotopic trapping. *Agron. J.* **86**, 782–788.

Chitwood, D.J. (1992). Nematicidal compounds from plants. In: *Phytochemical Resources for Medicine and Agriculture* (Ed. by H.N. Nigg and D. Seigler), pp. 185–204. Plenum Press, New York.

Chung, H.-H., and Kramer, P.J. 1975. Absorption of water and ^{32}P through suberized and unsuberized roots of loblolly pine. *Can. J. For. Res.* **5**, 229–235.

Clarkson, D.T., Sanderson, J. and Russell, R.S. (1968). Ion uptake and root age. *Nature* **220**, 805–806.

Coley, P.D. (1988). Effects of plant growth rate and leaf lifetime on the amount and type of anti-herbivore defense. *Oecologia* **74**, 531–536.

Coley, P.D., Bryant, J.P. and Chapin, III, F.S. (1985). Resource availability and plant anti-herbivore defense. *Science* **230**, 895–899.

Colpaert, J.V., van Assche, J.A., and Luijtens, K. (1992). The growth of the extra-matrical mycelium of ectomycorrhizal fungi and the growth response of *Pinus sylvestris* L. *New Phytol.* **120**, 127–135.

Crider, F.J. (1955). Root-growth stoppage resulting from defoliation of grass. *USDA Tech. Bull.* **1102**. Washington D.C., 23 pp.

Cropper, W.P., Jr and Gholz, H.L. (1991). *In situ* needle and fine root respiration in mature slash pine (*Pinus elliottii*) trees. *Can. J. For. Res.* **21**, 1589–1595.

Culvenor, R.A., Davidson, I.A. and Simpson, R.J. (1989). Regrowth by swards of subterranean clover after defoliation. 2. Carbon exchange in shoot, root and nodule. *Ann. Bot.* **64**, 557–567.

Dawes, C.J. and Bowler, E. (1959). Light and electron microscope studies of the cell wall structure of the root hairs of *Raphanus sativus*. *Am. J. Bot.* **46**, 561–565.

de Visser, R. (1985). Efficiency of respiration and energy requirements of N assimilation in roots of *Pisium sativum*. *Physiol. Plant.* **65**, 209–218.

Dillenburg, L., Whigham, D.F., Teramura, A.H. and Forseth, I.N. (1993). Effects of below- and aboveground competition for the vines *Lonicera japonica* and *Parthenocissus quinquefolia* on the growth of the tree host *Liquidambar straciflua*. *Oecologia* **93**, 48–54.

Dodd, J.L. (1980). The role of plant stresses in development of corn stalk rots. *Plant Dis.* **64**, 533–537.

Donald, C.M. (1958). The interaction of competition for light and nutrients. *Aust. J. Agr. Res.* **9**, 421–435.

Dubach, M., Russelle, M.P. (1994). Forage legume roots and nodules and their role in nitrogen transfer. *Agron. J.* **86**, 259–266.

Ehleringer, J. (1985). Annuals and perennials of warm deserts. In: *Physiological Ecology of North American Plant Communities* (Ed. by B.F. Chabot and H.A. Mooney), pp. 162–180, Chapman and Hall, New York.

Eissenstat, D.M. (1991). On the relationship between specific root length and the rate of root proliferation: a field study using citrus rootstocks. *New Phytologist* **118**, 63–68.

Eissenstat, D.M. (1992). Costs and benefits of constructing roots of small diameter. *J. Plant Nutr.* **15**, 763–782.

Eissenstat, D.M. (1997). Trade-offs in root form and function. In: *Agricultural Ecology* (Ed. by L.E. Jackson), Academic Press, Sand Diego (in press).

Eissenstat, D.M. and Duncan, L.W. (1992). Root growth and carbohydrate responses in bearing citrus trees following partial canopy removal. *Tree Physiol.* **10**, 245–257.

Eissenstat, D.M. and Van Rees, K.C.J. (1994). The growth and function of pine roots. *Ecol. Bull.* **43**, 76–91.

Eissenstat, D.M., Graham, J.H., Syvertsen, J.P. and Drouillard, D.L. (1993). Carbon economy of sour orange in relation to mycorrhizal colonization and phosphorus status. *Annals of Botany* **71**, 1–10.

Espeleta, J.F. (1995). The fate of citrus roots in surface dry soil: effects of tree juvenility and mycorrhizal status. M.S. Thesis, University of Florida, Gainesville.

Etherington, J.R. (1987). Penetration of dry soil by roots of *Dactylis glomerata* L. clones derived from well-drained and poorly drained soils. *Funct. Ecol.* **1**, 19–23.

Fabião, A., Persson, H.A. and Steen, E. (1985). Growth dynamics of superficial roots in Portuguese plantations of *Eucalyptus globulus* Labill. studied with a mesh bag technique. *Plant Soil* **83**, 233–242.

Fahey, T.J. (1992). Mycorrhizal and forest ecosystems. *Mycorrhiza* **1**, 83–89.

Fahey, T.J. and Hughes, J.W. (1994). Fine root dynamics in a northern hardwood forest ecosystem, Hubbard Brook Experimental Forest, NH. *J. Ecol.* **82**, 533–548.

Fernandez, O.A. and M.M. Caldwell. (1975). Phenology and dynamics of root growth of three cool semi-desert shrubs under field conditions. *J. Ecol.* **63**, 703–714,

Fitter, A.H. (1985). Functional significance of root morphology and root system architecture. In: *Ecological Interactions in Soil: Plant, Microbes, and Animals* (Ed. by A.H. Fitter, D. Atkinson, D.J. Read, and M.B. Usher), pp. 87–106. Blackwell Scientific Publ., London.

Fitter, A.H. (1991). Characteristics and functions of root systems. In: *Plant Roots: The Hidden Half* (Ed. by Y. Waisel, A. Eshel and U. Kafkafi), pp. 3–25. Marcel Dekker, Inc., New York.

Fitter, A.H. (1994). Architecture and biomass allocation as components of the plastic response of root systems to soil heterogeneity. In: *Exploitation of Environmental Heterogeneity of Plants: Ecophysiological Processes Above- and Belowground* (Ed. by M.M. Caldwell and R.W. Pearcy), pp. 305–323. Academic Press, New York.

Fusseder, A. (1987). The longevity and activity of the primary root of maize. *Plant Soil* **101**, 257–265.

Gange, A.C., Brown, V.K. and Sinclair, G.S. (1994). Reduction of black vine weevil larval growth by vesicular-arbuscular mycorrhizal infection. *Entomol. Exp. Applic.* **70**, 115–119.

Gardner, W.K., Barber, D.A., and Parbery, D.G. (1983). The acquisition of phosphorus by *Lupinus albus* L. III. The probable mechanism by which phosphorus movement in the soil/root interface is enhanced. *Plant Soil* **70**, 107–124.

Garnier, E. and Laurent, G. (1994). Leaf anatomy, specific mass and water content in congeneric annual and perennial grass species. *New Phytologist* **128**, 725–736.

Gholz, H.L., Hendry, L.C. and Cropper, W.P. Jr. (1986). Organic matter dynamics of fine roots in plantations of slash pine (*Pinus elliottii*) in north Florida. *Can. J. For. Res.* **16**, 529–538.

Goins, G.D. and Russelle, M.P. (1996). Fine root demography in alfalfa (*Medicago sativa* L.). *Plant Soil* (in press).

Goodchild, D.J. and Myers, L.F. (1987). Rhizosheaths – a neglected phenomenon in Australian agriculture. *Australian J. Agric. Res.* **38**, 559–563.

Gower, S.T., Gholz, H.L., Nakane, K. and Baldwin, V.C. (1994). Production and carbon allocation patterns of pine forests. *Ecol. Bull. (Copenh.)* **43**, 115–135.

Graham, J.H. (1995). Root regeneration and tolerance of citrus rootstocks to root rot caused by *Phytophthora nicotianae*. *Phytopathology* **85**, 111–117.

Graham, J.H., Leonard, R.T. and Menge, J.A. (1981). Membrane-mediated decrease in root exudation responsible for phosphorus inhibition of vesicular-arbuscular mycorrhizae formation. *Plant Physiol.* **68**, 548–552.

Graham, J.H., Brylansky, R.H., Timmer, L.W. and Lee, R.F. (1985). Comparison of citrus tree declines with necrosis of major roots and their association with *Fusarium solani. Plant Dis.* **69**, 1055–1058.

Greaves, M.P. and Darbyshire, J.F. (1972). The ultrastructure of the mucilaginous layer on plant roots. *Soil Biol. Biochem.* **4**, 443–449.

Grime, J.P. (1977). Evidence for the existence of three primary strategies in plants and its relevance to ecological and evolutionary theory. *Amer. Naturalist* **111**, 1169–1194.

Grime, J.P. (1994). The role of plasticity in exploiting environmental heterogeneity. In: *Exploitation of Environmental Heterogeneity of Plants: Ecophysiological Processes Above- and Belowground* (Ed. by M.M. Caldwell and R.W. Pearcy), pp. 1–20. Academic Press, New York.

Grime, J.P., Crick, J.C. and Rincon, J.E. (1986). The ecological significance of plasticity. In: *Plasticity in Plants* (Ed. by D.H. Jennings and A.J. Trewavas), pp. 5–29. Cambridge University Press, New York.

Gross, K.L., Peters, A., and Pregitzer, K.S. (1993). Fine root growth and demographic responses to nutrient patches in four old-field plant species. *Oecologia* **95**, 61–64.

Harley, J.L. and Smith, S.E. (1983). *Mycorrhizal Symbiosis.* Academic Press, New York.

Harper, J.L. (1977). *Population Biology of Plants.* Academic Press, New York.

Harper, J.L. (1989). The value of a leaf. *Oecologia* **80**, 53–58.

Harper, J.L., Jones, M. and Sackville Hamilton, N.R. (1991). The evolution of roots and the problems of analyzing their behaviour. In: *Plant Root Growth: An Ecological Perspective* (Ed. by D. Atkinson), pp. 3–22. Blackwell Scientific Publications, Oxford.

Hayes, D.C. and Seastedt, T.R. (1987). Root dynamics of tallgrass prairie in wet and dry years. *Can. J. Bot.* **65**, 787–791.

Head, G.C. (1966). Estimating seasonal changes in the quantity of white unsuberized root on fruit trees. *J. Hort. Sci.* **41**, 197–206.

Head, G.C. (1969). The effects of fruiting and defoliation on seasonal trends in new root production on apple trees. *J. Hort. Sci.* **44**, 175–181.

Head, G.C. (1973). Shedding of roots. In: *Shedding of Plant Parts* (Ed. by T.T. Kozlowski), pp. 237–293. Academic Press, New York.

Hendrick, R.L. and Pregitzer, K.S. (1992). The demography of fine roots in a northern hardwood forest. *Ecology* **73**, 1094–1104.

Hendrick, R.L. and Pregitzer, K.S. (1993). Patterns of fine root mortality in two sugar maple forests. *Nature* **361**, 59–61.

Hendricks, J.J., Nadelhoffer, K.J. and Aber, J.D. (1993). Assessing the role of fine roots in carbon and nutrient cycling. *Trends. Ecol. Evol.* **8**, 174–178.

Henry, Ch.M. and Deacon, J.W. (1981). Natural (non-pathogenic) death of the cortex of wheat and barley seminal roots, as evidenced by nuclear staining with acridine orange. *Plant Soil* **60**, 255–274.

Holden, J. (1975). Use of nuclear staining to assess rates of cell death in cortices of cereal roots. *Soil Biol. Biochem.* **7**, 333–334.

Hoogenboom, G., Huck, M.G. and Peterson, C.M. (1987). Root growth rate of soybean as affected by drought stress. *Agron. J.* **79**, 607–614.

Hooker, J.E., Black, K.E., Perry, R.L. and Atkinson, D. (1995). Arbuscular mycorrhizal fungi induced alteration to root longevity of poplar. *Plant Soil* **172**, 327–329.

Huang, B. and Nobel, P.S. (1992). Hydraulic conductivity and anatomy for lateral roots of *Agave deserti* during root growth and drought-induced abscission. *Journal of Experimental Botany* **43**, 1441–1449.

Huck, M.G., Hoogenboom, G. and Peterson, C.M. (1987). Soybean root senescence under drought stress. In: *Minirhizotron Observation Tubes: Methods and Applications for Measuring Rhizosphere Dynamics* (Ed. by H.M. Taylor), pp. 109–121. ASA Special Publ. **50**, Agronomy Society of America, Madison.

Ingestad, T. and Ågren, G.I. (1991). The influence of plant nutrition on biomass allocation. *Ecolog. Appl.* **1**, 168–174.

Itoh, S. and Barber, S.A. (1993a). A numerical solution of whole plant nutrient uptake for soil-root systems with root hairs. *Plant Soil* **70**, 403–413.

Itoh, S. and Barber, S.A. (1993b). Phosphorus uptake by six plant species as related to root hairs. *Agron. J.* **75**, 457–461.

Johnson, I.R. (1990). Plant respiration in relation to growth, maintenance, ion uptake, and nitrogen assimilation. *Plant Cell Environ.* **13**, 139–328.

Jones, M.D., Durall, D.M. and Tinker, P.B. (1990). Phosphorus relationships and production of extramatrical hyphae by two types of willow ecto-mycorrhizas. *New Phytol.* **115**, 259–267.

Jupp, A.P. and Newman, E.I. (1987) Morphological and anatomical effects of severe drought on the roots of *Lolium perenne* L. *New Phytol.* **105**, 393–402.

Klepper, B., Taylor, H.M., Huck, M.G. and Fiscus, E.L. (1973). Water relations and growth of cotton in drying soils. *Agron. J.* **54**, 307–310.

Koide, R.T. and Mooney, H.A. (1987). Spatial variation in inoculum potential of vesicular-arbuscular mycorrhizal fungi caused by formation of gopher mounds. *New Phytol.* **107**, 173–182.

Kosola, K.R. and Eissenstat, D.M. (1994). The fate of citrus seedlings in dry soil. *J. Exp. Bot.* **45**, 1639–1645.

Kosola, K.R., Eissenstat, D.M., and Graham, J.H. (1995). Root demography of mature citrus trees: the influence of *Phytophthora nicotianae*. *Plant Soil* **171**, 283–288.

Krauss, U. and Deacon, J.W. (1994). Root turnover of groundnut (*Arachis hypogaea* L.) in soil tubes. *Plant Soil* **166**, 259–270.

Kroehler, C.J. and Linkins, A.E. (1988). The root surface phosphatases of *Eriophorum vaginatum*: effects of temperature, pH, substrate concentration and inorganic phosphorus. *Plant Soil* **105**, 3–10.

Kummerow, J., Krause, D. and Jow, W. (1978). Seasonal changes in fine root density in the southern California chaparral. *Oecologia* **37**, 201–212.

Kuppers, M. (1994). Canopy gaps: competitive light interception and economic space filling – a matter of whole-plant allocation. In: *Exploitation of Environmental Heterogeneity of Plants: Ecophysiological Processes Above- and Belowground* (Ed. by M.M. Caldwell and R.W. Pearcy), pp. 111–144. Academic Press, New York.

Lambers, H. (1987). Growth, respiration, exudation and symbiotic associations: the fates of carbon translocated to the roots. In *Root Development and Function* (Ed. by P.J. Gregory, J.V. Lake, and D.A. Rose), pp. 125–145. New York: Cambridge University Press.

Lamont, B.B. (1995). Mineral nutrient relations in Mediterranean regions of California, Chile and Australia. In *Ecology and Biogeography of Mediterranean Ecosystems of Chile, California, and Australia* (Ed. by M.T.K. Arroyo, P.H. Zedler and M.D. Fox), pp. 211–238. New York: Springer-Verlag.

Ledig, F.T. (1983). The influence of genotype and environment on dry matter distribution in plants. In: *Plant Research in Agroforestry. International Council for Research in Agroforestry* (Ed. by P.H. Huxley), pp. 427–454. Nairobi, Kenya.

Lerdau, M. (1992). Future discounts and resource allocation in plants. *Functional Ecology* **6**, 371–375.

Lynch, J.M. and Whipps, J.M. (1991). Substrate flow in the rhizosphere. In: *The Rhizosphere and Plant Growth* (Ed. by D.L. Keister and P.B. Cregan), pp. 15–24. Kluwer Acad. Publ., Boston.
Lyr, H. and Hoffman, G. (1967). Growth rates and growth periodicity of tree roots. *Int. Rev. Forest. Res.* **2**, 181–206.
Marschner, H. (1986). *Mineral Nutrition of Higher Plants*. Academic Press, New York.
Marshall, J.D. (1986). Drought and shade interact to cause fine-root mortality in Douglas-fir seedlings. *Plant Soil* **91**, 51–60.
McCully, M.E. and Canny, M.J. (1985). Localisation of translocated ^{14}C in roots and root exudates of field-grown maize. *Physiol. Plant.* **65**, 380–392.
McCully, M.E. and Canny, M.J. (1988). Pathways and processes of water and nutrient movement in roots. *Plant Soil* **111**, 159–170.
McGonigle, T.P. and Fitter, A.H. (1988). Ecological consequences of arthropod grazing on VA mycorrhizal fungi. *Proc. Roy. Soc. Edin.* **94B**, 25–32.
McKay, H. and Coutts, M.P. (1989). Limitations placed on forestry production by the root system. *Aspect. Appl. Biol.* **22**, 245–254.
McMichael, B.L. and Taylor, H.M. (1987). Applications and limitations of rhizotrons and minirhizotrons. In: *Minirhizotron Observation Tubes: Methods and Applications for Measuring Rhizosphere Dynamics* (Ed. by H.M. Taylor), pp. 1–14. ASA Special Publ. **50**.
Meyer, W.S., Tan, C.S., Barrs, H.D. and Smith, R.C.G. (1990). Root growth and water uptake by wheat during drying of undisturbed and repacked soil in drainage lysimeters. *Aust. J. Agric. Res.* **41**, 253–265.
Milchunas, D.G. and Lauenroth, W.K. (1992). Carbon dynamics and estimates of primary production by harvest, ^{14}C dilution, and ^{14}C turnover. *Ecology* **73**, 593–607.
Milchunas, D.G., Lauenroth, W.K., Singh, J.S., Dole, C.V. and Hunt, H.W. (1985). Root turnover and production by ^{14}C dilution: implications of carbon partitioning in plants. *Plant Soil* **88**, 353–365.
Miller, R.M. and Jastrow, J.D. (1991). Extraradical hyphal development of vesicular-arbuscular mycorrhizal fungi in a chronosequence of prairie restorations. In: *Mycorrhizas in Ecosystems* (Ed. by D.J. Read, D.H. Lewis, A.H. Fitter and I.J. Alexander), pp. 171–176. CAB International, Wallingford, UK.
Miller, R.M., Reinhardt, D.R. and Jastrow, J.D. (1995). External hyphal production of vesicular-arbuscular mycorrhizal fungi in pasture and tallgrass prairie communities. *Oecologia* **103**, 17–23.
Millington, W.F. and Chaney, W.R. (1973). Shedding of shoots and branches. In: *Shedding of Plant Parts* (Ed. by T.T. Kozlowski), pp. 149–204. Academic Press, New York.
Molyneux, D.E. and Davies, W.J. (1983). Rooting pattern and water relations of three pasture grasses growing in drying soil. *Oecologia* **58**, 220–224.
Monk, C.D. (1966). An ecological significance of evergreeness. *Ecology* **47**, 504–505.
Nadelhoffer, K.J. and Raich, J.W. (1992). Fine root production estimates and belowground carbon allocation in forest ecosystems. *Ecology* **73**, 1139–1147.
Nadelhoffer, K.J., Aber, J.D. and Melilo, J.M. (1985). Fine roots, net primary production and nitrogen availability: a new hypothesis. *Ecology* **66**, 1377–1390.
Nambiar, E.K.S. (1987). Do nutrients retranslocate from fine roots? *Can. J. For. Res.* **17**, 913–918.
Nepstad, D.C., de Carvalho, C.R., Davidson, E.A., Jipp, P.H., Lefebvre, P.A., Negreiros, G.H., da Silva, E.D., Stone, T.A., Trumbore, S.E., and Vieira, S. (1994). The role of deep roots in the hydrological and carbon cycles of Amazonian forests and pastures. *Nature* **372**, 666–669.

Newman, E.I. (1974). Root and soil water relations. In: *The Plant Root and its Environment* (Ed. by E.W. Carson). University Press of Virginia, Charlottesville.

Newman, E.I. (1985). The rhizosphere: carbon sources and microbial populations. In: *Ecological Interactions in Soil* (Ed. by A.H. Fittter, D. Atkinson, D.J. Read and M. Busher), pp. 107–121. Blackwell Scientific Publications, Oxford.

Newman, E.I. (1988). Mycorrhizal links between plants: their functioning and ecological significance. *Adv. Ecol. Res.* **18**, 243–270.

Newman, E.I. and Reddell, P. (1988). The distribution of mycorrhizas among families of vascular plants. *New Phytol.* **106**, 745–751.

Newsham, K.K., Fitter, A.H., and Watkinson, A.R. (1995). Arbuscular mycorrhiza protect an annual grass from root pathogenic fungi in the field. *J. Ecol.* **83**, 991–1000.

Nobel, P.S., Alm, D.M. and Cavelier, J. (1992). Growth respiration, maintenance respiration and structural-carbon costs for roots of three desert succulents. *Funct. Ecol.* **6**, 79–85.

North, G.B., Huang, B., and Nobel, P.S. (1993). Changes in the structure and hydraulic conductivity of desert succulents as soil water status varies. *Botan. Acta* **106**, 126–135.

Northup, R.R., Yu, Z., Dahlgren, R.A. and Vogt, K.A. (1995). Polyphenol control of nitrogen release from pine litter. *Nature* **377**, 227–229.

Nye, P.H. (1992). Towards the quantitative control of crop production and quality. III. Some recent developments in research into the root–soil interface. *J. Plant Nutr.* **15**, 1175–1192.

Nye, P.H. and Tinker, P.B. (1977). *Solute Movement in the Soil–Root System.* Blackwell Scientific Publications, Oxford

Palta, J.A. and Nobel, P.S. (1989). Influences of water status, temperature, and root age on daily patterns of root respiration for two cactus species. *Ann. Bot.* **63**, 651–662.

Pearson, J.N. and Jakobsen, I. (1993a). Symbiotic exchange of carbon and phosphorus between cucumber and three arbuscular mycorrhizal fungi. *New Phytol.* **124**, 481–488.

Pearson, J.N. and Jakobsen, I. (1993b). The relative contribution of hyphae and roots to phosphorus uptake by arbuscular mycorrhizal plants, measured by dual labelling with ^{32}P and ^{33}P. *New Phytol.* **124**, 489–494.

Peng, S., Eissenstat, D.M., Graham, J.H., Williams, K., and Hodges, N.C. (1993). Growth depression of mycorrhizal citrus at high phosphorus supply: analysis of carbon costs. *Plant Physiol.* **101**, 1063–1071.

Poorter, H. (1994). Construction costs and payback time of biomass: a whole plant perspective. In: *A Whole Plant Perspective on Carbon-Nitrogen Interactions* (Ed. by J. Roy and E. Garnier), pp. 111–127.

Poorter, H., van der Werf, A., Atkin, O.K., and Lambers, H. (1991). Respiratory energy requirements of roots vary with the potential growth rate of a plant species. *Physiol. Planta* **83**, 469–475.

Portas, C.A.M. and Taylor, H.M. (1976). Growth and survival of young plant roots in dry soil. *Soil Sci.* **121**, 170–175.

Pregitzer, K.S., Hendrick, R.L. and Fogel, R. (1993). The demography of fine roots in response to patches of water and nitrogen. *New Phytol.* **125**, 575–580.

Pregitzer, K.S., Zak, D.R., Curtis, P.S., Kubiske, M.E., Teeri, J.A. and Vogel, C.S. (1995). Atmospheric CO_2, soil nitrogen and fine root turnover. *New Phytol.* **129**, 579–585.

Queen, W.H. (1967). Radial movement of water and ³²P through suberized and unsuberized roots of grape. Ph.D. Dissertation. Duke University, Durham, North Carolina.

Read, D.J. (1993). Mycorrhiza in plant communities. In: *Advances in Plant Pathology: Mycorrhiza Synthesis.* (Ed. by I.C. Tommerup), pp. 1–31. Academic Press, New York.

Reich, P.B., Uhl, C. Walters, M.B. and Ellsworth, D.S. (1991). Leaf lifespan as a determinant of leaf structure and function among 23 amazonian tree species. *Oecologia* **86**, 16–24.

Reich, P.B., Walters, M.B. and Ellsworth, D.S. (1992). Leaf life-span in relation to leaf, plant, and stand characteristics among diverse ecosystems. *Ecolog. Monogr.* **62**, 365–392.

Reid, J.B., Sorensen, I. and Petrie, R.A. (1993). Root demography of kiwifruit (*Actinidia deliciosa*). *Plant Cell Environ.* **16**, 949–957.

Richards, J.H. (1984). Root growth response to defoliation in two *Agropyron* bunchgrasses: field observations with an improved root periscope. *Oecologia* **64**, 21–25.

Ryan, M.G. (1991). Effects of climate change on plant respiration. *Ecolog. Appl.* **1**, 157–167.

Rygiewicz, P.T. and Andersen, C.P. (1994). Mycorrhizae alter quality and quantity of carbon allocated below ground. *Nature* **369**, 58–60.

Ryser, P. and Lambers, H. (1995). Root and leaf attributes accounting for the performance of fast- and slow-growing grasses at different nutrient supply. *Plant Soil* **170**, 251–265.

Ryser, P. (1996). The importance of tissue density for growth and life span of leaves and roots: a comparison of five ecologically contrasting grasses. *Functional Ecol.* (in press).

Schläpfer, B and Ryser, P. (1996). Leaf and root turnover of three ecologically contrasting grass species in relation to their performance along a productivity gradient. *Oikos* **75**, 398–406.

Schoettle, A.W. and Fahey, T.J. (1994). Foliage and fine root longevity in pines. *Ecolog. Bull.* **43**, 136–153.

Schwab, S.M., Menge, J.A. and Tinker, P.B. (1991). Regulation of nutrient transfer between host and fungus in vesicular-arbuscular mycorrhizas. *New Phytol.* **117**, 387–398.

Self, G.K., Brown, T.K., Graves, J. and Fitter, A.H. (1995). Longevity and respiration rates of roots of upland grassland species in relation to temperature and atmospheric CO_2 concentration (Abstract). *J. Exp. Bot.* **46** (Suppl.), 25.

Setälä, H. (1995). Growth of birch and pine seedlings in relation to grazing by soil fauna on ectomycorrhizal fungi. *Ecology* **76**, 844–1851.

Shaver, G.R. and Billings, W.D. (1975). Root production and root turnover in a wet tundra ecosystem, Barrow, Alaska. *Ecology* **56**, 401–409.

Simard, S.W., Perry, D.A., Jones, M.D., and Durall, D.M. (1995). Shading influences net carbon transfer among ectomycorrhizal tree seedlings in the field. *Dynamics of Physiological Processes in Roots*, 8–11 Oct. 1995, Cornell University, NY (Abstract).

Smith, P.F. (1976). Collapse of 'Murcott' tangerine trees. *J. Amer. Soc. Hort. Sci.* **101**, 23–25.

Smith, S.E. and Gianinazzi-Pearson, V. (1988). Physiological interactions between symbionts in vesicular arbuscular mycorrhizal plants. *Ann. Rev. Plant Physiol. Molec. Biol.* **39**, 221–244.

Smith, S.E., Dickson, S., Morris, C. and Smith, F.A. (1994). Transfer of phosphate from fungus to plant in VA mycorrhizas: calculation of the area of symbiotic interface and of fluxes of P from two different fungi to *Allium porrum* L. *New Phytol.* **127**, 93–99.

Snapp, S.S. and Lynch, J. (1996). Phosphorus distribution patterns in mature bean plants: root retention and leaf remobilization to pods. *Crop Sci.* **36**, 929–935.

Spaeth, S.C. and Cortes, P.H. (1995). Root cortex death and subsequent initiation and growth of lateral roots from bare steles of chickpeas. *Can. J. Bot.* **73**, 253–261.

Sprent, J.I. (1975). Adherence of sand particles to soybean roots under water stress. *New Phytol.* **74**, 461–463.

Stanton, N.L. (1988). The underground in grasslands. *Annu. Rev. Ecol. System.* **19**, 573–589.

Stasovski, E. and Peterson, C.A. (1991). The effects of drought and subsequent rehydration on the structure and vitality of *Zea mays* seedling roots. *Can. J. Bot.* **69**, 1170–1178.

Sylvia, D.M. (1988). Activity of external hyphae of vesicular-arbuscular mycorrhizal fungi. *Soil Biol. Biochem.* **20**, 39–43.

Sylvia, D.M. (1990). Distribution, structure and function of external hyphae of vesicular-arbuscular mycorrhizal fungi. In: *Rhizosphere Dynamics* (Ed. by J.E. Bo Jr, and L.C. Hammond), pp 144–167. Boulder, Colorado: Westview Press.

Szaniawski, R.K. and Kielkiewicz, M. (1982). Maintenance and growth respiration in shoots and roots of sunflower plants grown at different root temperatures. *Physiol. Planta* **54**, 500–504.

Taylor, H.M. and Klepper, B. (1973). Rooting density and water extraction patterns for corn (*Zea mays* L.). *Agron. J.* **65**, 965–968.

Toth, R. and Miller, R.M. (1984). Dynamics of arbuscule development and degeneration in a *Zea mays* mycorrhiza. *Am. J. Bot.* **71**, 449–460.

Toth, R., Miller, R.M., Jarstfer, A.G., Alexander, T. and Bennett, E.L. (1991). The calculation of intraradical fungal biomass from percent colonization in vesicular-arbuscular mycorrhizae. *Mycologia* **83**, 553–558.

van der Werf, A., Kooijman, A., Welschen, R. and Lambers, H. (1988). Respiratory energy costs for the maintenance of biomass for growth and for ion uptake in roots of *Carex diandra* and *Carex acutiformis*. *Physiol. Planta* **72**, 483–491.

van Rees, K.C.J. and Comerford, N.B. (1990). The role of woody roots of slash pine seedlings in water and potassium absorption. *Can. J. For. Res.* **20**, 1183–1191.

van Rees, K.C.J., Comerford, N.B. and Rao, P.S.C. (1990). Defining soil buffer power: implications for ion diffusion and nutrient uptake modeling. *Soil Sci. Soc. Am. J.* **54**, 1505–1507.

Veech, J.A. (1982). Phytoalexins and their role in the resistance of plants to nematodes. *J. Nematol.* **14**, 2–9.

Veen, B.W. (1981). Relation between root respiration and root activity. *Plant Soil* **63**, 73–76.

Vitousek, P. (1982). Nutrient cycling and nutrient use efficiency. *Amer. Naturalist* **119**, 553–572.

Vogt, K.A., Grier, C.C. and Vogt, D.J. (1986). Production, turnover and nutritional dynamics of above- and belowground detritus of world forests. *Adv. Ecol. Res.* **15**, 303–307.

Wang, Z., Burch, W.H., Mou, P., Jones, R.H., and Mitchell, R.J. (1995). Accuracy of visible and ultraviolet light for estimating live root proportions with minirhizotrons. *Ecology* **76**, 2330–2334.

Watkins, N.K., Fitter, A.H., Graves, J.D., and Robinson, D. (1996). Quantification using stable carbon isotopes of carbon transfer between C3 and C4 plants linked by a common mycorrhizal network. *Soil Biol. Biochem.* (in press).

Weste, G. (1986). Vegetation changes associated with invasion by *Phytophthora cinnamomi* of defined plots in the Brisbane Ranges, Victoria, 1975–1985. *Aust. J. Bot.* **34**, 633–648.

Whaley, E.L. (1995). *Uptake of Phosphorus by Citrus Roots in Dry Surface Soil.* M.S. Thesis, University of Florida, Gainesville, Florida.

Williams, M. and Yanai, R. (1996). Multi-dimensional sensitivity analysis and ecological implications of a nutrient uptake model. *Plant Soil* (in press).

Wilson, J.B. (1988). Shoot competition and root competition. *J. Appl. Ecol.* **25**, 279–296.

Xu, J.G. and Juma, N.G. (1994). Relations of shoot C, root C and root length with root-released C of two barley cultivars and the decomposition of root-released C in soil. *Can. J. Soil Sci.* **74**, 17–22.

Yanai, R.D. (1994). A steady-state model of nutrient uptake improved to account for newly-grown roots. *Soil Sci. Soc. Am. J.* **58**, 1562–1571.

Yanai, R.D., Fahey, T.J. and Miller, S.L. (1995). Efficiency of nutrient acquisition by fine roots and mycorrhizae. In: *Resource Physiology of Conifers* (Ed. by W.K. Smith and T.M. Hinckley), pp. 75–103. Academic Press, New York.

Appendix

Uptake equations

Solute uptake at the root surface depends on the concentration of solute in solution at the root surface. This concentration will differ from the average concentration in solution, due to gradients created by solute uptake by the root and movement of solute by diffusion and solution flow. The concentration profile around the root can be described by assuming that a steady state is reached. The concentration at the root surface can then be described as a function of the average concentration in solution.

$$C_0 = C_{av}v_0\left[\alpha + (v_0 - \alpha)\left(\frac{2}{2-\gamma}\right)\frac{(r_x/r_0)^{2-\gamma} - 1}{(r_x/r_0)^2 - 1}\right]^{-1} \tag{3}$$

where C_0 = concentration of substance at the root surface (mol cm^{-3}). C_{av} = the average solution concentration (mol cm^{-3}), α = root absorbing power (cm s^{-1}). v_0 = inward radial velocity of water at the root surface (cm), r_x = average radial distance to the next root's zone of influence (cm), r_0 = radius of the root (cm), and $\gamma = r_0v_0/(Db)$ (dimensionless), where D = diffusion coefficient (cm^2 s^{-1}) and b = buffer capacity of the soil, or the ratio between exchangeable and dissolved nutrient (dimensionless).

This expression for C_0 uses a linear representation of nutrient uptake kinetics, which is appropriate only at low concentrations. To modify the model to allow carrier saturation at high concentrations, we substitute Michaelis–Menten kinetics:

$$\alpha = V_{max} / (K_m - C_0) \tag{4}$$

where V_{max} = maximum rate of uptake (mol cm^{-2} s^{-1}), and K_m = concentration at the root surface at half of V_{max} (mol cm^{-3}). In addition, a C_{min} can be specified, such that uptake does not occur when $C_0 < C_{max}$.

When solute concentration at the root surface is obtained, solute uptake can be calculated from the root surface area, $2\pi r_0 L$, and the uptake kinetics.

$$UPTAKE = \lambda 2\pi r_0 \alpha C_0 \Delta t \tag{5}$$

where λ = specific root length (cm (g root)$^{-1}$) and Δt = the model timestep (s). Equations (3) and (5) are presented by Baldwin et al. (1973) and Nye and Tinker (1977) and were derived by Yanai (1994), along with equation (4). Calculations of b and D can be made sensitive to the volumetric soil water content, θ (van Rees et al., 1990):

$$b = \theta + \rho K_d \tag{6}$$

where K_d is the solid-liquid partitioning coefficient.

$$D = D_l \theta f / b \tag{7}$$

where D_l is the diffusion coefficient in water (cm^2 s^{-1}) and f is the impedance factor, which describes soil tortuosity.

Nocturnal Insect Migration: Effects of Local Winds

P.J.A. BURT AND D.E. PEDGLEY

I. SUMMARY

Many insect species undertake migratory flights at night, sometimes over tens or even hundreds of kilometres. Such flights are usually dominated by the wind because the insects climb to heights where the wind speed exceeds their own flying speed through the air. Consequently, the distance moved is essentially the product of wind speed and duration of flight.

The wind systems encountered by night-flying insects occur over a range of sizes up to thousands of kilometres and can be modified by topography. Larger, or broad-scale, systems may be inferred from the well-known synoptic weather maps, but local effects (with scales of tens of kilometres or less) may not be revealed by such charts. Because there is increasing evidence of

ADVANCES IN ECOLOGICAL RESEARCH VOL. 27
ISBN 0–12–013927–8

the effects of these smaller systems on both the movement and concentration of flying insects, an extensive search of the entomological and meteorological literature, particularly from the last 10 to 15 years, has been made to advance our understanding of such effects. Discussion is confined to the lowest kilometre or two of the atmosphere, where almost all insect migration takes place.

Several types of wind system are considered: low-level jet streams, coastal winds (sea and land breezes), rainstorm outflows, down-slope and down-valley winds, and a variety of local winds caused by barrier effects of mountains – these include rotors, spillage, jets and wakes. For some local wind systems no studies have been made of their likely effects on nocturnal insect migration. For insect pest species that have been well studied, the influences of these local wind systems on movement and concentration of flying adults have become clearer, but for most species, whether pests or not, little or nothing is known. It is likely, however, that comparable influences do occur and they should be taken into account in field studies of night flight intended to improve our understanding of the population ecology of migrant species.

II. INTRODUCTION

The principal aim of this paper is to review the evidence of the effects of local wind systems on insect migration at night. The results throw light on the occurrence of clouds of flying adults and on the clumped distribution of populations of migrants on the ground (including dense pest infestations). Much of the evidence comes from field studies of a few pest species, but it is likely that many other species are similarly affected.

Since about the middle of the twentieth century the idea of long-distance migration has become widely accepted as an integral part of the life-history strategy of insects (Johnson, 1969). By that time long-distance insect migration was found to be dominated by the behaviour of the atmosphere (for reviews, see Pedgley, 1982, 1983; Drake and Farrow, 1988; McManus, 1988; Drake and Gatehouse, 1995). A migrating insect is generally displaced downwind because it usually climbs above its flight boundary layer, i.e. to where the wind speed exceeds the insect's own speed through the air (Taylor, 1974). The distance travelled is consequently determined largely by wind speed and duration of flight. Many species fly at night and there is increasing evidence that individuals can travel hundreds, even thousands, of kilometres in a few nights.

Long-distance migrations are often within single, large, wind systems that are easily recognizable on weather maps. Wind systems, however, also occur on smaller scales, and not only can directions and speeds differ from those derived from such maps, but local areas of concentration of insects can occur whilst in flight (Drake and Farrow, 1989; Pedgley, 1990). This may lead to unexpectedly dense infestations which, for pest species, may be difficult to manage (e.g. African armyworm; Rose et al., 1995). The effects of such small-scale wind systems on migration have been less studied but they are becoming clearer.

Although there have been relatively few studies of the effects of local wind systems on insect flight at night, there have been many studies of the wind systems themselves because of their importance in, for example, the operation of aircraft and the spread of air pollution. Nevertheless, it is clear from this review that there are gaps in the descriptions, let alone the understanding, of some of the local wind systems which affect nocturnal insect flight.

This review draws attention to our current understanding of the structure of the lowest kilometre or two of the atmosphere, in order to encourage entomologists to consider the ways in which the atmosphere might affect insect flight at night. A separate bibliography of sources in the meteorological literature is appended as a guide to studies. Within each section, a simple outline of the relevant meteorology is given, although detailed meteorological discussion is avoided, and the known effects of weather conditions on insect movement is discussed.

III. VERTICAL PROFILES

The vertical distribution of flying insects is affected by the vertical distribution of wind and temperature. Normally, atmospheric temperature decreases with height above the ground. In open, flat, country, on days with little or no wind and cloud, however, a temperature inversion may develop towards dusk once the ground has become cooler than the air in contact with it. The top of the inversion builds upwards during the night by radiational transfer of heat from the air to the ground and may reach several hundred metres by dawn. Temperatures below the inversion level decrease during the night, and near the ground the air may become too cool for insect flight. After sunrise, as the ground begins to warm and atmospheric convection starts to build upwards, the inversion is progressively eroded until it is destroyed, usually by late morning.

Layers of insects centred up to some hundreds of metres above the ground, and even up to 1–2 km, have been found in radar studies of a number of species (Reid et al., 1979; Drake, 1984b, 1985b; Drake and Farrow, 1985; Mueller and Larkin, 1985; Drake and Rochester, 1994). They tend to develop during the night, evolving from the earlier upward decline in aerial density, apparently in association with changes in the vertical profiles of temperature and wind. For example, layering is common up to 500 m in Oriental Armyworm moths *Mythimna separata* (Lepidoptera: Noctuidae) migrating in China (Drake and Farrow, 1985). In Canada, layering of spruce budworm moths *Choristoneura fumiferana* (Lepidoptera: Tortricidae), is usually centred at 100–200 m

(Schaefer, 1976) and is particularly strong on nights with a temperature inversion near the ground (Greenbank *et al.*, 1980). Similar layering has been reported in a vast cloud of fall armyworm moths *Spodoptera frugiperda*, and corn earworm moths *Helicoverpa zea* (both noctuids) streaming downwind from a source in irrigated maize in the Lower Rio Grande Valley, Texas, USA (Wolf *et al.*, 1990). The leading edge of this cloud moved more than 400 km during the night. On another occasion, an airborne radar has shown *H. zea* layered at the top of a 800 m deep storm outflow (Westbrook *et al.*, 1987). An earlier study (Callahan *et al.*, 1972) used a vertical profile of light traps to reveal layers of *H. zea* in the stably stratified nocturnal boundary layer (to a height of 320 m). In Australia, Drake (1984b, 1985b) found layering of insects, probably moths, at the top of the temperature inversion layer, with the lower boundary sometimes very sharply defined, suggesting that the insects are sensitive to their surroundings. Sometimes, moths of the African armyworm *Spodoptera exempta* (Lepidoptera: Noctuidae) become concentrated into layers aloft: some layers centred at 300–400 m have been detected in Kenya (D.R. Reynolds, unpublished data). Radar observations in Kenya (Riley *et al.*, 1981, 1983) have shown moths taking-off after dusk and climbing at about 0.4–0.5 m s^{-1} to heights up to 1000 m, judged by the inclination of the flight ceiling and the maximum altitude reached by moths coming from the furthest part of an outbreak site (about 5 km upwind) as they displaced at 9 m s^{-1}. There was evidence of levelling out when they passed over a second radar 14 km downwind, after flying for 20–30 min.

Radar studies of the Senegalese grasshopper *Oedaleus senegalensis* (Orthoptera: Acrididae) in Mali (Riley and Reynolds, 1979, 1983) showed behaviour similar to that of the moths. They climbed at about 0.5 m s^{-1} and by midnight had often layered at a height of a few hundred metres – at or just above the level of highest temperature. Again in Mali, Reynolds and Riley (1988) found strong layering of mixed grasshopper species towards midnight at about 300 m. The Australian plague locust *Chortoicetes terminifera* (Orthoptera: Acrididae) has been observed to behave similarly (Drake and Farrow, 1983). The rate of climb was again about 0.5 m s^{-1} and, by mid-evening, layers were centred near 200 m as the inversion developed. Reid *et al.* (1979) found that some layers reached 1.8 km. A radar study of the rice brown planthopper *Nilaparvata lugens* (Homoptera: Delphacidae) at Nanjing, China, showed that there was often a dense layer between 400 and 1000 m, with a well-defined ceiling at the level where the temperature was 16°C (Riley *et al.*, 1991). Insect layers generally become less dense during the night, presumably as short-distance fliers land, but aphids have been found in layers even at dawn by aerial sampling from a helicopter (Isard *et al.*, 1990).

It seems from these records that layers often tend to be centred near the level of warmest air at the top of the night-time temperature inversion. In other cases, the insects probably occur at heights at which temperatures are the lowest for flapping flight (the ceiling layer).

On some nights with a temperature inversion, the level of warmest air is sometimes close to the level of strongest wind – the so-called nocturnal low-level jet. The jet forms when the drag (slowing-down effect) of daytime convective mixing is removed as the night-time inversion develops. The jet can flow faster because it is not being slowed by the lighter winds or the ground below. Jet maxima have been reported between 200 and 1500 m and wind speeds of 5–10 m s^{-1} are not unusual, while, in contrast, wind speeds near the ground may weaken to only 1–2 m s^{-1}. Sometimes the temperature inversion inhibits vertical mixing so strongly that the frictional drag of the ground is largely confined to the lowest few tens of metres of the atmosphere, and the weak surface flow can be unrelated to the wind aloft.

Wind shear in the vertical (the rapid change of wind speed with height) leads to incorrect estimation of insect displacement speeds and directions based only on surface measurements. Wind speeds (and directions) at insect flight height are best estimated from weather maps, using either direct measurements at that height or the known relationship between patterns of wind and atmospheric pressure. This relationship, however, breaks down when the wind at jet altitudes accelerates from about dusk onwards, reaching a maximum during the night (later in the night at lower latitudes). As the jet strengthens, the wind shear intensifies, sometimes leading to an instability and the onset of turbulent mixing which can build downwards to the ground, where it is recognized by the occurrence of spells of gusty wind in an otherwise calm night.

In the United States, the nocturnal low-level jet has been associated with the transport of mostly unidentified insects, but including variegated cutworm moths *Peridroma saucia* (Hubner) (Lepidoptera: Noctuidae) (Browning and Atlas, 1966; Beerwinkle *et al.*, 1994). An Australian case in which it was inferred that the migrating insects were moths has also been reported (Drake, 1985b).

It may be difficult for an observer on the ground to identify the cause of a low-level jet in open country, although weather charts or atmospheric soundings should indicate the presence of such a front, or the temperature profile described above, from which the origin of the low-level jet could be deduced. Pre-frontal low-level jets have been shown to transport insects in the USA, often over several thousands of kilometres. Such transport is known for various noctuid moths, such as the black cutworm *Agrotis ipsilon* (Showers *et al.*, 1989, 1993; Smelser *et al.*, 1991), *Peridroma saucia* (Buntin *et al.*, 1990), the green cloverworm *Plathypena scabra* (Wolf *et al.*, 1987), *Spodoptera frugiperda* (Johnson, 1987) and the American armyworm *Pseudaletia unipuncta* (McNeill, 1987).

IV. COASTAL WINDS

In warm, sunny weather along coastlines, on-shore winds (sea breezes), tend to develop during the day. Often, such winds persist into the evening before dying out. Also, if the land becomes cooler than the sea during the night, the wind direction reverses and a land breeze blows off-shore. The sea breeze may augment or be in opposition to the broad-scale wind flow. Strong onshore breezes occur if the two combine.

Sea and land breezes are forms of density current, in which cooler air undercuts warmer. Sea breezes have the following properties (Figure 1a).

- They start during the morning and reach greatest speed in the afternoon (5–10 m s^{-1}), but can persist well into the evening before fading away.
- Their depth is typically 0.5–1.0 km.
- The leading edge, the sea breeze front, is sharply defined (Figure 1a, F) and moves progressively inland, often tens of kilometres by late evening and sometimes 100 km or more.
- The passage of a sea breeze front is marked by a sudden onset of a wind from the sea and by falling temperatures.
- Just behind the front there may be a rotor – a large closed eddy with a horizontal axis (Figure 1a, R) with an updraught zone about 1 km wide and greatest upcurrents 1–3 m s^{-1}.
- Just ahead of the front there is another narrow zone of upcurrents.
- The seaward extent of the onshore wind increases during the day to some tens of kilometres but can exceed 100 km.
- A broad-scale wind blowing from the land intensifies but slows down the sea breeze front, or even prevents it from moving inland, whereas a broad-scale wind blowing from the sea enhances inland spread, although the front may be more diffuse.
- They are most strongly developed at low latitudes, and in warmer months at middle latitudes.

Sea breeze fronts are known to concentrate flying insects and there is an observed tendency for them to become concentrated in a band about 1 km broad along the front (Figure 1a, C). Detailed studies of night flight by the spruce budworm Moth *Choristoneura fumiferana* in New Brunswick, using an aircraft equipped with a downward-looking radar, showed the vertical structure of the radar echoes from the insects associated with a front (Schaefer, 1979; Greenbank *et al.*, 1980; Neumann and Mukammal, 1981). Another sea breeze front there had a similar echo (Dickison, 1990). The sudden appearance of aphids at coastal holiday areas in England has been attributed to concentration on sea breeze fronts (Cochrane, 1980), and radar studies of insect-trawling birds have been used to infer the concentration of insects there (Simpson, 1964, 1967; Simpson *et al.*, 1977; Puhakka *et al.*,

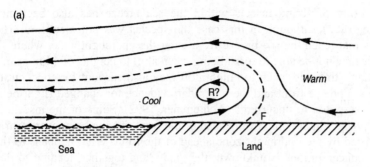

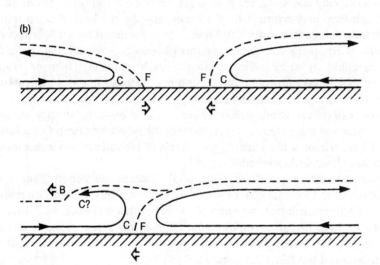

Fig. 1. (a) Airflow (shown by arrows) and sea breeze structure (F denotes sea breeze front, R? denotes position of rotor if present). (b) Colliding sea breezes, leading to generation of bore (B), showing possible insect concentration zone (C).

1986). Because the sea breeze is deeper than the height to which most insects climb, those that take off around dusk, even if they are in flight before the sea breeze front arrives, are likely to be taken inland 10 km or more, at speeds around 5–10 m s^{-1}. Repeated occurrence of forest defoliation by larvae of the gypsy moth *Lymantria dispar* (Lepidoptera: Lymantriidae) in New Jersey 8–10 km inland in a band 2–3 km wide, is consistent with accumulation in, and deposition from, the sea breeze front of first-instar larvae drifting on silk threads (Mason and McManus, 1980).

Layering of flying insects within the sea breeze has also been noted (Drake, 1982), although in this case the species was not identified. Those insects that reach the sea breeze front enter the updraught zone, where they may or may not be able to resist being taken aloft to heights of 1 km or more. Resistance to being taken too high aloft may also account for concentration in the sea breeze front and there is radar evidence for changed flight behaviour in response to upcurrents (Achtemeier, 1992). Some of the insects in a frontal band might come from ahead of the front, having been overtaken and then lifted by the updraught zone ahead of the front. What happens to these insect concentrations is unknown; they may land together, leading to dense populations on the ground, or they may land progressively as the sea breeze decays, and so be spread more widely. In either case, ground densities are likely to be greater than in the absence of the sea breeze.

Insects flying above the sea breeze can be taken out to sea in an off-shore wind, but they may return if they descend into the sea breeze. Such movements have been reported for day flight in both the desert locust *Schistocerca gregaria* (Orthoptera: Acrididae) swarms (Rainey and Waloff, 1948) and the cowpea aphid *Aphis craccivora* (Homoptera: Aphididae) (Johnson, 1957) and, although undocumented, there is no reason to assume that such movements could not persist into the evening.

Lakes can induce winds similar to sea and land breezes, but they are usually weaker and less extensive, even those of the North American Great Lakes and of Lake Victoria, in Africa. Again, effects of lake breezes on insect movements have been little studied.

A sea breeze may collide with a lake breeze, as occurs over Lake Okeechobee, in Florida, USA, or it may replace it entirely. Lake Kinneret (the Sea of Galilee), in Israel, provides an example that has been well studied, where the lake breeze is replaced in summer by a sea breeze from the Mediterranean. This sea breeze continues into the evening and enhances the development of the lake's land breeze.

Land breezes have been studied much less than sea breezes. They are generally weaker, shallower and less extensive, but the leading edge (the land breeze front) can be sharply defined. There has been little attempt to study the effects of land breezes on flying insects, although one investigation (Davey, 1959) suggested that convergence into the middle Niger flood plains of Mali by the migratory locust *Locusta migratoria* (Orthoptera: Acrididae) is related to the development of land breezes in the evening over the seasonal inland delta.

V. BORES AND SOLITARY WAVES

Sea breezes from opposite coasts of an island or peninsula can collide, as can land breeze fronts from opposite shores over a lake. Such collisions appear able to induce a bore in the upper surface of the layer of cool, marine air that has spread inland (Figure 1b, B). An atmospheric bore is a sudden and sustained rise in this surface, like the bore in a tidal estuary. The most-studied atmospheric bore is the 'Morning Glory' of northern Australia.

Atmospheric bores can travel hundreds of kilometres through the night. Sudden deepening of the cool layer may be indicated by a striking line, or a series of parallel lines, of cloud and a sudden strengthening of the wind: air temperatures near the ground may rise if a nocturnal inversion ahead of the bore is modified sufficiently by turbulence generated by the bore's passage. It appears that bores can also be induced in the nocturnal boundary layer by intrusive density currents other than sea breezes, such as drainage winds and rainstorm outflows.

Whether bores can concentrate flying insects (Figure 1b, C) is uncertain; they might be the cause of some night-time radar band echoes in West Africa (Schaefer, 1969, 1976; Riley and Reynolds, 1983). Instead of a bore, a solitary wave may be induced in the top of the cool layer. This waveform has a single crest, and such waves may be the cause of some radar band echoes from insect concentrations seen in Australia (Drake, 1984a, 1985a). A solitary wave was inferred to have perturbed, but not disrupted, layers of *Nilaparvata lugens*, as observed by radar in China (Riley *et al.*, 1991).

VI. RAINSTORM OUTFLOWS

Another form of density current is the rainstorm outflow: it has a similar structure to a sea breeze and its effects on flying insects are similar.

A storm outflow is caused by a cool downdraught in a rain shower striking the ground and spreading out as a density current undercutting the surrounding warmer air (Figure 2a). Cooling, by the partial evaporation of falling raindrops below the cloud base, produces a shaft of denser, sinking air, the downward motion being increased by frictional drag on the air by the falling drops. Downdraught speeds are typically a few metres a second but can exceed 10 m s^{-1}. Downdraughts in individual showers have diameters up to about 10 km but they can be larger. Some are weak and dissipate before reaching the ground, others reach the ground even though all raindrops have evaporated aloft.

When a downdraught strikes the ground it spreads radially. Small but powerful downdraughts (called microbursts) spread explosively, causing damage to vegetation, crops and buildings, and leading to loss of life in aircraft accidents. The more usual storm outflows are less violent and surface wind speeds are typically 10–20 m s^{-1} although they can reach 30–40 m s^{-1}. Speeds tend to decrease away from the storm origin. Depths are typically about 1 km although they can exceed 2 km. A typical gust front, the leading edge of the storm outflow (Figure 2a, G) has a speed a few m s^{-1} less than the speed of the outflow, consistent with the presence of an overturning rotor (Figure 2a, R) in

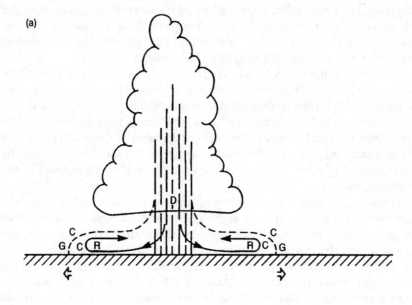

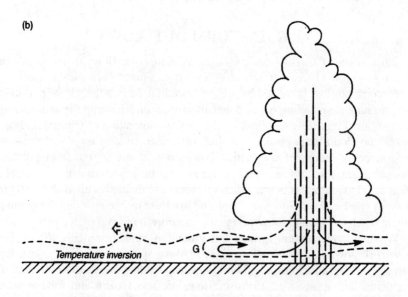

Fig. 2. (a) Thunderstorm outflow structure, showing downdraught (D) striking ground and spreading outwards, generating a gust front (G) and rotor (R). The potential zones of insect concentration are just behind the gust front. Note that this is a three-dimensional flow (there is also outflow into and out of the plane of the page). (b) Generation of gravity wave (W) when a downdraught strikes a temperature inversion.

the 'head' just behind the gust front. There may also be smaller circulations on the leading edge of the gust front. The strongest gusts at the ground are approximately equal to the speed of the outflow. Ahead of the front there is a strip of upcurrents, up to a few kilometres wide and strongest nearest the front, with upward speeds as much as 10 m s^{-1}. Outflow speed and depth generally decrease with time as the downdraught formation decays and the already cooled air spreads tens of kilometres, even more than 100 km, from the parent storm.

Passage of a gust front is indicated by the onset of a squall and by rapidly falling temperatures. Sometimes there is a striking line of cloud in the zone of upcurrents, or a wall of dust if the ground is dry and sparsely vegetated: the Arabic word 'haboob' for such a dust wall is used widely. Dust walls reveal detailed structures in the gust front: an over-hanging 'nose' (cf. F in Figure 1a and G in Figure 2) and a scalloped leading edge, both of which may be present, if difficult to detect, in sea breezes. Behind a gust front the surface wind direction changes to blow more or less outwards from the storm centre. Outflows can come in pulses, either because the originating downdraught is unsteady or because there are downdraughts from adjacent storms, each with its own life cycle.

Although storm outflows resemble sea breezes they are usually stronger and more localised. Moreover, because storms tend not to be isolated, outflows from neighbouring storms can collide, producing complex wind patterns.

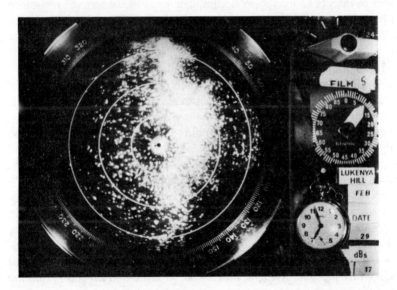

Fig. 3. The concentration of flying insects, mostly African armyworm moths (*Spodoptera exempta*), by a rainstorm outflow near Lukenya Hill, 35 km southeast of Nairobi, 29 February 1980. The band of dense insect echoes was formed by the convergence between the westerly outflow from the storm and the broad-scale wind from the northeast. Reproduced from Pedgley *et al.*, 1982, with permission.

Gust fronts have been associated with sudden mass arrivals of insects at night, for example, spruce budworm moths *Choristoneura fumiferana*, in New Brunswick (Greenbank *et al.*, 1980). Radar studies have linked gust fronts with insect concentrations (Figure 2, C). During a study of insect night flight in Kenya (Pedgley *et al.*, 1982), a concentration of insects (almost certainly dominantly *Spodoptera exempta*) was seen near a rainstorm soon after sunset (Figure 3). This concentration was revealed as a curved band of radar echo lying approximately north–south to the east of the rainstorm and sloping up towards it. The echo band moved eastwards at 5 m s^{-1} and, as it passed over the radar site, there was a change of surface wind from eastnortheast 4 m s^{-1} to west 5 m s^{-1}, accompanied by a rapid fall in temperature. Echoes from individual moths to the east of the band were moving from the northeast at 6–8 m s^{-1}, whereas those to the west were coming from the west at 8–9 m s^{-1}. It is clear, therefore, that both winds and moths were converging on the windshift line. As a result, there was a narrow zone of upcurrents that can be calculated to have reached 2 m s^{-1} at 200 m and that took insects upwards to a maximum height of 750 m. If the insects had been simply carried aloft there would have been no concentration and they would then have been swept westwards as the upcurrents levelled out and flowed over the top of the westerlies. For concentration to have occurred there must have been some resistance to being carried aloft. Some evidence for this resistance was the absence of a layer of insects above the westerlies. The most likely form of resistance would have been active flight downwards, perhaps induced by falling temperature in the rising air. Schaefer (1976) gives a very similar example from Sudan: a curved echo band probably caused mostly by the Sudan plague locust *Aiolopus simulatrix* (Orthoptera: Acrididae).

In an early study, Harper (1960) drew attention to daytime arcs parallel to the edges of showers, but several kilometres distant and sloping up towards them. Direct observations showed that these arcs were caused by swifts, almost certainly feeding on insects. Reynolds and Riley (1988) and Riley and Reynolds (1990) give two examples of band echoes at night in Mali, caused by various species of grasshoppers, that resulted from convergence at the leading edges of outflows, probably from distant rainstorms.

There is evidence that concentration of *Spodoptera exempta* by storm outflows is followed by mass egg-laying and then dense outbreaks of larvae (Blair, 1972; Rose and Law, 1976; Tucker, 1983; Tucker and Pedgley, 1983). Similar infestations have been reported in a number of moth species in temperate latitudes – e.g. the noctuids *Spodoptera exigua*, *Spodoptera littoralis*, *Agrotis ipsilon*, *Agrotis segetum*, *Autographa gamma*, *Xestia c-nigrum* and *Peridroma saucia*, and the pyralid *Margaritia sticticalis*. These infestations may have resulted from moths concentrated by wind convergence near rainstorms.

Insect take-off may be inhibited by the strength of some outflows, but insects do fly in outflows and can also form layers aloft (Westbrook *et al.*,

1987). Flying insects overtaken by an outflow can become concentrated near the gust front to produce bands similar to those observed at sea breeze fronts. An outflow can be likened to an increasing circular ring in a sheet of flying insects, expanding across country and perhaps encountering other rings, but eventually allowing the insects to be deposited as with a sea breeze front. The net effect will be a more clumped population on the ground than would have been expected in the absence of the storms.

Downdraughts at night are sometimes unable to displace the colder air within the inversion near the ground (Figure 2b), with the result that the effects of the outflow are hardly felt at ground level, although vertical oscillations, in the form of gravity waves, may be induced at the boundary of the inversion and the air above. Powerful downdraughts, however, may be able to displace the colder air, leading to temporary rises of temperature at the ground.

VII. MOUNTAIN WINDS: DRAINAGE WINDS

When a nocturnal temperature inversion forms over sloping terrain (for example the sides of a hill, valley or valley bottom) the air nearest the ground is cooler than air at the same height away from the slope and it tends to drain down-slope under the influence of gravity. The resulting breeze is variously called a drainage, down-slope, gravity or katabatic wind. Conventionally, researchers distinguish these types of flow from those caused by wind acceleration through valleys or gaps between mountains. Although sometimes misleadingly referred to in the literature as 'drainage' winds, such accelerated flows are mechanical in origin, as are other winds which are caused by cold air spilling over mountains (fall winds) and by the cascading of storm outflows down slopes. These phenomena are discussed in section VIII.

Although drainage winds are common among hills and mountains during clear, quiet weather, there appear to be few studies of their effects on night-flying insects, perhaps because of the difficulties of field work. It is not difficult, however, to visualize possible effects of hill-side drainage winds based on the following known properties.

- Minimum slope needed seems to be very gentle – about 1:150 or 1:100 (say 0.5°).
- Onset is usually a few hours after local sunset, which is before astronomical sunset, where slope is towards the east and therefore sheltered from the setting sun.
- Wind speed increases with height, away from ground friction, reaching a maximum (usually less then 2 m s^{-1}) at heights varying from a few metres to around 100 m. Above the maximum wind there is a zone of mixing with the broad-scale wind.

- Speed increases down the slope, but it may be slowed by vegetation (and even dammed by a forest edge) or accelerated at a sudden increase in slope.
- Downslope windspeed is controlled by slope angle: a downslope wind over a 10° slope is twice as strong as one over a 1° slope, under the same external conditions.
- On an exposed slope, the drainage wind can be temporarily replaced by the broad-scale wind.
- The leading edge of the downslope wind can be sharply defined: it resembles the head of a storm outflow, but on a smaller scale. The passage of the leading edge marks the sudden onset of the down-slope wind and is accompanied by a falling temperature.
- There is evidence that a broad-scale wind of 3 m s^{-1} or above inhibits downslope wind formation.

It may be deduced from these properties that all but the strongest fliers, dispersing over hill slopes within a few tens of metres of the ground on clear, quiet nights, will tend to be taken downslope and perhaps become concentrated in valleys, or even in the windshifts at leading edges of the slope winds. Indeed, some radar band echoes have been attributed to such windshift lines, although without supporting evidence (Schaefer, 1976; Riley and Reynolds, 1983). A sloping night-time concentration seen in the vertical plane by radar over California (Richter *et al.*, 1973) may have corresponded with the leading edge of a slope wind. In light slope winds, *upwind* flight would tend to take insects to high altitude. Such flight has been suggested as a mechanism for larch bud moths *Zeiraphera diniana* (Lepidoptera: Tortricidae) to reach the broad-scale wind aloft (Baltensweiler and Fischlin, 1979).

Although valley-bottom drainage winds can develop independently they are usually a result of the channelling of drainage winds from valley walls. They have been studied more than those down hillsides because of their relevance to the transport and trapping of pollutants from industrial sites on valley floors. The following properties have been discovered.

- Direction is more or less parallel to the valley axis, but convergence of slope winds from opposite walls may induce an overturning of the air and hence a helical down-valley flow. In the absence of a sufficient gathering area the down-valley wind can be weak.
- The whole depth of a valley may be filled but the greatest speeds are in a jet at a height of about a quarter or a third of the valley depth (i.e. about half the depth of the temperature inversion). Jets can meander down valleys, possibly causing oscillations, or pulses, with periods of some fraction of an hour at a fixed point, and heights tend to be lower over steeper valley floors. On meeting a sharp bend in a valley the jet may bounce off the valley wall to produce downstream oscillations.

- Speeds are stronger than hill-slope winds: jet speeds of about 5 m s^{-1} are common, and 9 m s^{-1} has been reported.
- Speed and depth tend to increase down-valley, but there is much variability. If the flow is weak, pools of stagnant cold air can form as the nocturnal temperature inversion develops or obstructions on the valley floor cause blockages. See-saw oscillations of this ponded cold air have been reported, like a seiche in a lake.
- Above the down-valley wind is a return flow at about ridge height, and above this is a transition layer to the broad-scale flow. The speed and depth of the down-valley wind can be affected by the strength of the broad-scale wind and by its direction relative to the valley axis: sometimes it is completely inhibited.
- Onset can be sudden and accompanied by a temperature fall: the leading edge appears to behave like a density current.
- In broken country, the pattern of down-valley winds can be very complex and we may suppose that jets become superimposed at valley junctions, where cross-valley flows may occur.

We may deduce from these properties that insects flying on clear nights in valley bottoms will tend to be taken down-valley: they may become concentrated and stranded in stagnant areas of air where valleys meet or where wind shifts at leading edges are halted. Down-valley winds spreading out over plains and weakening may deposit insects at particular sites, perhaps accounting for clumping of populations.

VIII. MOUNTAIN WINDS: BARRIER WINDS

Mountains act as barriers to the wind, deflecting and accelerating it in much the same way as buildings do, but on larger scales of space and time. The distortions of the airflow are complicated: they vary with height and are controlled by the shape and size of the mountain and also by the undisturbed (upstream) vertical profiles of temperature (i.e. lapse rate) and wind (i.e. shear).

In general, there is a greater tendency for the wind to flow around rather than over an obstacle, especially at night when the lapse rate is often less than 10°C km^{-1}. Deflection of the wind results in localised regions of acceleration and deceleration, and hence to regions of wind shear that cause waves and eddies to form, with vertical and horizontal axes. The positions of similar eddies around a building are revealed by blowing smoke and drifts of leaves or snow. In broken country, the resulting complexity of barrier winds can be appreciated by comparison with winds observed in a city centre, or with the flow of river over and around boulders in its bed. Flow around isolated mountains and across ridges (or mountain chains) has been studied most, both in the field and particularly in the laboratory by the use of simulation models. In the following sections, the nature of such flows is summarised and some further descriptions of how other forms of topography modify airflow are presented. When reading the following sections it may be useful to bear in mind that many of the phenomena described do not occur in isolation. For example, not only is the behaviour of the air flowing over a mountain ridge affected as it crosses the ridge, there are also modifications of the flow in valleys behind that ridge.

A. Effects of Isolated Mountains

Winds blowing around isolated hills and mountains have been studied main-ly by simulation in the laboratory using flow tanks, but the few available field studies support the laboratory results. Detailed tank experiments with stably stratified flow gave results summarised in Figure 4. Four regions are distin-guished:

- flow undeflected well above the summit
- flow with lee waves and rotor near summit height
- flow nearly horizontal, with deflection and acceleration around the sides
- flow with downstream wake and vortices.

The positions of these regions are to some extent variable and will change with the properties of the approaching airstream, especially temperature lapse rate (static stability) or vertical shear, as the night progresses or owing to large-scale events.

  The nature of air flow around an isolated hill is complex and its effects on flight and landing of insects do not appear to have been examined. It is poss-ible, however, on the basis of flow investigations, to suggest where insects might be concentrated.

  With strong atmospheric stability (small lapse rate or even an inversion), only the wind near summit level blows across the hill, whilst that below sum-mit level accelerates around the sides, separated from a wake zone of light winds (wind shadow). Take-off will be hindered in zones of acceleration, but insects which are already airborne are likely to be transported rapidly. In con-trast, insects are more likely to be concentrated, or taken in directions against the general wind (e.g. up leeward slopes), or provided opportunities to land, where speeds are light (in the area of wind shadow). Moreover, just as a sta-tionary zone of turbulent overturning with an abrupt rise in water level down-stream can occur at the foot of a weir or the lee of a submerged boulder, so a similar atmospheric overturning can appear at the foot of a leeward slope. This is known as a hydraulic jump, Figure 5d, H: it is a wave continually breaking upstream. A hill wake may contain a pair of quasi-stationary vor-tices with more or less vertical axes and upstream, upslope, flow along the centreline (Figure 4c). Beyond the wake the divided flow comes together

(a)

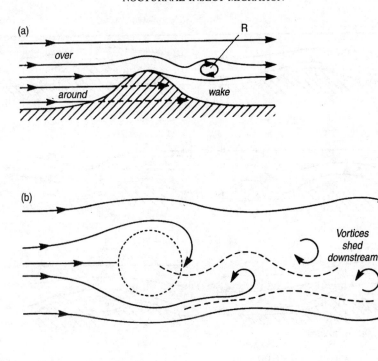

(b)

(c)

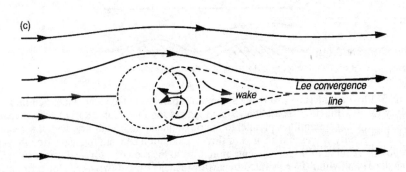

Fig. 4. (a) Side view of air flow around an isolated hill, with the generation of a rotor (R). (b) Shedding of wake vortices downstream of an isolated hill. (c) Quasi-stationary vortices and zone of lee convergence, downstream of an isolated hill.

again along a convergence line. Insects caught in the air forming these regions of circulation are likely to be trapped.

With lesser stability (greater lapse rate), although the depth of flow passing over the hill increases, the hydraulic jump strengthens and moves downstream and the size and depth of the wake decrease, until eventually the whole

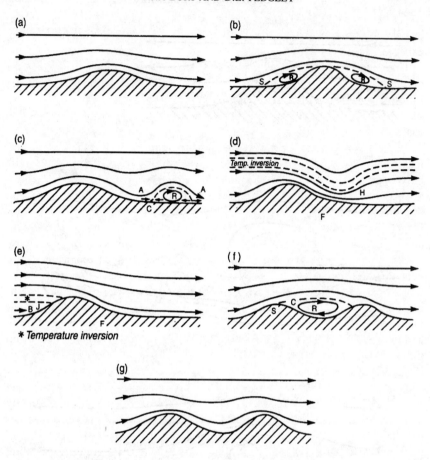

Fig. 5. Flow over a ridge, showing streamlines parallel to the ridge under neutral stability (a, b) with flow separation (S) and generation of a rotor (R). (c) Flow over a ridge, showing streamlines parallel to the ridge under static stability, with the generation of gravity waves and an area of flow acceleration beneath the wave trough (A). A possible region of insect concentration (C) and location of a rotor (R) are also indicated. (d) Fall wind (F) generated by airflow accelerating down a lee slope, forming a hydraulic jump (H) downstream. (e) Generation of a Föhn wind (F) by blocking (B) of airflow due to the presence of an inversion below the crest of an obstacle, and the generation of a topographic jet (J). (f) Generation of a rotor (R) in a leeward valley and possible zone of insect concentration (C). (g) Sweeping flow across a valley.

depth of the approaching flow is able to pass over the hill and both the lee jump and separated wake disappear. Thus, under such conditions it may be expected that fewer insects will be concentrated in the wake region, with most being swept past the hill as the lapse rate becomes greater.

As neutral stability is approached, a separated wake reforms but flow in it is turbulent. Again, some insect concentration in such a region might be expected.

Under most circumstances, therefore, in complex terrain the distribution of populations on the ground would become clustered. The same would be true for other windborne organisms, including fungal spores. As a result, there are likely to be marked geographical variations in the incidence of pests and diseases in crops near hills and on valley floors and sides.

Wake vortices are sometimes shed downstream and periodically reform, leading to a trail consisting of two rows, or streets (Figure 4b). Such trails have been recognised on satellite images of clouds in the lee sides of mountainous islands (Aleutians, Cape Verde, Canaries, Cheju, Guadeloupe, Jan Mayen, La Reunion, Madeira and Mauritius). Aircraft surveys have confirmed their structure in the lee of Hawaii. Street lengths range up to 800 km, and individual vortices last up to 20 h. Vortex street formation is favoured by an inversion layer based below summit level, with near neutral stability beneath, together with little variation of wind speed or direction with height to well above summit level. Vortex streets sometimes occur over land but single vortices have been more often reported in the lee of ridge ends (near Denver, Melbourne and Tokyo, and along the coasts of Oregon, California and Morocco). These vortices can be separated by a convergence line from the accelerated flow (jet) passing around the ridge end. Their effects on flying insects are unknown but influences may extend far downwind.

Outbreaks of *Spodoptera exempta* larvae in Kenya tend to occur to the west of hills (Pedgley and Rose, 1982). Other reports of insect concentrations by isolated hills are sparse and are often complicated by the presence of other topography. For example, although Noda and Kiritani (1989) found, during their nationwide survey of Japan to identify the characteristics of landing places of migrating planthoppers (Homoptera: Delphacidae) *Nilaparvata lugens* and *Sogatella furcifera*, that some of these insects occurred in the shelter of hills, others were at the heads of windward valleys, where eddies can also form.

A valley lying across the wind can be filled with an eddy, and insects may be taken from ridge top to ridge top, as has been reported for larvae of the gypsy moth *Lymantria dispar*, drifting on silk threads (Cameron *et al.*, 1979).

B. Ridges

A neutrally stable airstream readily crosses a ridge, with acceleration over the crest (Figure 5a) depending on height, cross-section shape and steepness of the upstream slope, as well as the nature of the ground and vegetation. Flying insects will find fewer opportunities to land and will be swept across the crest. Steeper upstream slopes do not necessarily produce stronger crest winds, because an upstream rotor may develop (Figure 5b, R), effectively increasing the width of the ridge and therefore reducing its steepness. The flow leaves the ground upstream of the ridge if the crest is sharp (an event called separation, Figure 5b, S) and makes contact with the ground again only near the crest. A similar rotor can form on the downstream side, particularly after a sudden increase in steepness of the lee slope.

Beneath a rotor there is reversed flow; on the upstream side this flow meets the undisturbed wind at a line of convergence. Flying insects can be expected to concentrate there, as occurred with the *Spodoptera exempta* moths in Kenya (Pedgley *et al.*, 1982), and it is likely that similar concentration in upstream rotors allowed deposition of the planthoppers in Japan, reported by Noda and Kiritani (1989), at the heads of valleys where slope increased rapidly. A weak rotor provides an area of light winds in which flying insects can land. Where the wind crosses a ridge at an angle the rotor has a helical motion, and beyond the ridge end this can stream away downwind, as happens at Gibraltar. These observations may be compared with those on a much smaller scale in the lee of windbreaks (Lewis, 1965a, b, 1966, 1969a, b, 1970).

Radar has detected flying moths temporarily concentrated over periods of about an hour in eddies downwind of an escarpment in Kenya (Pedgley *et al.*, 1982). A similar concentration may have been responsible for the appearance of colorado beetles *Leptinotarsa decemlineata* (Coleoptera: Chrysomelidae) on the ground in strips parallel to the Öre Mountains in former Czechoslovakia (Forchtgott, 1950). An association is not unreasonable if there are topographically determined zones of convergent winds that might concentrate flying moths, or if there are zones of light winds or calms that would provide preferred landing sites. The grass webworm *Eudonia sabulosella* (Lepidoptera: Pyraustidae) in New Zealand aggregates on leeward slopes when wind speeds exceed 10 km h^{-1} (Cowley, 1987). Similar sheltering, or wind-shadow effects might be expected to occur widely but there appear to be few reports.

A statically stable airstream (lapse rate <10°C km^{-1}) may be able to cross a ridge if the approach speed is great enough, but the lifted air becomes cooler, by expansion, than unlifted air at the same height and the induced negative buoyancy resists the lifting. On the leeward side, sinking of the cooled air is possible and it can overshoot, with the result that vertical oscillations are induced in the form of gravity waves (Figure 5c). Sometimes these waves propagate upstream at a speed equal to but opposite that of the wind so that they are more or less stationary relative to the ridge. Their presence is sometimes indicated by the formation of more or less stationary clouds caused by the cooling air as it ascends into the wave crests. Beneath a crest there is sometimes an overturning, in the form of a rotor, once again leading to a strip in which winds are opposed to the general wind direction, which again allows concentration of flying insects (Figure 5c, C). Beneath a wave trough (Figure 5c, A) there is an acceleration in wind speed, sometimes revealed in dry country by blowing dust.

C. Fall Winds

Acceleration over a rounded crest sometimes continues down the leeward slope, producing a zone of strong wind, called a fall wind (Figure 5d, F). In the lee of some large mountains sustained speeds of 20 m s^{-1} or more occur. They result from spillage over the ridge and can be particularly strong when the approaching airstream has a temperature inversion based just above crest height. The inversion hinders lifting so that the cold air beneath it accelerates, first over the crest and then down the lee slope. (On a clear night, a drainage component adds to the downslope flow). Many mountainous areas have well-known cold fall winds of this type and their flow is similar to that of a river over a weir. Insects entering the very turbulent winds, having been swept down a leeside slope in a fall wind, would probably have great difficulty in landing.

An inversion based *below* crest height may prevent lifting and cause stagnation in much of the approaching airstream (known as blocking, see Figure 5e, B). Flying insects entering such a stagnant area would have increased opportunities to land. Winds above the inversion can cross the ridge and, on descending the leeward slope, become warmed by compression so that, level for level, air on the leeward side is warmer than that on the windward side. Such a warm fall wind is one type of föhn, a warm, dry wind blowing from the mountains (Figure 5e, F). Many mountainous areas have well-known föhn winds. High-flying insects may then be taken over a ridge even though the flow at lower altitudes is blocked. Slopes in the blocked layer on the windward side can develop downslope breezes at night if skies are clear. Insects taking off from those slopes could then be taken in a direction opposite to that of the broad-scale wind.

Winds within a leeward valley may be dominated by either separation (with a rotor filling much of the valley; Figure 5f) or sweeping flow (Figure 5g), depending on whether or not buoyancy can suppress inertia in the airstream. A rotor-filled valley may account for the ridge-top distribution of windborne larvae of *Lymantria dispar* (Cameron et al., 1979).

D. Topographic jets

Where a stably stratified airstream approaches a ridge at an angle, instead of blocking there can be acceleration parallel to the ridge axis in the form of a jet centred over the upstream slope (Figure 5e, J). Such jets have been found in large mountain chains in various parts of the world: the Alps, the Antarctic Peninsula, the northern slope of the Brooks Range (Alaska), the eastern slopes of the Rockies and of the Appalachians, and the eastern slopes of the highlands of East Africa. Such jets are likely to promote rapid transport of insects caught within them, providing few opportunities for landing, or for the take-off of further individuals from the slope into the airflow. It may be reasonable to expect insects to accumulate further downstream of such a site as the wind slows or transported insects are cast into a zone of wind shadow (depending on the topography).

E. Gap Winds

Stably stratified airstreams, resistant to lifting, accelerate through gaps or passes in mountain chains. Downstream, the strong winds either persist as a jet, even with lateral vortices, or they fan out.

Wind acceleration as a result of channelling through passes or along valleys may induce faster and greater flights than might otherwise occur. An example is the warm down-canyon winds carrying beet leafhoppers, *Circulifer tenellus* (Hom: Cicadellidae) into the plains of the San Joachin valley of California, where evening cooling then prevents flight (Lawson *et al.*, 1951). Winds blowing persistently from particular directions through passes may account for reports of flying insects apparently 'choosing' their routes across mountains.

F. Mountains and Sea Breezes

When a sea breeze advancing inland meets an isolated mountain, it is likely to behave in ways already described (see section IV). If it meets a mountain range it will be channelled up those valleys that face the coast, and deflected by higher ground. Because of the stabler stratification (smaller lapse rate) resulting from the low-level intrusion of a layer of cool sea air, there is resistance to flow over mountains. Consequently, either blocking or lateral deflection is possible, e.g. turning to blow parallel to the mountains. Coastal mountains, combined with a temperature inversion over the sea below ridge height, restrict a sea breeze to the layer below the inversion. Where spillage of the intruding sea breeze air over the mountain occurs, there can be both a slowing of the sea breeze as the mountains are approached and a subsequent strengthening of the wind speed down the lee slope, sometimes with the formation of a hydraulic jump in the leeside valley (as can happen with fall winds). This may lead to insect deposition on the upstream side of the mountains, and their rapid transport (with little opportunity for landing) down the lee side, with possible effects caused by the hydraulic jump described in section VIIIA.

The movement of sea breezes inland can also be blocked if they encounter pools of cold air trapped within mountainous country. Such areas may be favourable zones for insect accumulation.

Where a sea breeze is channelled through mountain gaps the resulting flow pattern on the leeward side may be very complex, with the formation of local convergence zones. These have been well described for California. Annual reappearance of the peach potato aphid *Myzus persicae* (Homoptera: Aphididae) each autumn in the desert valleys of interior California has been attributed to windborne spread from the coast through gaps in coastal mountains (Dickison and Laird, 1967).

A sea breeze reaching mountains in the evening may encounter, and be undercut by, hill-slope winds and down-valley winds that have already started to develop. If the lifted airflow contains insects, they might avoid the zones of concentration suggested in section VII. Katabatic winds may enhance a land breeze where mountains are sufficiently close to the coast, leading to a greater air flow speed and reducing the likelihood of insects landing.

IX. CONCLUSIONS

Our review has shown that, although a number of field studies have revealed much valuable information on both the movements of insects and the small-scale wind systems in which they fly, there is still considerable ignorance about the effects of a variety of such wind systems on the movement and concentration of most night-flying species. A primary cause of this ignorance is inadequate examination of wind systems as integral parts of field studies on nocturnal insect flight. Moreover, meteorologists have yet to describe fully, let alone understand, some of these systems. Nevertheless, we can summarize known effects and speculate on others.

The presence of a nocturnal temperature inversion often leads to a layering of flying insects. The high altitude of flight of many nocturnal migrants has contributed to the slow realization of the numbers flying and the persistence of their flight. Development of a nocturnal jet can lead to more rapid insect movement than might be inferred from weather maps, and much more rapid than from surface winds often experienced at night by entomologists in the field. The mechanisms used by flying insects to select their height of flight on a given night are poorly understood; no doubt they involve sensing both warmth and zones of turbulence in the air.

Sea breezes persisting into the evening may well sweep inland any night-flying insects from near the coast. Concentration can occur on the sea breeze front, which can sometimes go surprisingly far inland. Whether bores and solitary waves can concentrate flying insects is unclear – perhaps only those with rotors can – but their structures and behaviour, as atmospheric phenomena, require much more research. It is possible that they are of importance, particularly over the vast subtropical plains of Africa and Australia. The pattern of night-time winds is likely to be complex over land with many lakes.

A night with rainstorms is likely to be have complicated wind patterns because outflows are often stronger than the broad-scale wind on which they are superimposed. They will cause considerable redistribution of flying insects over distances of tens of kilometres, and sometimes much more. Individual insects can be expected to be moved along convoluted trajectories, and to become involved in one or more concentrating zones. As a result, there is likely to be a clumping of an originally more dispersed population, with potentially important implications for pest management. Where that clumping is likely to be greatest may be very difficult to deduce except, perhaps, where a single storm has produced a quasi-stationary gust front. Satellites may be an increasingly valuable aid for more precisely locating storms that concentrate flying insects: as has been investigated for African armyworm (Tucker, 1995).

Drainage winds are highly dependent on local topography. Almost nothing is known about their effects on night-flying insects. Provided that temperatures are not so low as to inhibit flight we may suppose that drainage winds redistribute insects in hilly country on a scale of kilometres, and sometimes tens of kilometres, particularly if there is any associated down-valley jet. Some insects may be carried from valley mouths across nearby plains. Local obstructions, such as forest edges and rocky knolls, may cause ponding, stagnation of air, and opportunities for insects to land, particularly the slower fliers. Where a drainage wind develops a sharply-defined leading edge, concentration of flying insects is possible but this has not been shown conclusively, probably because such drainage wind fronts appear to be rare.

Although half of the land surface may be described as hilly or mountainous, often little or nothing is known about the influence of local wind systems on night-flying insects. This is partly because most entomological radars require fairly flat sites. The principal modifications to the broad-scale wind are local zones of acceleration, deceleration, and convergence all of which affect speed, direction of movement, aerial concentration and landing. There is considerable scope for investigating the consequences of mountain-modified winds on the horizontal distribution of insects, both in the air and on the ground and vegetation. It is possible, for example, that places well-known locally for insect outbreaks result from such winds, particularly where rotors form or a wind shadow allows landing. In contrast, acceleration, particularly in jets, may move insects much faster and further than would otherwise be expected.

A question likely to be asked is where and when mountain-induced jets, rotors and wind shadows are likely to appear in a particular hilly region. The answer will depend in part on knowledge of the vertical profiles of temperature and wind upstream at the time. These will often be unmeasured, but it should be possible for local meteorologists to interpolate them from routine observations. Then, using a topographic map it becomes possible to estimate where concentration, for example, may occur. Forecasters of insect flight at night should be provided with such estimates, or at least they should be aware of the complexities of small-scale wind systems. There is scope for the development of numerical models, perhaps personal computer-based, that can simulate these wind systems, given local topography (perhaps in the form of digital elevation models) and vertical structure of the airstream as inputs. Some such work has already been done in the USA (Seem and Russo, 1995).

We emphasise the following as needing further research.

(1) Controls on the altitude of insect layers.
(2) Effects of drainage winds on insect movement, and the role of frontal structure in concentration.
(3) Barrier effects of topography on movement and concentration.
(4) Frequency and horizontal extent of sea and lake breezes sweeping insects inland from coastal zones during the evening and concentrating them at the sea breeze front.
(5) The roles of bores and solitary waves as concentrating mechanisms.
(6) The mechanisms by which flying insects land at high densities after concentration within wind convergence zones.

If our understanding of the influences of local wind systems on insect flight is to progress, field studies need to put emphasis on atmospheric behaviour through supplementary observations of winds and temperatures in the lowest kilometre or two of the atmosphere.

ACKNOWLEDGEMENTS

We are grateful for the assistance from our colleagues in the library of the Natural Resources Institute (Kate Nicholas, Mike Scanlon, and Mel Williams), who seldom failed to provide references in what must have seemed at times an endless succession of requests. We also thank the following colleagues for their comments on earlier drafts of this paper: Anthea Cook, Charles Dewhurst, Margaret Haggis, Joyce Magor, Tessema Megenasa, Peter Odiyo, Bill Page, Don Reynolds, Jane Rosenberg and Mike Tucker. We are most grateful to Martyn Rothery for producing the figures.

REFERENCES

Achtemeier, G.L. (1992). Grasshopper response to rapid vertical displacements within a 'clear air' boundary layer as observed by Doppler radar. *Environ. Entomol.* **21**, 921–938.

Baltensweiler, W. and Fischlin, A. (1979). The role of migration for the population dynamics of the larch bud moth *Zeiraphera dianiana* Gn (Lep: Tortricidae). *Bull. Soc. Entomol. Suisse* **52**, 259–271.

Beerwinkle, K.R., Lopez, J.D., Witz, J.A., Schleider, P.G., Eyster, R.S. and Lingren, P.D. (1994). Seasonal radar and meteorological observations associated with nocturnal insect flight at altitudes to 900 meters. *Environ. Entomol.* **23**, 676–683,

Blair, B.W. (1972). An outbreak of African Armyworm, *Spodoptera exempta* (Walker) (Lep.: Noctuidae), in Rhodesia during December 1971 and January 1972. *Rhodesian J. Agric. Res.* **10**, 159–168.

Browning, K.A. and Atlas, D. (1966). Velocity characteristics of some clear-air dot angels. *J. Atmos. Sci.* **23**, 592–604.

Buntin, G.D., Pedigo, L.P. and Showers, W.B. (1990). Temporal occurrence of the variegated cutworm Lepidoptera Noctuidae adults in Iowa USA with evidence for migration. *Environ. Entomol.* **19**, 603–608.

Callahan, P.S., Sparks, A.N., Snow, J.W. and Copeland, W.W. (1972). Corn Earworm Moth: vertical distribution in nocturnal flight. *Environ. Entomol.* **1**, 497–503.

Cameron, E.A., McManus, M.L. and Mason, C.J. (1979). Dispersal and its impact on the population dynamics of the Gypsy Moth in the United States of America. *Bull. Soc. Entomol. Suisse* **52**, 169–179.

Cochrane, J. (1980). Meteorological aspects of the numbers and distribution of the Rose-Grain Aphid, *Metopolophium dirhodum* (Wlk.), over south-east England in July 1979. *Plant Pathol.* **29**, 1–8.

Cowley, J.M. (1987). Oviposition site selection and effect of meteorological conditions on flight of *Eudonia sabulosella* (Lep.: Scopariinae) with implications for pasture damage. *N. Z. J. Zool.* **14**, 527–533.

Davey, J.T. (1959). The ecology of *Locusta* in the semi-arid lands and seasonal movements of populations. *Locusta*, **7** (International African Migratory Locust Organisation). 180 pp.

Dickson, R.C. and Laird Jr, E.F. (1967). Fall dispersal of Green Peach Aphids to desert valleys. *Ann. Ent. Soc. Amer.* **60**, 1088–1091.

Dickison, R.B.B. (1990). Detection of mesoscale synoptic features associated with dispersal of Spruce Budworm moths in eastern Canada. *Phil. Trans. R. Soc. Lond.*, **B328**, 607–617.

Drake, V.A. (1982). Insects in the sea-breeze front at Canberra: a radar study. *Weather* **37**, 134–143.

Drake, V.A. (1984a). A solitary wave disturbance of the marine boundary layer over Spencer Gulf revealed by radar observations of migrating insects. *Aust. Meteorol. Mag.* **32**, 131–135.

Drake, V.A. (1984b). The vertical distribution of macro-insects migrating in the nocturnal boundary layer: A radar study. *Boundary-Layer Meteorol.* **28**, 353–374.

Drake, V.A. (1985a). Solitary wave disturbances of the nocturnal boundary layer revealed by radar observations of migrating insects. *Boundary-Layer Meteorol.* **31**, 269–286.

Drake, V.A. (1985b). Radar observations of moths migrating in a nocturnal low-level jet. *Ecol. Entomol.* **10**, 259–265.

Drake, V.A. and Farrow, R. (1983). The nocturnal migration of the Australian Plague Locust, *Chortoicetes terminifera* (Walker) (Orthoptera: Acrididae): quantitative radar observations of a series of northward flights. *Bull. Ent. Res.* **73**, 567–585.

Drake, V.A. and Farrow, R.A. (1985). A radar and aerial-trapping study of an early spring migration of moths (Lepidoptera) in inland New South Wales. *Aust. J. Ecol.* **10**, 223–235.

Drake, V.A. and Farrow, R.A. (1988). The influence of atmospheric structure and motions on insect migration. *Annu. Rev. Entomol.* **33**, 183–210.

Drake, V.A. and Farrow, R.A. (1989). The 'aerial plankton' and atmospheric convergence. *Trends Evol. Ecol.* **4**, 281–285.

Drake, V.A. and Gatehouse, A.G. (eds) (1995). *Insect Migration: Tracking Resources Through Space and Time.* Cambridge University Press, Cambridge, UK.

Drake, V.A. and Rochester, W.A. (1994). The formation of layer concentrations by migrating insects. *11th Conf. Biomet. Aerobiol., San Diego, Amer. Meterol. Soc.,* pp. 411–414.

Forchtgott, J. (1950). The transport of microparticles or insects over the Erzgebirge. *Met. Zpr. Prague* **4**, 14–16. (In Czech.)

Greenbank, D.O., Schaefer, G.W. and Rainey, R.C. (1980). Spruce Budworm (Lepidoptera: Tortricidae) moth flight and dispersal: new understanding from canopy observations, radar and aircraft. *Mem. Entomol. Soc. Can.* 110.

Harper, W.G. (1960). An unusual indicator of convection. *Marine Observer* **30**, 36–40.

Isard, S.A., Irwin, M.E. and Holinger, S.E. (1990). Vertical distribution of aphids (Homoptera: Aphididae) in the planetary boundary layer. *Environ. Entomol.* **19**, 1473–1484.

Johnson, B. (1957). Studies on the dispersal by upper winds of *Aphis craccivora* Koch in New South Wales. *Proc. Linn. Soc. NSW* **82**, 191–198.

Johnson, C.G. (1969). *The Migration and Dispersal of Insects by Flight.* Methuen and Co. Ltd, London.

Johnson, S.J. (1987). Migration and the life history strategy of the fall armyworm *Spodoptera frugiperda* in the western hemisphere. *Insect Sci. Appl.* **8**, 543–549.

Lawson, F.R., Chamberlain, J.C. and York, G.T. (1951). Dissemination of the Beet Leafhopper in California. *United States Department of Agriculture Tech. Bull. 1030.*

Lewis, T. (1965a). The effects of an artificial windbreak on the aerial distribution of flying insects. *Ann. Appl. Biol.* **55**, 503–512.

Lewis, T. (1965b). The effect of an artificial windbreak on the distribution of aphids in a lettuce crop. *Ann. Appl. Biol.* **55**, 513–558.

Lewis, T. (1966). Artificial windbreaks and the distribution of turnip mild yellows virus and *Scaptomyza apicalis* (Dip.) in a turnip crop. *Ann. Appl. Biol.* **58**, 371–376.

Lewis, T. (1969a). Factors affecting primary patterns of infestation. *Ann. Appl. Biol.* **63**, 315–317.

Lewis, T. (1969b). The distribution of flying insects near a long hedgerow. *J. Appl. Ecol.* **6**, 443–452.

Lewis, T. (1970). Patterns of distribution of insects near a windbreak of tall trees. *Ann. Appl. Biol.* **65**, 213–220.

McManus, M.L. (1988). Weather, behaviour and insect dispersal. *Mem. Entomol. Soc. Can.* **146**, 71–94.

McNeill, J.N. (1987). The true armyworm *Pseudaletia unipuncta.* A victim of the pied piper or a seasonal migrant? *Insect Sci. Appl.* **8**, 591–597.

Mason, C.J. and McManus, M.L. (1980). The role of dispersal in the natural spread of the Gypsy Moth. *Proc. 2nd IUFRO Conf., Dispersal of Forest Insects.* (Ed. by A.A. Berryman and L. Safranyik) pp. 94–115.

Mueller, E.A. and Larkin, R.P. (1985). Insects observed using dual-polarisation radar. *J. Atmos. Ocean. Techn.* **2**, 49–54.

Neumann, H.H. and Mukammal, E.I. (1981). Incidence of mesoscale convergence lines as input to Spruce Budworm control strategies. *Int. J. Biometeorol.* **25**, 175–187.

Noda, T. and Kiritani, K. (1989). Landing places of migratory planthoppers *Nilaparvata lugens* (Stål) and *Sogatella furcifera* (Horvath) (Hom.: Delphacidae) in Japan. *Appl. Entomol. Zool.* **24**, 59–65.

Pedgley, D.E. (1982). *Windborne Pests and Diseases: Meteorology of Airborne Organisms.* Ellis Horwood Ltd, Chichester.

Pedgley, D.E. (1983). Windborne spread of insect-transmitted diseases of animals and man. *Phil. Trans. R. Soc. Lond.* **B302**, 463–470.

Pedgley, D.E. (1990). Concentration of flying insects by the wind. *Phil. Trans. R. Soc. Lond.* **B328**, 631–653.

Pedgley, D.E., Reynolds, D.R., Riley, J.R., and Tucker, M.R. (1982). Flying insects reveal small-scale wind systems. *Weather* **37**, 295–306.

Pedgley, D.E. and Rose, D.J.W. (1982). International workshop on the control of the Armyworm and other migrant pests in East Africa, Arusha 1982. *Tropical Pest Management*, **28**, 437–440.

Puhakka, T., Koistinen, J. and Smith, P. (1986). Doppler radar observation of a sea-breeze front. *Proc. 23rd Conf. Radar Meteorology, Amer. Meterol. Soc.*, pp. 198–201.

Rainey, R.C. and Waloff, Z. (1948). Desert Locust migration and synoptic meteorology in the Gulf of Aden area. *J. Anim. Ecol.* **17**, 101–112.

Reid, D.G., Wardhaugh, K.G. and Roffey, J. (1979). Radar studies of insect flight at Benalla, Victoria, in February 1974. *CSIRO Div. Entomol., Tech. Paper No.* **16**, 1–21.

Reynolds, D.R. and Riley, J.R. (1988). A migration of grasshoppers, particularly *Diabolocatantops axillaris* (Thunberg) (Orthoptera: Acrididae), in the West African Sahel. *Bull. Ent. Res.* **78**, 251–271.

Richter, J.H., Jensen, D.R., Noonkester, V.R., Kreasky, J.B., Stimman, M.W. and Wolf, W.W. (1973). Remote radar sensing: atmospheric structure and insects. *Science* **180**, 1176–1178.

Riley, J.R. and Reynolds, D.R. (1979). Radar-based studies of the migratory flight of grasshoppers in the middle Niger area of Mali. *Proc. R. Soc. Lond.* **B204**, 67–82.

Riley, J.R. and Reynolds, D.R. (1983). A long-range migration of grasshoppers observed in the Sahelian zone of Mali by two radars. *J. Anim. Ecol.* **52**, 167–183.

Riley, J.R. and Reynolds, D.R. (1990). Nocturnal grasshopper migration in West Africa: transport and concentration by the wind, and the implications for air-to-air control. *Phil. Trans. R. Soc. Lond.* **B328**, 655–672.

Riley, J.R., Reynolds, D.R. and Farmery, M.J. (1981). Radar observations of *Spodoptera exempta*, Kenya, March–April 1979. *COPR Misc. Rep. No.* **54**.

Riley, J.R., Reynolds, D.R. and Farmery, M.J. (1983). Observations of the flight behaviour of the armyworm moth, *Spodoptera exempta*, at an emergence site using radar and infra-red optical techniques. *Ecol. Entomol.* **8**, 395–418.

Riley, J.R., Cheng, X.-N., Zhang, X.-X., Reynolds, D.R., Xu, G.-M., Smith, A.D., Cheng, J.-Y., Bao, A.-D. and Zhai, B.-P. (1991). The long-distance migration of *Nilaparvata lugens* (Stål) (Delphacidae) in China: radar observations of mass return flight in the autumn. *Ecol. Entomol.* **16**, 471–489.

Rose, D.J.W. and Law, A.B. (1976). The synoptic weather in relation to an outbreak of the African Armyworm, *Spodoptera exempta* (Wlk.). *J. Entomol. Soc. S.A.* **39**, 125–130.

Rose, D.J.W., Dewhurst, C.F. and Page, W.W. (1995). The bionomics of the African Armyworm *Spodoptera exempta* in relation to its status as a migrant pest. *Integ. Pest Manag. Rev.*, **1**, 49–64.

Schaefer, G.W. (1969). An airborne radar technique for the investigation and control of migrating pest species. *Phil. Trans. R. Soc. Lond.* **B287**, 459–465.

Schaefer, G.W. (1976). Radar observations of insect flight. *Proc. 7th Symp. Roy. Entomol. Soc.*, pp. 157–197.

Schaefer, G.W. (1979). Radar studies of locusts, moth and butterfly migration in the Sahara. *Proc. R. Entomol. Soc. Lond.* **(C)34**, 33, 39–40.

Seem, R.C. and Russo, J.M. (1995). Linking disease and weather forecast systems: evolution and resolution. *Proc. XIII Int. Pl. Prot. Congress, The Hague, The Netherlands, 2–7 July 1995.*

Showers, W.B., Smelser, R.B., Keaster, A.J., Whitford, F., Robinson, J.F., Lopez, J.D. and Taylor, S.E. (1989). Recapture of marked black cutworm (Lepidoptera: Noctuidae) males after long-range transport. *Environ. Entomol.* **18**, 447–458.

Showers, W.B., Keaster, A.J., Raulston, J.R., Hendrix, W.H., Derrick, M.E., McCorcle, M.D., Robinson, J.F., Way, M.O., Wallendorf, M.J. and Goodenough, J.L. (1993). Mechanism of southward migration of a nocturid moth [*Agrotis ipsilon* (Hufnagle)] a complete migrant. *Ecology* **74**, 2303–2314.

Simpson, J.E. (1964). Sea-breeze fronts in Hampshire. *Weather* **19**, 196–201.

Simpson, J.E. (1967). Aerial and radar observations of some sea-breeze fronts. *Weather* **22**, 306–327.

Smelser, R.B., Showers, W.B., Shaw, R.H. and Taylor, S.E. (1991). Atmospheric trajectory analysis to project long-range migration of black cutworm *Lepidoptera Noctuidae* adults. *J. Econ. Entomol.* **84**, 879–885.

Taylor, L.R. (1974). Insect migration, flight periodicity and the boundary layer. *J. Anim. Ecol.* **43**, 225–238.

Tucker, M.R. (1983). Light-trap catches of African Armyworm moth, *Spodoptera exempta* (Walker) (Lep.: Noctuidae), in relation to rain and wind. *Bull. Ent. Res.* **73**, 315–319.

Tucker, M.R. (1995). Epidemiology of the African Armyworm (*Spodoptera exempta*) intra- and interseasonal variability in outbreak severity in eastern Africa in relation to weather and moth migration. PhD Thesis, University of Wales, Bangor.

Tucker, M.R. and Pedgley, D.E. (1983). Rainfall and outbreaks of the African Armyworm, *Spodoptera exempta* (Walker) (Lep.: Noctuidae). *Bull. Ent. Res.* **73**, 195–199.

Westbrook, J.K., Wolf, W.W., Pair, S.D., Sparks, A.N. and Raulston, J.R. (1987). Empirical moth flight behavior in the nocturnal planetary boundary layer. *Proc. 18th Conf. Agric. For. Met.* pp. 263–264.

Wolf, W.W., Pedigo, L.P., Shaw, R.H. and Newsom, L.D. (1987). Migration-transport of the green cloverworm *Plathypena scabra* (F.) into Iowa as determined by synoptic-scale weather patterns. *Environ. Entomol.* **16**, 1169–1174.

Wolf, W.W., Westbrook, J.K., Raulston, J.R., Pair, S.D. and Hobbs, S.E. (1990). Recent airborne observations of migrant pests in the United States. *Phil. Trans. R. Soc. Lond.* **B328**, 619–630.

BIBLIOGRAPHY

A selection of references to descriptive meteorological studies of wind systems likely to be encountered by night-flying insects.

Vertical Profiles

Blackadar, A.K. (1957). Boundary layer wind maxima and their significance for the growth of nocturnal inversions. *Bull. Amer. Meteorol. Soc.* **38**, 283–290.

Bonner, W.D., Esbensen, S. and Greenberg, R. (1968). Kinematics of the low-level jet. *J. Appl. Meteor.* **7**, 339–347.

Clarke, R.H. (1983a). Fair weather nocturnal inland wind surges and atmospheric bores: Part I Nocturnal wind surges. *Aust. Meteorol. Mag.* **31**, 133–145.

Clarke, R.H. (1983b). Fair weather nocturnal inland wind surges and atmospheric bores: Part II Internal atmospheric bores in northern Australia. *Aust. Meteorol. Mag.* **31**, 147–160.

Finnigan, J.J., Einaudi, F. and Fua, D. (1984). The interaction between an internal gravity wave and turbulence in the stably-stratified nocturnal boundary layer. *J. Atmos. Sci.* **41**, 2409–2436.

Kinuthia, J.H. (1992). Horizontal and vertical structure of the Lake Turkana Jet. *J. Appl. Meteor.* **31**, 1248–1274.

Meyer, J.H. (1971). Radar observation of land breeze fronts. *J. Appl. Meteor.* **10**, 1224–1232.

Panofsky, H.A. (1974). The atmospheric boundary layer below 150 meters. *Annu. Rev. Fluid Mech.* **6**, 147–177.

Reynolds, R.M. and Gething, J.T. (1970). Acoustic sounding at Benalla, Victoria and Julia Creek, Queensland. *Project EAR, University of Melbourne Meteorology Dept, Report* **VII**.

Smith, R.K. (1988). Travelling waves and bores in the lower atmosphere: the morning glory and related phenomena. *Earth-Sci. Rev.* **25**, 267–290.

Thuillier, R.H. and Lappe, U.O. (1964). Wind and temperature profile characteristics from observations on a 1400 ft Tower. *J. Appl. Meteor.* **3**, 299–306.

Coastal Winds

Asculai, E., Doron, E. and Terliuc, B. (1984). Mesoscale flow over complex terrain – a field study in the Lake Kinneret area. *Boundary-Layer Meteorol.* **30**, 313–331.

Bitan, A. (1981). Lake Kinneret (Sea of Galilee) and its exceptional wind system. *Boundary-Layer Meteorol.* **21**, 477–487.

Estoque, M.A., Gross, J. and Lai, H.W. (1976). A lake breeze over southern Lake Ontario. *Mon. Weather Rev.* **104**, 386–396.

Fraedrich, K. (1972). A simple climatological model of the dynamics and energetics of the nocturnal circulation at Lake Victoria. *Quart. J. R. Meteorol. Soc.* **98**, 322–335.

Goldreich, Y., Druyan, L.M. and Berger, H. (1986). The interaction of valley/mountain winds with a diurnally veering sea/land breeze. *J. Climatol.* **6**, 551–561.

Hsu, S-A. (1970). Coastal air-circulation system: observations and empirical model. *Mon. Weather Rev.* **98**, 487–509.

Meyer, J.H. (1971). Radar observations of land breeze fronts. *J. Appl. Meteor.* **10**, 1224–1232.

Rider, G.C. and Simpson, J.E. (1968). Two crossing fronts on radar. *Meteorol. Mag.* **97**, 24–30.

Simpson, J.E. (1994). *Sea Breeze and Local Winds*. Cambridge University Press, Cambridge.

Simpson, J.E., Mansfield, D.A. and Milford, J.R. (1977). Inland penetration of sea-breeze fronts. *Quart. J. R. Meteorol. Soc.* **103**, 47–76.

Sumner, G.N. (1977). Sea breeze occurrence in hilly terrain. *Weather* **32**, 200–208.

Wakimoto, R.M. and Atkins, N.T. (1994). Observations of the sea-breeze front during CaPE. Part 1: Single doppler, satellite and cloud photogrammetry analysis. *Mon. Weather Review*, **122**, 1092–1114.

Wakimoto, R.M. and Kingsmill, D.E. (1995). Structure of an atmospheric undular bore generated from colliding boundaries during CaPE. *Mon. Weather Review*, **123**, 1374–1393.

Rainstorm Outflows

Goff, R.C. (1976). Vertical structure of thunderstorm outflows. *Mon. Weather Rev.* **104**, 1429–1440.

Intrieri, J.M., Bedard, A.J. & Hardesty, R.M. (1990). Details of colliding thunderstorm outflows as observed by doppler lidar. *J. Atmos. Sci.* **47**, 1081–1098.

Klingle, D.L., Smith, D.R. and Wolfson, M.M. (1987). Gust front characteristics as detected by doppler radar. *Mon. Weather Rev.* **115**, 905–918.

Mahoney, W.P. (1988). Gust front characteristics and the kinematics associated with interacting thunderstorm outflows. *Mon. Weather Rev.* **116**, 1474–1491.

Scott, R.W. and Ackermann, B. (1983). Surface signatures of a dry nocturnal gust front. *Mon. Weather Rev.* **111**, 197–204.

Simpson, J.E. (1987). *Gravity currents: In the Environment and the Laboratory.* Ellis Horwood Ltd, Chichester.

Sinclair, P.C., Purdom, J.F.W. and Dattore, R.E. (1988). Thunderstorm outflow structure. *15th Conf. Severe Local Storms,* Amer. Meteorol. Soc. pp. 233–239.

Sun, J. and Crook, A. (1994). Wind and thermodynamic retrieval from single-doppler measurements of a gust front observed during Phoenix II. *Mon. Weather Rev.* **122**, 1075–1091.

Drainage Winds

Allwine, K.J. (1993). Atmospheric dispersion and tracer ventilation in a deep mountain valley. *J. Appl. Meteorol.* **32**, 1017–1037.

Bell, R.C. and Thompson, R.O.R.Y. (1980). Valley ventilation by cross winds. *J. Fluid. Mech.* **96**, 757–767.

Clements, W.E., Archuleta, J.A. and Hoard, D.E. (1989). Mean structure of the nocturnal drainage flow in a deep valley. *J. Appl. Meteorol.* **28**, 457–462.

Doran, J.C. (1991). Effects of ambient winds on valley drainage flows. *Boundary-Layer Meteorol.* **55**, 177–189.

Grace, W. and Holton, I. (1990). Hydraulic jump signatures associated with Adelaide downslope winds. *Aust. Meteorol. Mag.* **38**, 43–52.

Gryning, S-E., Mahrt, L. and Larsen, S. (1985). Oscillating nocturnal slope flow in a coastal valley. *Tellus* **37A**, 196–203.

Horst, T.W. and Doran, J.C. (1986). Nocturnal drainage flow on simple slopes. *Boundary-Layer Meteorol.* **34**, 263–286.

McNider, R.T. (1982). A note on velocity fluctuations in drainage flows. *J. Atmos. Sci.* **39**, 1658–1660.

Mahrt, L. and Larsen, S. (1982). Small scale drainage front. *Tellus* **34**, 579–587.

Nappo, C.J. and Snodgrass, H.F. (1981). Observations of nighttime winds using pilot balloons in Anderson Creek valley, Geysers, California. *J. Appl. Meteorol.* **20**, 721–727.

Porch, W.M., Fritz, R.B., Coulter, R.L. and Gudiksen, P.H. (1989). Tributary, valley and sidewall air flow interactions in a deep valley. *J. Appl. Meteorol.* **28**, 578–589.

Post, M.J. and Neff, W.D. (1986). Doppler lidar measurements of winds in a narrow mountain valley. *Bull. Amer. Meteorol. Soc.* **67**, 274–281.

Barrier Winds

Barry, R.G. (1992). *Mountain Weather and Climate 2nd edn.* Methuen, London and New York.

Bell, G.D. and Bosart, L.F. (1988). Appalachian cold-air damming. *Mon. Weather Rev.* **116**, 137–161.

Brighton, P.W.M. (1978). Strongly stratified flow past three-dimensional obstacles. *Quart. J. R. Meteorol. Soc.* **104**, 289–307.

Bromwich, D.H. (1988). A satellite study of barrier-wind airflow around Ross Island. *Antarctic J.* **23**, 167–169.

Chen, W-D. and Smith, R.B. (1987). Blocking and deflection of airflow by the Alps. *Mon. Weather Rev.* **115**, 2578–2597.

Chiba, O., Naito, G., Kobayashi, F. and Toritani, H. (1994). Wave trains over the sea due to sea breezes. *Boundary-Layer Meteorol.* **70**, 329–340.

Etling, D. (1989). On atmospheric vortex streets in the wake of large islands. *Meteorol. Atm. Phys.* **41**, 157–164.

Finnigan, T.D., Vine, J.A., Jackson, P.L., Allen, S.E., Lawrence, G.A. and Steyn, D.G. (1994). Hydraulic physical modeling and observations of a severe gap wind. *Mon. Weather Rev.* **122**, 2677–2687.

Grant, A.L.M. (1994). Wind profiles in the stable boundary layer and the effect of low relief. *Quart. J. R. Meteorol. Soc.* **120**, 27–46.

Jenkins, G.J., Mason, P.J., Moores, W.H. and Sykes, R.I. (1981). Measurements of the flow structure around Ailsa Craig, a steep, three dimensional, isolated hill. *Quart. J. R. Meteorol. Soc.* **107**, 833–857.

Mason, P.J. and Sykes, R.I. (1979). Flow over an isolated hill of moderate slope. *Quart. J. R. Meteorol. Soc.* **105**, 383–396.

Mass, C.F. and Albright, M.D. (1989). A severe windstorm in the lee of the Cascade Mountains of Washington State. *Mon. Weather Rev.* **117**, 2406–2436.

Ramachandran, G., Rao, K.V. and Krishna, K. (1980). An observational study of the boundary-layer winds in the exit region of a mountain gap. *J. Atmos. Sci.* **19**, 881–888.

Faunal Activities and Soil Processes: Adaptive Strategies That Determine Ecosystem Function

P. LAVELLE

I. SUMMARY

Soils host an extremely diverse community of invertebrates that differ in their adaptive strategies and hence in the functions they fulfil in soils. Soil invertebrates have relatively limited abilities to digest soil resources and rely to a large extent on micro-organisms to derive the assimilates they need from soil

ADVANCES IN ECOLOGICAL RESEARCH VOL. 27
ISBN 0–12–013927–8

organic resources. Conversely, activation of microflora by larger organisms appears to be essential to the maintenance and enhancement of activity of microbial communities. A paradigm is proposed (the 'Sleeping Beauty Paradox') that explains the overall mechanism of these interactions. Three major groups of invertebrates may be defined based on the nature of the relationship that they develop with soil microflora. This classification sets special emphasis on the creation of structures by invertebrates, and on the nature of these structures. The microfauna comprises invertebrates of less than 0.2 mm on average that live in water filled soil pore space. They do not create structures and they make use of micro-organisms mainly through predation in micro-foodweb systems. Mesofauna (= invertebrates 0.2–2 mm in size) and large arthropods comprise the group of litter transformers that create purely organic structures (their faecal pellets) inside which a mutualist relationship ('external rumen digestion') develops. Earthworms, termites and, to a lesser extent, ants, are 'ecosystem engineers' that create diverse organo-mineral structures and interact with micro-organisms through an internal rumen type of digestion. Dynamics of organic matter are largely influenced by the nature and persistence in time of structures created by invertebrates. Holorganic structures made by litter transformers allow a two-phase dynamics of organic matter (i.e., acceleration of mineralisation in a short time scale followed by a significant reduction due to compaction of the structure). Organomineral structures have a comparable effect on soil organic matter dynamics, but they may last for much longer periods. Because of their diversity and occasional abundance, these structures may significantly influence organic matter dynamics and soil physical properties, especially porosity and aggregation that determine water storage and circulation, and hence, potential erosion.

It is suggested that the three systems defined operate at nested scales of time and space and have decreasing overall effects on the determination of soil function in the order micro-foodwebs < litter transformers < ecosystem engineers. Nonetheless, ecosystem engineers are less able to withstand high levels of natural or anthropogenic environment constraints and lower order groups may become predominant with significant differences in the function of the ecosystem. Modifications of soil fauna communities may lead to losses of diversity and result in losses of functions when specific structural patterns or regulation mechanisms are lost.

II. INTRODUCTION

Soils host an extremely high diversity of organisms. In one hectare of temperate forest, several hundred species of soil invertebrates may coexist (Schaefer and Schauerman, 1990; André et al., 1994). It has been – and it still is – an enormous task for zoologists to identify and classify all these species. However, the perception of the functional importance of soil invertebrates is as old as the interest for their classification. Aristotle called earthworms 'the intestine of the earth' and Darwin was the first of a long lineage of zoologists who have been fascinated by the unwearying and multifarious activity of earthworms. His book *The Formation of Vegetable Mould Through the Action of Worms with Observations on Their Habits* (1881) and the work of Müller on humus forms (1887) are still famous products of an active scientific community which created soil ecology (Bal, 1982). At the same time, Dokuchaev (1889) formulated the basic concepts of pedology, but soil biology and pedology have largely ignored each other for several decades while they devoted considerable efforts to classifications and the separate elucidation of basic processes.

In the 1970s, the International Biological Programme produced a large amount of quantitative data on the abundance of soil invertebrates and their participation to fluxes of energy in the ecosystem (see e.g., Petersen and Luxton, 1982). It was concluded that their direct participation to the release of CO_2 from soils was limited to a few percent of the total, micro-organisms being by far the major producers of CO_2. During the following decade, considerable efforts were devoted to the description and quantification of the direct and indirect effects of soil invertebrates on major processes of the soil function, i.e. regulation of microbial activities, the dynamics of soil organic matter and nutrient cycling, and the formation and maintenance of the physical structure (Lavelle, 1978; Coleman, 1985; Anderson et al., 1983; Eschenbrenner, 1986; Barois, 1987; Martin, 1991; Andren et al., 1990; Blanchart, 1992). Experiments in small laboratory designs called microcosms, and field observations and experimentations, mainly developed over short time periods, gave a large amount of results, sometimes quite contradictory (Anderson, 1987). The effects of soil invertebrate activities on plant growth have been considered in a number of studies, as a means ultimately to test their effects on fertility (Stockdill, 1959, 1982; Setälä and Huhta, 1991; Spain et al., 1992; Okello-Oloya and Spain, 1986; Pashanasi et al., 1992).

These experiments have demonstrated that (1) in small laboratory designs, soil invertebrates often have effects which are largely disproportionate to their abundance and biomass, especially on short-term processes such as the release of mineral-N and P from decomposing litter, or aggregation of soil; (2) in field conditions, these short term effects may not be significant over a greater scale of time and space since climate, soil characteristics or the chemical composition of organic matter may have a larger impact in the determination of soil processes than invertebrate activities. Under specific conditions, however, invertebrate pests, e.g., phytoparasitic nematodes or coleopteran larvae may have dramatic effects on crop production.

The rise of problems linked to soil management has recently stimulated the development of multifactorial experimental approaches which attempt to integrate biological, chemical and physical soil components and processes into comprehensive models of soil function. They have been applied to maintain the sustainability of soil fertility, monitor the storage and release of carbon and pollutants in a globally changing environment, and to quantify and understand the function of biodiversity (Coleman *et al.*, 1984a, 1989; Swift, 1986; Anderson and Flanagan, 1989; Lavelle *et al.*, 1993; Moore *et al.*, 1993).

Two major kinds of approaches may be distinguished: the foodweb approach that tries to describe and quantify fluxes of carbon and nutrients through trophic networks associating mainly Protozoa, nematodes and microarthropods to micro-organisms, and evaluate the impact of such interactions on soil–plant systems (Ingham *et al.*, 1986; Andren *et al.*, 1990; Moore and De Ruiter, 1991). The other approach considers 'Biological systems of regulation', i.e. large entities such as earthworms or termites or roots and the part of the soil that they influence, and tries to elucidate mechanisms of the soil function at the scale of these specific units (Lee and Wood, 1971; Coleman *et al.*, 1984b; Lavelle, 1984; Clarholm, 1985; Lee, 1985; Wood, 1988; Abbadie and Lepage, 1989; Lavelle *et al.*, 1992c).

This paper aims first at linking these diverse approaches into a general conceptual model of the relationships between soil invertebrates and the other animate and inanimate components of the soil system. A functional classification of soil organisms based on their adaptive strategies is then attempted. Three large functional groups (guilds) of soil invertebrates are defined and their effects on soil organic matter dynamics, nutrient cycling and the soil physical structure are detailed.

III. THE ROLE OF INVERTEBRATES IN THE SOIL SYSTEM

Soils receive energy that flows through the ecosystem in the form of (1) biological energy as dead organic matter, (2) physical energy, i.e. the convective and diffusive energies linked to water infiltration and drainage and temperature changes, and (3) chemical energy released by oxidations and hydrolyses.

Physical and chemical energies are largely dissipated through the processes of soil formation. These involve alteration and fragmentation of the bedrock at scales of time of a minimum of 10 000 years, and a minimum spatial scale of (e.g.) a water catchment. Biological energy received as decomposing residues is dissipated along the decomposer foodweb at much shorter scales of time and space. The energy flow based on the decomposition of organic resources creates structures and releases partially reduced chemical compounds which further participate in the above-mentioned processes of alteration by means of oxidative reactions (Berthelin *et al.*, 1979). Soil faunal activities may become significant determinants of pedogenesis, e.g. when they facilitate the stabilisation of acid organic compounds by mixing them to clay mineral elements, or when they accumulate organomineral pellets which are the component elements of aggregated structures. In some cases the participation of earthworm casts to the formation of soils is so evident that they have been called vermisols (Nye, 1955; Wielemaker, 1984, Eschenbrenner, 1986; Pop and Postolache, 1987).

Decomposition of organic residues is almost a purely biological process: 80–99.9% of CO_2 released by soils are issued from oxidative digestion processes (Seastedt *et al.*, 1987; Moorehead and Reynolds, 1989; Scharpenseel *et al.*, 1989). The rates at which this energy flows, and the forms that it takes, are determined by environmental factors and specific properties of soil organisms. Interactions with other organisms and the creation of specific microsites where these interactions occur, are characteristic features of the impact of organisms on soil processes.

Soil invertebrate activities are part of the multiple factors that determine microbial activities. Physical and chemical determinants, and the effects of soil invertebrates, operate at different scales of time and space; this network of interactions among all these determinants of microbial activities may be best represented by a hierarchical model in which climatic factors that operate at the largest scale of time and space are likely to constrain edaphic determinants (especially the abundance and mineralogy of clays and nutrient stocks), the quality of organic inputs, and finally activities of soil invertebrates and roots (Lavelle *et al.*, 1993; Figure 1).

Soil micro-organisms, roots and invertebrates have complementary adaptive strategies whereby they influence three major processes in soils, i.e. decomposition and the dynamics of soil organic matter, the formation and maintenance of the soil structure, and nutrient and water supply to plants.

IV. ADAPTIVE STRATEGIES OF SOIL ORGANISMS: THE SLEEPING BEAUTY AND THE ECOSYSTEM ENGINEERS

Soil organisms have evolved in an environment that imposes three major requirements, (1) to move in a compact environment with a loosely connected porosity, (2) to feed on low-quality resources and (3) to adapt to the occasional drying or flooding of the porous space (Lavelle, 1988). A continuum of adaptive strategies based on size is observed in soils, from micro-organisms to macroinvertebrates (Swift *et al.*, 1979). The apparent contradiction between the short potential generation time of micro-organisms (*ca* 20 h, Clarholm and Rosswall, 1980) and their slow turnover time (1–1.5 years, e.g. Jenkinson and Rayner, 1977; Chaussod *et al.* 1986) has been defined as the 'sleeping beauty paradox' based on evidence that soil invertebrates and root activities were necessary to stimulate this dormant microflora (Lavelle *et al.*, 1994). Microbial communities are both numerous and diverse, but they are largely dormant. Their inability to move in the compact soil environment limits their activity to the immediate microsite in which they reside. They are to

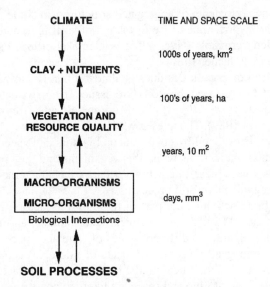

Fig. 1. An hierarchical model of the factors determining soil processes in terrestrial ecosystems (modified from Lavelle *et al.*, 1993).

a great extent dependent on larger organisms, roots and soil fauna, for access to new substrates. Among the organisms thus acting as 'Prince Charmings', special attention is paid to the 'ecosystem engineers', i.e. those that are able to modify the soil environment through their mechanical activities (Stork and Eggleton, 1992).

Soil invertebrates have a continuum of strategies from the smallest microfauna that colonize the water-filled pore space in the same way as do microorganisms, to macrofauna that alter the soil environment to their own needs. The size of invertebrates and the aquatic or aerial nature of their respiration, reflect their way of adapting to spatial constraints. Three groups have been distinguished (Bachelier, 1978; Swift *et al.*, 1979), i.e.

(1) *Microfauna*, which comprise aquatic invertebrates living in the water-filled soil porosity. They are small, less than 0.2 mm, on average, and mainly include Protozoa and Nematodes, plus other groups of lesser importance like tardigrades, rotifers

(2) *Mesofauna*, which comprise microarthropods (mainly collembolans and acarids) and the small Oligochaeta, Enchytraeidae, which have an average size of 0.2–2 mm and live in the air-filled pore space of the soil and litter

(3) *Macrofauna*, which include invertebrates larger than 2 mm, on average. Termites, earthworms and large arthropods are the main components of this group. They have the ability to dig the soil and create specific

structures for their movements and living activities (e.g. burrows, galleries, nests and chambers) plus casts and faecal pellets resulting from their feeding activities. These organisms have also been called 'ecosystem engineers' for their ability to profoundly affect the soil structure and hence major soil processes via the structures that they build (Stork and Eggleton, 1992).

Soil invertebrates have limited ability to digest the complex organic substrates of the soil and litter, but many have developed interactions with microflora that permit them to exploit soil resources. With increasing size, the relationship between microflora and fauna gradually shifts from predation to mutualisms of increasing efficiency. The concept of a foodweb becomes more and more difficult to apply as the trophic structure becomes increasingly, 'fluid and interactive with individual species operating on several levels which might be distinguished as trophically different' (Swift et al., 1979). Excrement of invertebrates is of the utmost importance in the evolution of organic matter (Martin and Marinissen, 1993), in the formation and maintenance of soil structure and, over long periods of time, to specific pedological processes called 'zoological ripening of soils' (Bal, 1982).

Three major guilds of soil invertebrates may be distinguished on the basis of the relationship that they have with soil micro-organisms and the kind of excrement that they produce (Figure 2).

Micro-foodwebs mainly comprise microfauna that are predators of bacteria and fungi, and their predators. Microfauna do not appear to produce recognizable solid excrements and hence, the effect of these invertebrates on soil organic matter dynamics is not prolonged in structures that are stable for some time after deposition. They have, however, a significant impact on population dynamics of micro-organisms and the release of nutrients immobilised in microbial biomass (Trofymow and Coleman, 1982; Clarholm, 1985). This process is especially developed in the rhizosphere. Predatory Acarina or Collembola, and even larger invertebrates (earthworms) may extend this foodweb over several trophic levels.

Litter transformers mainly comprise mesofauna and large arthropods which normally ingest purely organic material and develop an external ('exhabitational' *sensu* Lewis, 1985) mutualism with microflora based on the 'external rumen' (Swift et al., 1979). Litter arthropods may digest part of the microbial biomass or develop mutualistic interactions in their faecal pellets: in these structures, organic resources which have been fragmented and moistened during the gut transit, are actively digested by microflora. After some days of incubation, arthropods often reingest their pellets and absorb the assimilable organic compounds that have been released by microbial activity, and occasionally, part of the microbial biomass (Hassal and Rushton, 1982). This specific type of exhabitational mutualism is called the external rumen type of digestion (Swift et al., 1979). Inside this general adaptive strategy, a

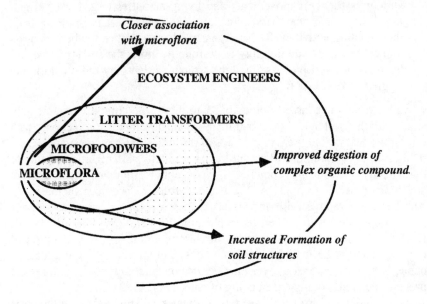

Fig. 2. Regulation of microbial activities by three guilds of soil invertebrates operating at increasing scales of time and space.

large diversity of behaviours may be identified (Vannier, 1985). Some insects such as Diptera (Sciaridae) which have comparatively efficient enzymatic machinery produce liquid faeces that they never reingest (Deleporte and Rouland, 1991). Their digestion system is not able to digest tannin–protein compounds and highly polymerized polysaccharides (lignin) efficiently and these accumulate and hamper the progress of decomposition (Minderman, 1968; Gourbière, 1982; Toutain, 1987). The acid organic compounds released in the course of decomposition are not flocculated in the presence of mineral particles; they behave as aggressive compounds that may leach and actively participate in the weathering of minerals, thus favouring processes like podzolisation (Berthelin et al., 1979; Pedro, 1989).

Finally, *the ecosystem engineers*, comprise macrofauna, mainly earthworms and termites, that are large enough to develop mutualistic relationships with microflora inside their gut proper. These interactions may involve obligate (such as the protozoa contained in the posterior pouch of lower termites) or facultative symbionts; the latter occur in the gut of higher termites and also in earthworms (Barois et al., 1987; Breznak, 1984). These organisms usually ingest a mixture of organic and mineral elements. Organic acids produced by digestion, and the subsequent incubation of organic matter in casts, are normally flocculated in the presence of clay minerals and a high microbial activity. Digestion is efficient and complex organic compounds like cellulose, lignin and tannin–protein complexes are assimilated (Butler and Buckerfield, 1979; Breznak, 1984; Toutain, 1987; Rouland et al., 1990). Faecal pellets which are large (in the range 0.1–2 cm and more) may be the component elements of macro-aggregate structures, and participate prominently in the formation of stable structures through the regulation of porosity, aggregation, bulk density and surface features (Bal, 1982). These organisms also build large structures, such as mounds and networks of galleries and chambers, which have significant impacts on the evolution of soils at medium time scales. Ants may be considered as part of this group, although the vast majority of them use soil only as an habitat and have a limited impact on soil organic matter dynamics. They will not be considered in this paper because of the small number of studies devoted to their effects on soil processes.

Whenever conditions are suitable for their activities, macrofauna, and especially earthworms and termites, become major regulators of microbial activities within their sphere of influence (i.e., the termitosphere of termites and the drilosphere of earthworms, Lavelle, 1984) in which they also determine the abundance and activities of smaller groups of soil fauna (Dash et al., 1980; Yeates, 1981). These 'Biological Systems of Regulation' include the rhizosphere in which roots are the major determinant (Lavelle et al., 1993).

The apparent looseness of the soil trophic structure may be due to the juxtaposition of two fundamentally opposite types of relationships between invertebrates and micro-organisms, i.e. (1) a 'classical' foodweb in which organisms of a given size feed on organisms of a lower size, at a lower level in the foodweb, and are eaten by larger organisms which comprise the upper level; and (2) biological systems of regulation (BSRs) which are systems of interactions based on the mutualism between micro-organisms and invertebrates of different sizes.

The overall structure of the foodweb is further complicated by the effect of interactions between BSRs; in some places, for example, favourable conditions may promote high levels of activity of termites or anecic earthworms and, hence, significantly reduce resources available for the development of litter transformers and micropredators.

V. MICROFAUNA: MICROBIAL REGULATION IN MICRO-FOODWEBS

A. The Composition of Micro-foodwebs

Micro-foodwebs are defined as the part of the foodweb that links micro-organisms to their predators; they are clearly distinguished from systems organised by litter arthropods and the ecosystems engineers that are based on mutualistic relationships of larger organisms with microflora.

The soil micropredator foodweb includes micro-organisms, mainly bacteria and fungi, Protozoa, nematodes and some predacious Acarina. Several levels of complexity exist since the system includes microbial grazers, predators and superpredators. Nonetheless, the foodchain length is normally limited to three or four transfers (Moore *et al.*, 1993).

The structure of micro-foodwebs has been established in a number of sites (Hendrix *et al.*, 1986; Elliott *et al.*, 1988; Ingham *et al.*, 1989) and models have been built to simulate the effects of such assemblages on soil processes (Hunt *et al.*, 1987). The composition of foodwebs is largely influenced by the predominance of either fungi or bacteria, which in turn is determined by abiotic factors (Whitford, 1989) and management practices (Hendrix *et al.*, 1986).

Microflora are generally divided into bacteria and fungi, which perform distinct functions and often colonise distinct microsites. Small populations of predatory bacteria and fungi may be found. They are bacterial predators of bacteria, e.g., of the genus *Bdellovibrio* (Casida, 1988) or nematode-trapping fungi (Cooke and Godfrey, 1964; Mitsui, 1985). Other fungi of the genus *Dactylella* have been reported to capture and consume amoebae.

Mycophagy by protozoa has rarely been studied quantitatively but the apparent ease with which mycophagous forms are isolated suggests that it is not rare (Petz *et al.*, 1986). They are voracious consumers: a single heterotrophic nanoflagellate may ingest three to nine bacteria per hour and 8300 bacteria are consumed for each new amoeba produced in soil microcosms at 23°C (Bloem *et al.*, 1989). Stout and Heal (1967) estimated that Protozoa consume 150–900 g bacteria m^{-2} $year^{-1}$. Bamforth (1988) distinguishes two kinds of Protozoa: a group which may serve as food for a large variety of organisms and others which are endosymbionts (commensals or parasites). The latter are regularly found in the gut of some termites and earthworms (Astomata) and may have endosymbiotic bacteria in their cytoplasm. Protozoa of the former group are often considered as transformers of bacterial protoplasm into higher trophic levels. Mycophagous amoebae and colpodid ciliates may also ingest fungal spores and thus play a role in the suppression of phytopathogenic fungi and control of fungal populations (Chakraborty *et al.*, 1983; Coûteaux, 1985a).

Nematodes are another constant component of microfoodwebs, at the same level, or one step above Protozoa. A few adult nematodes may actually ingest flagellates and some amoebae (Elliott *et al.*, 1980) although a vast majority are either bacterial- or fungal-feeders. Nematodes are attracted to their food sources through a number of stimuli including pheromones, CO_2 and temperature gradients (Freckman, 1988). Bacterial-feeders ingest up to 5000 cells min^{-1} or 6.5 times their own weight per day. Overall consumption may be as much as 800 kg bacteria ha^{-1} $year^{-1}$ and the amount of nitrogen turned over in the range of 20–130 kg (review by Coleman *et al.*, 1984). Finally, mycophagous nematodes have been reported to suppress plant diseases by ingesting pathogenic fungi (Curl, 1988).

Some components of mesofauna, i.e. Acarina, Collembola and Enchytraeidae, may also be part of micropredator foodwebs, although the majority of them probably rely more on mutualistic relationships than predation (Moore, 1988). Predatory mites can feed on nematodes (e.g., Martikainen and Huhta, 1990). In North American desert systems, they have been reported to regulate nematode density (Whitford, 1989).

The vast majority of macrofauna rely on mutualistic relationships with microflora for their digestion and, hence, do not participate in microfoodwebs. Nevertheless, some degree of predation exists, especially on Protozoa. For example, earthworms can digest ciliates (Piearce and Phillips, 1980) and amoebae (Rouelle, 1983). On the other hand, Protozoa may participate to the function of the external rumen type of digestion in association with earthworms: incubation in faeces results in the excystment and development of protozoan populations which are further digested by invertebrates. In a microcosm experiment, earthworms were introduced into sterilized soil. After 5 days, the density of Protozoa in the soil was $3–70 \times 10^6$ g^{-1} soil with a predominance of amoebae (Shaw and Pawluk, 1986).

B. Distribution and Dispersal of Microsites

The activity of soil microfoodwebs mainly occurs in the water-filled soil porespace and water films which cover solid particles. These microsites have a defined distribution in time and space. They depend largely on soil moisture content and porosity, which in turn depend on texture and the overall biological activity. Finally, the presence of carbon resources determines the accumulation of a large microbial biomass which feeds micropredators. In predetermined conditions of substrate availability, the size of pores determines composition of the communities at the microsite level. Flagellates and small amoebae colonize pores of a minimum diameter of 8 μm; larger Protozoa and nematodes live in larger pores and most nematodes live in the external medium, outside the aggregates (Hattori, 1988). Soil texture and porosity are therefore critical determinants of microfoodwebs. The importance of porosity was demonstrated by an experiment in which bulk density of humus was increased from 0.25 to 0.41 g cm^{-3}, the abundance of pores larger than 10 μm decreased, and growth of ciliates and thecamoebae was inhibited (Coûteaux, 1985b).

The availability of organic resources and transportation of invertebrates across soil microsites are critical to the multiplication of microsites favourable to micropredator foodweb activities. As a result, such microsites are regularly found in the rhizosphere where root growth provides energy for bacteria and their micropredators and the growth of roots consequently disseminates communities in the volume of soil explored by the root system. In the presence of guilds of larger invertebrates, microsites may be included in the structures created by these invertebrates, i.e. the termitosphere of termites, or simply disappear.

C. Evidence for the Functional Importance of Micro-foodwebs: Effects on Soil Organic Matter (SOM) Dynamics

The functional importance of micro-foodwebs has been demonstrated several times in small-scale laboratory designs or microcosms. Effects on carbon and nutrient cycles have been observed and quantified.

In the carbon (C) cycle, respiration of microfauna only represents a few per cent of overall soil respiration (0.6–2% for nematodes according to Freckman, 1988). Nonetheless, their effect is disproportionate to their size and they significantly stimulate the growth and turnover of microbial populations, thus promoting faster rates of mineralisation, decomposition and nutrient turnover (Hendrix et al., 1986; Hunt et al., 1987; Sohlenius et al., 1988; Setälä et al., 1990, 1991).

The nitrogen (N) cycle is particularly affected by micro-foodweb interactions since bacteria have C:N ratios of 6:1 on average, close to that of Protozoa (5:1 according to Reich, 1948, in Stout and Heal, 1967) and slightly inferior to that of nematodes (10:1). Predation of micro-organisms thus results in the release of mineral-N that may be further used by plants. In the presence of nematodes for example, ammonification by bacteria and nematodes is greatly increased (Ingham et al., 1985). N fixation may be enhanced through the maintenance of a young population, provision of stimulating compounds or decreased O_2 concentration (Darbyshire and Greaves, 1973).

The phosphorus (P) cycle is also affected, although contrasting results have been produced. Coleman et al. (1984a) observed that, in the presence of amoebae, a significant part of P accumulated in microbial biomass is transformed into bicarbonate-extractable P. Nonetheless, the flux of P affected by this transformation may be limited; this would explain why few effects on P cycle have been noted, even in microcosms (Baath et al., 1980).

Positive effects of microfoodweb activities on plant growth have been measured in microcosms: when wheat plants were grown in sterilised soil into which both bacteria and Protozoa had been added, their production was increased by 80% and N mineralisation from soil organic matter was increased by 59%. In a control soil without plants, no increase in inorganic nitrogen was observed, which emphasises the role of root-derived carbon as a source of energy in the system (Clarholm, 1984). In another 35-day microcosm-reduced experiment, bacterial biomass was eight-fold and N uptake by plants was increased by 20% by protozoan activity. In the presence of Protozoa, shoot:root ratio of the plants increased and 65% more bacterial ^{15}N was taken up by plants (Kuikman and van Veen, 1989).

VI. SAPROPHAGOUS ARTHROPODA:
THE LITTER TRANSFORMERS

A. Niche

Mesofauna and non-social large litter arthropods live in the air-filled pore-space of soil and litter; the smallest live inside the soil in cracks and crevices, the largest in the looser strata of the A_0 horizon. They are big enough to create distinct structures in which microbial activities are regulated: these are the faecal pellets that accumulate in the litter layers and in the rhizosphere. However, they do not ingest significant amounts of mineral soil and have a limited ability to dig the soil and create galleries or aggregates. The kind of mutualism that they develop with microflora seems to be mostly exhabitational (i.e., in structures created outside the bodies of both components of the association) and no examples of inhabitational mutualist relationships are known.

These invertebrates, which do not possess a proper cellulase, mainly rely on the *external rumen* type of digestion to digest polysaccharides: they periodically reingest their faeces (or faeces produced by other invertebrates) and thus take advantage of the release of assimilable compounds resulting from microbial incubation (Swift *et al.*, 1979). They comminute, fractionate and moisten the ingested material thereby enhancing microbial activity. Most litter-feeding epigeic arthropods use this digestive system. Macrotermitine termites or ants of the tribe Attini, which cultivate fungi on especially elaborated wood or leaf material, belong to this group, although they also develop inhabitational relationships with microflora. Epigeic earthworms and Mollusca which are restricted to the litter layers also belong to the guild of litter transformers.

B. Structures

Litter transformers influence soil processes by (1) disseminating fungal spores and mycelium and (2) accumulating faecal pellets which are basic structural units of the holorganic layers (AH) and microsites for enhanced microbial activity.

The size and shape of faecal pellets produced by arthropods are highly variable (Bal, 1982). However, the vast majority of them are small (<100 μm) and unstable aggregates made up of untransformed organic matter. They have a low proportion of mineral elements and the rate of decomposition, which is often enhanced in fresh pellets, may be lower in ageing excrements than in the surrounding litter (Reyes and Tiedje, 1976; Hanlon, 1981; Griffiths *et al.*, 1989) due to compaction and accumulation of chemical compounds that are resistant to decomposition.

C. Effects on Physical Parameters of Soil

Litter invertebrates normally ingest small amounts of mineral particles, if any at all. Their contribution to physical soil processes is therefore limited, although some components, such as Enchytraeidae, may occasionally produce intensive burrow systems in selected microsites (Jegen, 1920, in Didden, 1990; Wolters, 1991). They are active agents of fragmentation and transfer of litter material to deeper strata where they accumulate and gradually form the amorphous organic matter of the H layer. Some groups, however, may ingest significant proportions of mineral elements and play some role in the development of the soil structure. Didden (1990) observed that 21–35% of the enchytraeids of a Dutch agro-ecosystem contained mineral particles and that the annual transfer of soil due to Enchytraeids amounted to 10–75 g m^2 of soil of the upper 40 cm, i.e. 0.001 to 0.01% of the total soil. The proportions of aggregates corresponding to the size of their faecal pellets and of the pores of a size corresponding to their body size, and air permeability were significantly increased. Nonetheless, differences were limited to a few per cent despite favourable experimental conditions (sieved soil and exclusion of other soil fauna components).

D. SOM Dynamics

Litter arthropods may ingest litter at different stages of decomposition although they prefer to ingest decayed litter when given a choice (Hassall *et al.*, 1987); Sciaridae (Diptera) which possess cellulase (Deleporte and Rouland, 1991), in contrast to the other arthropods, actively feed on leaves of the L litter layer and assimilate fresh better than previously decomposed material (Deleporte, 1987). Assimilation rates greatly vary among taxonomic groups and also depend on the quality of the litter ingested: they may vary from 9 to 92% in the Diptera larva *Bradysia confinis*, although most values are in the range 5–20% (Petersen and Luxton, 1982). Assimilation rates of different chemical components of litter show great differences: Bocock (1963) stated that the millipede *Glomeris marginata* utilises 6–10.5% of the dry matter contained in its food (*Fraxinus* litter), 43% of crude fat, 28% of holocellulose, 29% of soluble carbohydrates, and 0.3–0.4% of the nitrogen. Another part of the nitrogen is mineralised and appears as ammonium in faeces.

Microbial communities are modified in faecal pellets as a result of the digestion, compaction and fragmentation of the litter material. In isopod faeces derived from freshly fallen leaves of *Betula pendula*, densities of fungal and bacterial propagules, respectively, were increased by 3.2 and 126 times (Hassall *et al.*, 1987). Similar effects were observed by Hanlon (1981) with the millipede *Glomeris marginata* and the isopod *Oniscus asellus* fed oak litter: bacterial standing crop was multiplied by *ca* 10 and pH increased from 4.4 to 6.3–6.6. Differences in the relative response of bacteria and fungi appeared to be a result of the compaction of the litter material in faecal pellets: in artificially compacted samples consisting of small particles of litter (<0.2 µm) fungal growth was reduced whereas bacterial growth was not changed (Hanlon, 1981). Similar effects occurring in faecal pellets which mainly comprise fine particles would explain this difference.

The introduction of isopods into microcosms containing litter resulted first in a significant increase of CO_2 evolution, which was doubled with the optimal number of introduced isopods after 14 days (Hanlon and Anderson, 1980). Microbial respiration then decreased to levels which were still significantly higher than the control after 40 days with optimal numbers of arthropods; in the treatment with a higher density of isopods respiration was lower than in the control. These results suggest that invertebrates promote, in their excrement, an active respiration which is further decreased, possibly owing to changes in the composition of the microflora (decrease of fungal densities) and the compaction of the litter material (Figure 3).

In a comparable study, Hassall et al. (1987) concluded that the overall effect of arthropods on chemical and physical parameters of litter does not necessarily accelerate decomposition very significantly. The major process whereby they accelerate decomposition is the deposition of their faecal pellets into more humid microsites, deeper in the soil profile.

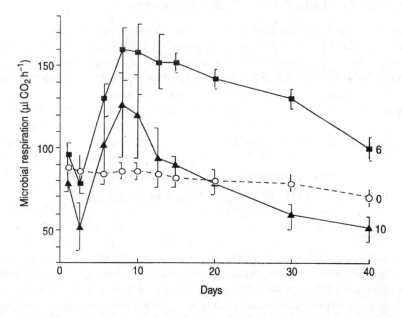

Fig. 3. Effects on microbial respiration of the isopod *Oniscus asellus* feeding on 1g of oak leaf litter in microcosms. 0, 6, 10: number of individuals introduced (Hanlon and Anderson, 1980).

One major limitation of the guild of litter arthropods seems to be their inability to digest complex tannin–protein compounds which normally block a large proportion of proteins from the cytoplasm of fresh leaves. In such conditions, decomposition of these compounds is slow and only proceeds through weak hydrolyses which slowly unwind the complex phenolic chemical structures of litter (Toutain et al., 1982).

Another major characteristic of litter invertebrate communities is their ability to stimulate nitrogen mineralisation and release ammonium, which may be further nitrified and taken up by plants, or temporarily accumulated before it is lost through volatilisation or immobilised in microbial biomass (Verhoeff and De Goede, 1985). In laboratory experiments, Anderson et al. (1985) established a correlation between the biomass of litter invertebrates and mineral-N release. In a field experiment using microlysimeters, Anderson et al. (1983) showed that during the first 32 weeks, from spring to leaf fall, the activity of soil fauna (13 g fresh weight m^{-2}) increased the total nitrogen losses by 11.4 kg ha^{-1} (5.2 kg NO-N + 6.2 kg NH_4-N), which represents about one third of the annual nitrogen input through litter. After leaf fall, less mineral-N was released in the presence than in the absence of fauna. Roots when present, absorbed about half of the mineral nitrogen.

E. Effects on Plant Growth in Microcosms

Effects of the guild of litter transformers on chemical and physical characteristics of the litter system may have positive effects on plant growth. This effect has been demonstrated several times in laboratory microcosms and larger field designs that simulated conditions of the real forest. Setälä and Huhta (1991) observed significant effects of litter invertebrates on the growth of seedlings of Betula pendula: after two years, leaf, stems and root biomass were, respectively, 70%, 53% and 38% greater and respective N and P contents were 3 and 1.5 times higher.

VII. TERMITES AND EARTHWORMS: THE ECOSYSTEM ENGINEERS

A. Roles

Termites and earthworms are major components of the guild of 'ecosystem engineers'. Ants and some large arthropods may to some extent be included in this group. When present, these organisms play a dominant role in the regulation of soil processes because (1) they ingest a mixture of organic and mineral particles and their faecal pellets are large and contain stabilised organic matter; (2) they have efficient symbiotic digestion systems in association with soil microflora which enable them to digest the most complex substrates, i.e.

tannin–protein complexes, lignin and humic compounds; and (3) they create diverse and abundant structures in soil and hence, have a dramatic effect on the soil physical structure and pedogenesis at different scales of time and space (Figure 4). The abundance and texture of these structures are determined by the consumption rates and selective ingestion of mineral and organic particles. Behavioural characteristics of species, and, to some extent, local edaphic and climatic conditions determine the type of structures created, which in turn affect all soil processes.

B. Effects on Physical Structure

1. Selective Ingestion of Particles

Soil invertebrates often select the particles that they ingest for their size and quality (organic or mineral). The behaviour of endogeic earthworms is highly variable and depends on the species, the size of the individuals and soil type. For example, when kept in an uncompacted, sieved soil in laboratory cultures, the common pantropical species *Pontoscolex corethrurus* preferentially ingested small, 0–20 μm, particles in a coarse-textured regosol from Mexico, but coarser particles in a fine-textured ultisol (Barois, 1992 in Lavelle *et al.*, 1992a). There is some evidence that earthworms may select independently for mineral and organic particles and, when given the choice ingest large organic and small mineral particles. However, in a natural soil, endogeic earthworms ingest the soil to feed and move into this compact environment, which does not leave much possibility for a true selection of particles. It is likely that any selection observed results from the selective exploitation of micro-environments and selective behaviours that are still unknown.

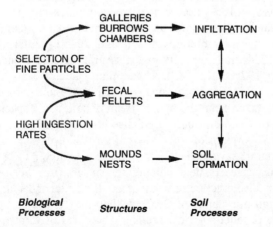

Fig. 4. Effects of macro-invertebrates on soil structure.

Termite distribution and activity are both affected by, and affect soil texture. The mound building termites are largely excluded from heavy clay, or excessively sandy soils which are not convenient for the building of their nests (Lee and Wood, 1971; Holt and Coventry, 1982). Termites generally select the smaller particles from within the soil profile and bring to the surface significant amounts of clay materials (Williams, 1968; Boyer, 1982). Species-specific effects and variations depending on the soil type have been often mentioned, although a comprehensive understanding of these effects is still lacking (Lee and Wood, 1971; Butler and Buckerfield, 1979; Rajagopal *et al.*, 1982; Spain *et al.*, 1983; Anderson and Wood, 1984; Lòpez-Hernàndez *et al.*, 1984; Garnier-Sillam *et al.*, 1987; Okwakol, 1987; Cox *et al.*, 1989).

2. Ingestion Rates and Bioturbation

Earthworms and termites may ingest large amounts of soil and litter, and hence actively participate to the process of bioturbation and become major regulators of the dynamics of litter and SOM in the ecosystem. Anecic earthworm populations, that live in subvertical burrows and feed on a mixture of leaf litter and soil, may incorporate the total annual litter production in as little as 2–3 months when they are abundant (Lee, 1985). Their activity however, is limited by food availability and the environmental determinants of their activity. In a pasture in southern France, Bouché *et al.* (1983) estimated that earthworms had assimilated 30% of the carbon available in litter in 17 weeks. Such a figure is only acceptable if earthworms reinvest the litter contained in their casts several times. In regularly burned humid savannahs in the Côte d'Ivoire, anecic earthworms incorporate 180–510 kg of dry litter to the soil annually. This represents *ca* 30% of the weight of litter annually decomposed (i.e. mineralised or incorporated into the soil instead of being destroyed by fire), but less than 10% of the annual production (Lavelle, 1978). Endogeic earthworms which mainly feed on soil from the A_1 horizon, and anecic species which ingest a mixture of soil and 20–50% litter (Bouché and Kretzschmar, 1974) may annually ingest several hundred tonnes of dry soil per ha (Lavelle, 1978; Lee, 1985). A maximum value of 1250 Mg ha^{-1} year^{-1} has been measured in humid savannahs at Lamto (Côte d'Ivoire) where this soil is normally taken from the upper 10 cm of soil; some species, however, particularly in the humid tropics, specialise in the ingestion of soil from the deep (20–50 cm) horizons of soils. Most of this soil is egested inside the soil profile: in the humid savannahs of Lamto (Côte d'Ivoire), 3 to 18% of the soil is egested as surface casts, depending on the species (Lavelle, 1978).

Termites may also ingest significant amounts of litter and soil: in low-lying parts of Sahelian grasslands of Sénégal, they may consume up to 49% of annual herbage production and in semiarid pastures of Kenya, they have a similar impact to grazing mammals with an annual consumption estimated at

1 Mg ha^{-1} (Lepage, 1974, 1981). Several estimates indicate rates of consumption in the range 1–16% of annual primary production in tropical forests and savannahs (Lee and Wood, 1971; Collins, 1983). Soil brought to the surface in surface nests and sheathings may represent up to 2400 Mg ha^{-1} and cover 10% of the soil surface (Meyer, 1960).

3. Structures

Macro-invertebrate activities create three types of structures: nests and chambers that host individuals or parts of a society; casts which may be deposited at the soil surface or inside the profile; and voids, pores and galleries that result from the movements of the invertebrates in soil or at the surface (like surface galleries of termites).

Nests may be compact and somewhat complex structures made of selected materials and/or faecal pellets; they may remain intact for several decades (as long as the colony is alive). At the death of the colony, they start to decay. This is a slow process that may leave visible mounds, often representing islands of soils with a relatively high fertility on which the density of trees is high (Lavelle et al., 1992c). Some termites and ants may have diffuse nests made of small subterranean chambers of a few centimetres connected by a dense network of galleries. Termites also produce surface sheathings which cover the soil and litter when they go foraging.

Earthworms produce large amounts of casts; the majority are round-shaped and compact structures which are isolated from the rest of the soil by a thin cortex of clay minerals and organic colloids (Blanchart et al., 1993). Their size and shape depend on the size of the species, the soil texture and feeding habits of the species. Lee (1985) thus distinguish globular casts made of the fusion of oval-shaped pellets one to several millimetres in diameter into 'paste-like slurries', and granular casts made of an accumulation of small and fragile fine-textured pellets.

Galleries are a frequent product of activities of the 'ecosystem engineers': they may be subhorizontal and connect chambers of a nest, or subvertical and connect deep soil horizons (where anecic earthworms find suitable conditions of temperature and moisture, out of the reach of predators) to the surface, where they come at night to ingest litter. Endogeic geophagous earthworms also dig subhorizontal galleries that they immediately refill with their casts as they eat their way through the soil. Gallery networks may be extremely dense: in temperate pastures in France, Kretzschmar (1982) found 4000–29 800 galleries m^{-2} with a total length of 142–890 mm^{-2} depending on the season. 16.3% of these galleries were found deeper than 60 cm depth and the overall surface of gallery walls was 1.6–12 m^2 m^{-2}. In a large survey in several regions of France, Lopes-Assad (1987) examined 24 different soil profiles and found 60–290 active (i.e., opened to the surface) galleries m^{-2}; their total length ranged from 6.2 to 66.6 m m^{-2} and they comprised *ca* 0.6% of the total volume of soil. Galleries are cylindrical and their walls are regularly coated with cutaneous mucus each time that the worm passes through. Cast deposits and iron oxide depositions are regular features of these galleries (Jeanson, 1979; Kretzschmar, 1987; Lamparski *et al.*, 1987).

Termites may also dig extensive networks of galleries; the diameter of termite galleries is in the range 1–20 mm and their networks may comprise up to 7.5 km ha^{-1} (Darlington, 1982; Grassé, 1984; MacKay *et al.*, 1985, 1988; Wood, 1988). Some Macrotermitinae may open holes at the surface to use to go foraging at night; the overall surface of these openings has been estimated at 2–4 m^2 ha^{-1} (Lepage, 1981).

4. Effects on Soil Physical Parameters and Pedogenesis

The activities of earthworms and termites mainly affect surface roughness, porosity and aggregation with significant effects on hydraulic properties of soil, especially rates of water infiltration.

Infiltration of water and gases. Termite galleries and earthworm burrows have a significant effect on water infiltration despite their low contribution (<1%) to the soil volume (see review by Lee, 1985; for earthworms, Lepage, 1979). Experiments in arable land have demonstrated that when direct-drilling is substituted for ploughing, anecic earthworms are favoured. As a result of their enhanced activities, hydraulic conductivity at the interface between A and B horizons may be doubled (Urbanek and Dolezal, 1992). In West Africa, Casenave and Valentin (1988) found a significant relationship between the presence of earthworm casts and termite sheathings and water infiltration rates: infiltration increased with the percentage of soil covered by earthworm casts and termite sheathings and maximum values were obtained when more than 20% and 30% of the soil surface were covered with earthworm casts and termite sheathings, respectively. Kladivko *et al.* (1986) also concluded that earthworms could diminish surface crusting.

The increase of infiltration stems mainly from water retention owing to an increase of surface roughness and the presence of macropores and galleries below surface structures. Similar effects have been mentioned by several authors (Lal, 1988; Aina, 1984; Lavelle, 1992a; review by Lee, 1985). However, species-specific effects on water infiltration have been observed in earthworms: In pot experiments at Lamto (Côte d'Ivoire), Derouard *et al.* (1996) clearly distinguished species which compact the soil and decrease infiltration rates (e.g., *Millsonia anomala*) and species which decompact the soil and increase water infiltration (*Hyperiodrilus africanus* and *Chuniodrilus zielae*) whereas associations of species might give intermediate results (Figure 5). In the same experiment, there was evidence that the effect of earthworms on soil physical parameters might change, depending on the plant that was present in the experimental unit: under similar experimental conditions soil compaction and surface crusting were maximum in pots cropped to rice (*Oriza sativa*) and minimum in pots with peanut (*Arachis*).

Gas diffusion is affected in the same way as water infiltration, although diffusion around galleries is more important than direct diffusion along the burrows (Kretzschmar, 1989).

Similar effects have been observed with termites and their exclusion from the soil of a North American desert significantly increased the bulk density of soil and erosion while water infiltration was decreased (Elkins *et al.*, 1986).

Porosity. Earthworms and termites generally contribute to the maintenance of a relatively high porosity by digging galleries. Deposition of earthworm casts at the soil surface is a mechanism that regulates soil porosity, as a volume of voids equivalent to that of casts is created inside the soil. This may represent volumes of tens of m³ of voids ha⁻¹ year⁻¹ which preferably are created at times and in sites where compaction is occurring. Seasonal variations of soil porosity have been sometimes related to variations of the abundance of earthworm populations (Hopp, 1973). Correlations between soil porosity and colonisation of new soils by earthworms always results in a significant increase of porosity which may be doubled (Hoeksema and Jongerius, 1959).

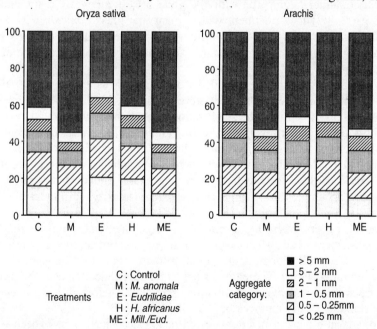

Fig. 5. Effect of different earthworm species or assemblages of species on the aggregation of a savanna alfisol in pot experiments over 80 days (Derouard, 1993). Blocks of a same category with different letters are significantly different (*P*<0.05). Control with plant but no worms

Soil aggregation. Endogeic earthworms especially affect the soil structure by promoting macro aggregation, i.e. the combination of soil particles into stable compound structures larger than 2 mm diameter. These earthworms may annually ingest between a few hundreds to more than 1000 Mg dry soil ha^{-1} (Lavelle, 1978). In four arable soils with unfertilised and fertilised barley, a grass plot and a lucerne layer, the annual cast production of *Aporrectodea caliginosa* was estimated at 36–108 Mg ha^{-1}, of which 20–50% was deposited at the soil surface. In humid tropical grasslands, endogeic earthworms may ingest daily 5–6 (*Pontoscolex corethrurus*) and up to 25–30 (*Millsonia anomala*) times their own weight of soil (Lavelle, 1978; Lavelle *et al.*, 1987). As a result, the overall annual production of casts may be as high as several hundreds, and up to 1250 Mg ha^{-1}. A small proportion of these is deposited at the soil surface and subterranean casts are the component units of stable macro-aggregate structures.

Experiments have demonstrated that such structures may be built in surprisingly short periods of time, thus pointing to the fact that factors other than the simple production of globular casts participate in aggregation (Blanchart *et al.*, 1989): in 33 days, the activity of 5 g of *Millsonia anomala* resulted in the formation of 2883 g of aggregates (i.e., 42.4% of the soil in the experimental container), compared with 906 g (13.3%) in a control treatment and 1075 g (15.8%) in a treatment with a plant, the grass *Panicum maximum.* Production of casts by the earthworms was estimated at 1815 g, i.e. 63% of the aggregates formed; the remaining 37% might be due to proliferation of fungal hyphae.

When earthworms were excluded from a soil in which they had built a macro-aggregate structure, this structure remained stable for a long time (at least several years) due to the stabilisation of aggregates with time. However, the introduction of earthworms producing granular casts causes destruction of the structure as large aggregates are split into much smaller and fragile ones (Blanchart *et al.*, 1996) As a result, there is a regulation of soil aggregation with small earthworms breaking large aggregates and thus preventing an excessive accumulation of large compact aggregates which otherwise may negatively affect plant growth (Rose and Wood, 1980).

Faecal pellets of some termites are organomineral micro-aggregates (Garnier-Sillam *et al.*, 1987) which form the walls of termite mounds and even surface soil horizons of some tropical soils (Wielemaker, 1984; Eschenbrenner, 1986). These effects are largely dependent on adaptive strategies of species and more information is required before a comprehensive understanding of these processes is available.

Long-term effects of invertebrates on pedogenesis have not yet been comprehensively considered and studies that specifically address this issue are needed. A few studies have indicated that termites and earthworms may be significant determinants of pedogenetic processes in the upper 30–60 cm of soils

(Lee and Wood, 1971; Pop and Postolache, 1987). Vermic soils with high earthworm activity have been recognised as a separate entity in the USDA '7th approximation'. Pop and Postolache (1987) state that 'by definition, a normally developed soil must be vermic (because) when defining mull, Kubiena states 'Practically all aggregates are earthworm casts or residues of them'. There is also some evidence that termites may influence pedogenesis in deeper soil strata but more studies are needed (Wielemaker, 1984; Eschenbrenner, 1986).

C. Effects on SOM dynamics

1. Digestion of SOM: Importance of Priming Effects

Termites and earthworms have highly efficient digestion systems which allow them to digest woody material or live on the highly dispersed and low-quality soil organic matter.

Earthworms digest soil organic matter with enzymes that are produced partly by the worm itself and partly through a mutualist association with the ingested microflora. They add 80–120% water to the ingested soil and 5–38% of the dry weight of soil as mucus, a readily assimilable substrate that they energetically mix in the anterior part of the gut. The addition of mucus to the soil triggers a priming effect (*sensu* Jenkinson, 1966) on soil organic carbon: microbial activity is greatly enhanced. Although the earthworm is able to produce some of the enzymes that are found in the gut content, such important enzymes as cellulase and mannanase (which degrades an important component of root material) are actually released by micro-organisms in the gut lumen (Zhang *et al.*, 1993). In the second part of the gut, when mucus has disappeared, micro-organisms digest soil organic matter and digestion products are then partly absorbed by the worm (Barois and Lavelle, 1986; Martin *et al.*, 1987; Trigo and Lavelle, 1992; Lavelle and Gilot, 1994).

Priming effects have not been described in the gut of termites. These insects seem to be able to produce a large variety of enzymes which interact with enzymes produced by the associated microflora to digest cellulose, lignin and other components of SOM (see, e.g. Rouland *et al.*, 1988, 1991). Some termites have protozoan symbionts whereas fungus growers rely on a mixture of inhabitational mutualism with microflora in their gut, and an external rumen with fungi cultivated in fungus gardens. Assimilation rates in the termites are high and published values range between 54% and 93% of the food eaten (Wood, 1978). Endogeic earthworms which feed on soil organic matter may live on soils with only 1% organic matter with assimilation rates of a few per cent (Martin and Lavelle, 1992).

2. Short- and Long-term Effects

Earthworms and termites have contrasting effects on SOM dynamics depending on the scale (Figure 6).

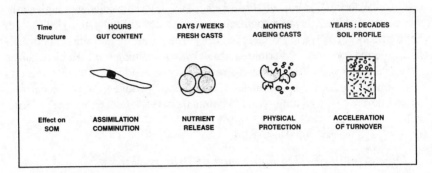

Fig. 6. Effects of earthworms on SOM (soil organic matter) dynamics at different scales of time and space.

At the short scale (hours and mg to cg of soil) of a gut transit, mineralisation of SOM is strongly increased. In a few hours (and even 30 min in the case of some tropical earthworms) part of the ingested SOM will have been assimilated. Detailed studies on temperate and tropical earthworms with bomb radiocarbon and natural ^{13}C labelling demonstrate that they mostly assimilate young organic matter from the labile fractions (Scharpenseel *et al.*, 1989). However, *Millsonia anomala*, an endogeic African earthworm is able to digest all particle-size fractions of a savannah alfisol with the same efficiency, which would mean that they are able to digest some resistant organic matter (Martin *et al.*, 1991). Earthworm digestion results in the fragmentation of the ingested organic debris and the release of significant amounts of mineral nitrogen and phosphorus (Sharpley and Syers, 1976; Barois *et al.*, 1987; Lavelle *et al.*, 1992b). The ability of earthworms to release mineral N and P from resources with high C:N and C:P ratios is a result of (1) their high efficiency at digesting compounds – that they excrete as intestinal and cutaneous mucus – which leaves nutrients in excess; (2) the rapid turnover of nitrogen in biomass (Ferrière and Bouché, 1985; Barois *et al.*, 1987; Cortez *et al.*, 1989), and (3) the continuation of mineralisation for some hours in freshly deposited casts resulting from enhanced microbial activity. Effects on phosphorus are especially interesting since part of the pool which is normally adsorbed in soil may be desorbed after a transit through earthworm guts (Lopez-Hernandez *et al.*, 1993; Brossard *et al.*, 1996). Estimates of the overall release of assimilable N range from 25 to 150 kg mineral N ha⁻¹ in tropical grasslands (Barois *et al.*, 1987; Lavelle *et al.*, 1992b) whereas Brossard *et al.* (1996) estimated the amount of mineral-P-released in a vertisol in Martinique covered with pasture at 50 kg ha⁻¹.

Although termites have a highly efficient digestion, they do not appear to release important amounts of assimilable nutrients in their faeces. They have developed highly efficient mechanisms for N conservation, and N-fixing organisms have been found in their guts as in the guts of earthworms (Breznak, 1984). Another specific feature of their digestion is the release of methane by humivorous and, to a lesser extent, xylophagous, and fungus-growing termites.

At the scale of days to months, earthworms and termites affect the dynamics of SOM in the structures that they have created: in earthworm casts, mineralisation rates rapidly decrease after a few days following their deposition. In casts of *Millsonia anomala* fed with a poor sandy alfisol, concentrations of NH_4-N in casts decrease to control levels after 8 days (Lavelle et al., 1992b). In casts of European Lumbricidae (Syers *et al.*, 1979) and the pantropical species *P. corethrurus* fed with richer soils (Lavelle *et al.*, 1992b), the amount of NH_4-N and NO-N is still higher than in the control 2 weeks after the deposition of casts. Over a longer period, mineralisation is inhibited in the compact structure of casts and, after 1 year of incubation at field capacity and 28°C, the decomposition rate in casts of *M. anomala* was three times lower than in a control soil sieved at 2 mm. Interestingly, there was clear evidence in this experiment that the coarse organic fractions had been significantly protected from decomposition whereas the decomposition of the finest fractions was increased (Martin, 1991). Termites also have significant effects on the release of mineral-N in the structures that they create. The walls of chambers of fungus gardens of *Trinervitermes geminatus* have much higher potentials for N-mineralisation than the bulk soil in short-term incubations (Abbadie and Lepage, 1989).

In the long term, i.e. from a few years to decades, the overall effect of earthworms and termites is still uncertain because there are few long term experiments. In 3-year field experiments comparing low-input agricultural systems with and without earthworms, differences between treatments were limited: the introduction of earthworms had not impeded the rapid loss of organic matter following the conversion to annual cultures, and only slight differences were seen in the proportion of different particle size fractions (Gilot, 1994; Pashanasi *et al.*, 1996). However, after the fifth crop at Yurimaguas (Peruvian Amazonia), soil organic content decreased more slowly in earthworm treatments and a significant protection of SOM started to be apparent. In a similar experiment conducted at Lamto (Côte d'Ivoire) there was no evidence of a better conservation of SOM due to the presence of earthworms. However, the quality of SOM had changed because standard laboratory tests indicated that CO_2 evolution was significantly higher in the earthworm treatment than in the control. There are, therefore, some indications that the presence of earthworms may limit the losses of organic matter in disturbed systems such as annual crops and modify the quality of SOM by operating a relative protection of young organic matter and facilitating the assimilation of part of the old resistant organic matter. Long-term experiments

118   P. LAVELLE

based on natural ^{13}C labelling of organic matter will allow this hypothesis to be tested. It is likely that the effect of earthworms will depend in the end on a hierarchy of factors including soil characteristics (especially clay mineralogy and abundance), the quality and quantity of organic inputs and last, but not least, species-specific effects. The diet of earthworms and the shape and structure of their casts will probably be critical criteria to consider when looking for a comprehensive explanation of the process involved.

In the long-term, termites tend to accumulate the undigested organic matter of their faeces in the walls of termitaria; they are conserved as organomineral compounds which resist decomposition. This is, for example, the case for the humivorous termites of the species *Thoracotermes macrothorax*, a dominant species in the Mayombe forest in Congo (Garnier Sillam *et al.*, 1991) Generally, the organic matter content of walls of the termitaria may be up to 50% greater than in the surrounding soil. This organic matter has a low C:N ratio and forms stable complexes with clay minerals (Lee and Wood, 1971; Arshad and Schnitzer, 1987; Garnier-Sillam *et al.*, 1987; Okwakol, 1987). Some fungus-growing termites such as *Macrotermes mulleri* do not incorporate their faeces to the walls of their termitaria and, consequently, the walls have lower contents in organic matter than the surrounding soil. These termites do not appear to play any role in the long-term conservation of soil organic matter in the ecosystem (Garnier-Sillam *et al.*, 1987).

D. Effects on Plant Growth

The effects of macro-invertebrates on soil processes, at different scales of time and space, may considerably affect soil fertility and, ultimately, result in dramatic changes of plant growth. Small-scale experiments have given somewhat variable results owing to species-specific effects and large-scale field experiments are still scarce.

The effect of termites on plant growth and production is still largely ignored. An abundant literature describes the negative effects of termites on crops and wood products and the ways to get rid of them (Harris, 1969; Vinson, 1986; Cowie *et al.*, 1991; Kumarasinghe, 1991; Rajagopal *et al.*, 1991). Indirect effects *via* the modification of soil processes induced by termite activities do not appear to have been evaluated. Experiments have demonstrated that soil from the walls of nests of *Amitermes laurensis* may be much better for plant growth than the surrounding soil (Okello-Oloya and Spain, 1986). This effect was significant with soil taken within 1m from the termite mound and at a depth of up to 50 cm. This effect has long been recognised by indigenous populations and the use of soil from some termitaria as fertiliser is not an uncommon practice (e.g., Swift *et al.*, 1989). However, no comprehensive scientific study has ever evaluated this practice in the general context of a farming system.

The effects of earthworms on plant growth have been well documented, especially in short-term pot experiments. Most studies indicate clearly positive effects on plant growth (Stockdill, 1959, 1982; Rose and Wood, 1980; Hoogerkamp et al., 1983; Lee, 1985; Senapati et al., 1985a; Curry and Boyle, 1987; Tomati et al., 1988; Buse, 1990; Galli et al., 1990; Clements et al., 1991; Haimi et al., 1992; Pashanasi et al., 1992; Spain et al., 1992 for a complete review). Nonetheless, there is some evidence that not all plants respond equally to the activities of earthworms of a given species: Pashanasi et al., (1992) observed a 14–24-fold increase of biomass of seedlings of the tree *Bixa orellana* in 120-day pot experiment after the inoculation of *Pontoscolex corethrurus*, whereas the growth of *Eugenia stipitata* seedlings was increased only 1.6–2.5-fold and that of the palm-tree *Bactris gasipaes* was 1.8–2.7 lower. A likely explanation was that *Bixa*, having a dense root system made of fine roots, was much more receptive to earthworm activities than the little developed and coarse root system of *Bactris*. Other pot experiments have demonstrated that the effect of earthworms may considerably vary depending on the species used and biomass. Some combinations of earthworm and plant species may give negative results and an excessive, or insufficient biomass may also have zero or negative effects on plant growth (Derouard et al., 1996). Plants produced in the presence of earthworms may have higher nutrient contents (Spain et al., 1992).

Larger-scale field experiments have confirmed these observations, although these effects tend to be much lower owing to limitation of earthworm activities by climatic factors or organic resources provided by soils. Conversely, the introduction of earthworms may sometimes trigger a flush of activity which may not be sustained for a long time if the increase of production does not result in an increased ability of the soil to sustain large populations.

Introduction of earthworms in ecosystems which did not have them, or that were colonised by an inefficient population often gives spectacular results (Hoogerkamp et al., 1983). In New Zealand, the replacement of native grasses by European species resulted in better production but a thick mat of slowly decomposing litter accumulated at the surface. The introduction of European Lumbricidae accelerated the incorporation of litter to soil and pasture production was further improved (Stockdill, 1959; 1982). The introduction of earthworms is currently used in projects of land reclamation or creation of new soils as, for example, in polders (Hoogerkamp et al., 1983; Marinissen and van den Bosche, 1992).

Experiments have been carried out to test the effect of inoculation of earthworms in tropical soils from Mexico, Peru and Côte d'Ivoire submitted to slash-and-burn low-input agriculture (Barois et al., in Lavelle et al., 1992; Gilot, 1994; Pashanasi et al., 1996). Soils under that type of management generally have very small populations due to the lack of adaptable populations. A significant correlation was observed between the increase in plant production and earthworm biomass (Figure 7).

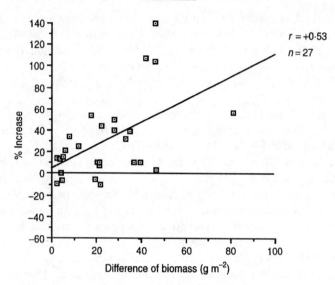

Fig. 7. Relationship between the difference of earthworm biomass measured between a control and a plot inoculated with earthworms and the increase of grain production (maize or rice) in field experiments carried out at three sites of the humid tropics (Lavelle and Gilot, 1993), each point represents one crop.

The major limitation to earthworm biomass was the ability of the system to provide fresh organic matter to feed the worms. The addition to the soil of organic residues which are normally not used in farming systems is envisaged as a means to increase earthworm biomass; coir, sawdust, coffee residues might be used for that purpose (Lavelle *et al.*, 1993).

The mechanisms whereby earthworms stimulate plant growth are still unclear. They are certainly manifold and include (1) mobilisation of nutrient pools that normally are not available (e.g., adsorbed phosphorus and nutrients contained in resistant organic fractions); (2) favourable changes in water and oxygen supply to roots; (3) more efficient use of nutrients based on an improved synchronisation and juxtaposition of nutrient release and nutrient uptake by plants (Swift, 1986; van Noordwig and De Willigen, 1986); and (4) a 'hormone-like' effect of earthworms (Tomati *et al.*, 1988). In experiments where ^{15}N-labelled plant material or microbial biomass had been added, the recovery of ^{15}N in plant biomass is always improved in the presence of earthworms (Lavelle *et al.*, 1992a; Spain *et al.*, 1992).

VIII. CONCLUSIONS

Significant advances have been made in the understanding of the role of soil fauna in soil function. Their relation to soil fertility is better understood although several detailed mechanisms are still unclear. It is now possible to revisit a few old questions and myths of soil science regarding soil zoology and propose novel research avenues.

Do the soil fauna create fertility, or is it fertility that sustains faunal communities? There is now clear evidence that faunal activities contribute to soil fertility since they play a large role in the transformations of soil organic matter and nutrients, at different scales of time and space, which influences their turnover and conservation, and probably improves the efficiency of the use of nutrients by plants. Digestive processes and physical structures created by soil invertebrates contribute significantly to these improvements. However, mechanical activities necessary to these processes have a cost in terms of carbon that the ecosystem must be able to provide. Faunal activities consume carbon whereas nutrients are only temporarily immobilised in their biomass. In humid savannahs of Côte d'Ivoire at Lamto, for example, earthworms which annually turn over *ca* 1000 Mg dry soil ha^{-1} annually assimilate 1.2 Mg ha^{-1} of soil organic matter containing some 700 kg carbon, of which 96% is used for maintenance and mechanical activities. Studies on soil fertility and sustainable use often put a major emphasis on nutrient cycling and stocks. Our synthesis supports the commonly accepted idea that SOM is the key element to sustain fertility; however, more attention should be paid specifically to the energetic role of SOM, which is poorly addressed. An optimised production (or importation) and allocation of C should be considered as an important objective, as much as nutrient supply and conservation of the soil structure. This might prevent cultivated soils having such deficient faunal activities, as is usually observed, and would probably improve the sustainability of their use. Carbon budgets should be considered at relatively large scales, from the watershed catchment (in which 'parcels' of different shapes with different treatments may interact, and exchange carbon and fauna), or region where transfers of carbon residues may be operated, to the farming system such as using sawdust or coir produced from industrial settlements to low-input agriculture. These include an accurate evaluation of the quantity and quality of the organic matter produced, the allocation of carbon and nutrients from this source to soil biota and the different compartments of soil organic matter

The contribution of soil invertebrates to pedogenesis. The role of biological processes in pedogenesis has rarely been directly addressed. The role of organic acids released by litter transformers in their faecal pellets, and the impact of the structures built – and transfers of elements operated by – large invertebrates (ecosystem engineers) should be quantified and incorporated into models of soil formation. This is particularly true of tropical soils in which termite and earthworm activities may have considerable impact on soil formation through bioturbation and the transit of enormous amounts of soil through the gut. Intestines of macro-invertebrates are powerful microbial and chemical reactors in which unexpected modifications of minerals and organomineral compounds are likely to occur. Changes in the crystallisation of clay minerals and desorption of P from resistant pools are examples of processes that are suspected to occur (Boyer, 1982; Lopez-Hernandez *et al.*, 1993; Brossard *et al.*, 1996).

Reaction to human activities and global environmental changes. Soil invertebrates should be considered as a resource that is highly sensitive to human impacts. High-input agriculture based on annual crops has a measurably negative impact on their communities through mechanical disturbance, depletion of assimilable carbon and non-target effects of chemicals. Attention should be paid to conserve biodiversity of soil invertebrates and assess the impact of land-use practices on their spatial distributions, at different scales, from that of a parcel to that of a watershed catchment and regional and biogeographical scales. Dynamics of communities during agricultural rotations or long-term disturbances is not known. Because of specific biogeographical features and slow rates of colonisation, the most efficient species in terms of stimulation of plant growth and soil conservation are not always present everywhere. This is especially true when human intervention creates novel ecosystems, such as a grassland in the middle of an old rainforest, or a fertilised pasture with high-quality grasses in a land that was always covered with low-quality grasses adapted to nutrient deficiencies. Under such circumstances, the management of soil fauna should be considered; pioneering introductions of European earthworms in New Zealand indicate that it is both feasible and useful. At a smaller scale, attention should be paid to the impact of the shape and relative localisation of parcels dedicated to different uses on the conservation and colonisation of soil fauna populations.

REFERENCES

Abbadie, L. and Lepage, M. (1989). The role of subterranean fungus comb chambers (Isoptera, Macrotermitinae) in soil nitrogen cycling in a preforest savannah (Côte d'Ivoire). *Soil Biol. Biochem,* 21 (8), 1067–1071.
Aina, P. O. (1984). Contribution of earthworms to porosity and water infiltration in a tropical soil under forest and long-term cultivation. *Pedobiologia* 26, 131–136.
Anderson, J. M. (1987). Interactions between invertebrates and microorganisms: noise or necessity for soil processes? In: *Ecology of Microbial Communities,* pp. 125–145. Cambridge University Press, Cambridge, UK.
Anderson J. M. and Flanagan P. (1989). Biological processes regulating organic matter dynamics in tropical soils. In: *Dynamics of Soil Organic Matter in Tropical Ecosystems* (Ed by T. Oades, G. Uheara and D.C. Coleman), pp. 97–125. NifTAL project, Univ. of Hawaii., Honolulu.
Anderson, J. M. and Wood, T. G. (1984). Mound composition and soil modification by two soil feeding termites (Termitinae, Termitidae) in a riparian Nigerian forest. *Pedobiologia* 26, 77–82.

Anderson J. M., Ineson, P. and Huish, S. A. (1983). The effects of animal feeding activities on element release from deciduous forest litter and soil organic matter. In: *New Trends in Soil Zoology*. (Ed. by Ph. Lebrun *et al.*), pp. 87–100. Dieu-Brichart, Ottignies-Louvain-la-Neuve, Belgium.

Anderson, J. M., Leonard, M. A., Ineson, P. and Huish, S. (1985). Faunal biomass: a key component of a general model of nitrogen mineralization. *Soil Biol. Biochem.* **17** (5), 735–737.

Andren, O., Lindberg, T., Boström, U., Clarholm, M., Hansson, A.C., Johansson, G., Lagerlöf, J., Paustian, K., Persson, J., Pettersson, R., Schnürer, J., Sohlenius, B. and Wivstad, M.A. (1990). 5. Organic carbon and nitrogen flows In: *Ecology of Arable Land – Organisms, Carbon and Nitrogen Cycling*, (Ed by O. Andren, T. Lindberg, K Paustian and T. Rosswall), pp. 84–125. Munksgaard, Copenhagen, Denmark.

Andre, H.M., Noti, M.I. and Lebrun, P. (1994). The soil fauna – the other last biotic frontier. *Biodiversity and Conservation*, **3**(1), 45–56.

Arshad, M. A. and Schnitzer M. (1987). The chemistry of a termite Fungus Comb. *Plant Soil* **98**, 247–256.

Baath, E., Berg, B., Lohm, U., Lundgren, B., Lundkvist, H., Rosswall, T., Söderström, B. and Wiren, A. (1980). Effects on experimental acidification and liming on soil organisms and decomposition in a Scots pine forest. *Pedobiologia* **20**, 85–100.

Bachelier, G (1978). *La Faune des sols, son Ecologie, son Action*. ORSTOM, Paris, France

Bal, L. (1982). *Zoological Ripening of Soils*. PUDOC, Wageningen, The Netherlands.

Bamforth, S.S. (1988). Interactions between protozoa and other organisms. *Agric. Ecosyst. Environ.* **24**, 229–234.

Barois, I. (1987). *Interactions entre les Vers de Terre (Oligochæta) tropicaux géophages et la microflore pour l'exploitation de la matière organique du sol*. PhD thesis, University of Paris VI. 301 pp.

Barois, I. and Lavelle, P. (1986). Changes in respiration rate and some physicochemical properties of a tropical soil during transit through *Pontoscolex corethrurus* (Glossoscolecidæ, Oligochæta). *Soil Biol. Biochem.* **18** (5), 539–541.

Barois, I., Verdier, B., Kaiser, P., Mariotti, A., Rangel, P. and Lavelle, P. (1987) .Influence of the tropical earthworm *Pontoscolex corethrurus* (Glossoscolecidæ) on the fixation and mineralization of nitrogen. In: *On Earthworms*, (Ed. by A.M. Bonvicini and P. Omodeo), pp. 151–158. Mucchi, Bologna, Italy.

Berthelin, J., Souchier, B. and Toutain, F. (1979). Intervention des phénomènes biologiques dans l'altération. *Sci. du Sol* **2** (3), 175–187.

Bocock, K.L. (1963). The digestion and assimilation of food by Glomeris. In: *Soil Organisms* (Ed. by Doeksen, J. and Van der Drift J.) pp. 86–91. North Holland, Amsterdam, The Netherlands

Blanchart, E. (1992). Restoration by earthworms (Megascolecidae) of the macroaggregate structure of a destructured savanna soil under field conditions. *Soil Biol. Biochem.* **24** (12), 1587–1594.

Blanchart, E., Lavelle, P. and Spain, A. (1989). Effects of two species of tropical earthworms (Oligochaeta: Eudrilidae) on the size distribution of aggregates in an african soil. *Rev. d'Ecol. Biol. du Sol* **26**, 417–425.

Blanchart, E., Bruand, A. and Lavelle, P. (1993). The physical structure of casts of *Millsonia anomala* (Oligochaeta: Megascolecidae) in shrub savanna soils (Côte d'Ivoire). *Geoderma*, **56** 119–132.

Blanchart, E., Lavelle, P., Brandean, E., Le Bissonais, Y. and Valentin, C. (1996). Regulation of soil structure by geophagous earthworm activities in humid savannahs of Côte d' Ivoire. *Soil. Biol. Biochem.* (in press).

124   P. LAVELLE

Bloem, J., EllenBroek, F.M., Bär-Gilissen, M.J.B. and Cappenberg, T.E. (1989). Protozoan grazing and bacterial production in stratified lake Vechten estimated with fluorescently labeled bacteria and by thymidine incorporation. *Appl. Environ. Microbiol.* **55** (7), 1787–1795.

Bouché, M.B. and Kretzschmar, A. (1974). Fonction des lombriciens. II. recherches méthodologiques pour l'analyse du sol ingéré (étude du peuplement de la station RCP 165/PBI). *Rev. d'Ecol. Biol. du Sol* **11** (1), 127–139.

Bouché, M.B., Rafidison, Z. and Toutain, F. (1983). Etude de l'alimentation et du brassage pédo intestinal du lombricien *Nicodrilus velox* (Annelida, Lombricidae) par l'analyse élémentaire. *Rev. d'Ecol. Biol. du Sol* **20** (1), 49–75.

Boyer, P. (1982). Quelques aspects de l'action des termites sur les argiles. *Clay Mineral.* **17**, 453–462.

Breznak, J.A. (1984). Biochemical aspects of symbiosis between termites and their intestinal microbiota, In: *Invertebrate–microbial interactions* (Ed by J.M. Anderson, A.D.M. Rayner and D.W.H. Walton), pp.173–24. Cambridge University Press, Cambridge, UK.

Brossard, M., Lavelle, P and Laurent, J.Y. (1996). Digestion of a vertisol by an endogeic earthworm (*Polypheretima elongata*, Megascolecidae) increases soil phosphate extractibility). *Euro. J. Soil Biol.* (in press).

Buse, A. (1990). Influence of earthworms on nitrogen fluxes and plant growth in cores taken from variously managed upland pastures. *Soil Biol. Biochem.* **22** (6), 775–780.

Butler, J.H.A. and Buckerfield, J.C. (1979). Digestion of lignin by termites. *Soil Biol. Biochem.* **11**, 507–513.

Casenave, A. and Valentin, C. (1989). *Les états de Surface de la Zone Sahélienne. Influence sur l'Infiltration.* ORSTOM, Paris, France.

Casida, L.E. Jr. (1988). Minireview: Nonobligate bacterial predation of bacteria in soil. *Microb. Ecol.* **15** 1–8.

Chakraborty, S., Theodorou, C. and Bowen, G. (1983). Amoebae from a take-all suppressive soil which feed on *Gaeumannomyces graminis tritici* and other soil fungi. *Soil Biol. Biochem.* **15**, 17–24.

Chaussod, R., Nicolardot, B., Catroux, G. and Chretien, J. (1986). Relations entre les caractéristiques physico-chimiques et microbiologiques de quelques sols cultivés. *Sci. du Sol* **2**, 213–226.

Clarholm, M. (1984). Microbes as predator or prey. Heterotrophic, free-living protozoa: neglected microorganisms with an important task in regulating bacteria populations. In: *Current Perspectives on Microbial Ecology* (Ed. by M.J. Klug and C.A. Reddy), pp.321–326. Washington DC, USA.

Clarholm, M. (1985). Possible roles for roots, bacteria, protozoa and fungi in supplying nitrogen to plants. In: *Ecological Interactions in Soil; Plants, Microbes and Animals* (Ed. by D. Atkinson, A.H. Fitter, D.J. Read and M.B. Usher), pp. 355–365. Blackwell Scientific Publications, Oxford, UK.

Clarholm, M. and Rosswall, T. (1980). Biomass and turnover of bacteria in a forest soil and a peat. *Soil Biol. Biochem.* **12**, 49–57.

Clements, R.O., Murray, P.J. and Sturdy, R.G. (1991). The impact of 20 years' absence of earthworms and three levels of N fertilizers on a grassland soil environment. *Agric. Ecosyst. Environ.* **36**, 75–85.

Coleman, D.C. (1985). Through a ped darkly: an ecological assessment of root-soil-microbial-faunal interactions. In: *Ecological Interactions in Soil; Plants, Microbes and Animals* (Ed. by D. Atkinson, A.H. Fitter, D.J. Read and M.B. Usher), pp. 1–21 Blackwell Scientific Publications, Oxford, UK.

Coleman, D.C., Anderson, R.V., Cole, C.V., McClellan, J.F., Woods, L.E., Trofymow, J.A. and Elliott, E.T. (1984a). Roles of protozoa and nematodes in nutrient cycling. In: *Microbial–plant Interaction* pp.17–28. ASA Special Publication No 47 Madison, E. Winsconsin.

Coleman, D.C., Ingham, R.E., McClellan, J.F. and Trofymow, J.A. (1984b). Soil nutrient transformations in the rhizosphere via animal-microbial interactions. In: *Invertebrates–microbial interactions.* (Ed. by J.M. Anderson, D.M. Rayner and D.W.H. Walton), pp. 35–58. Cambridge University Press, Cambridge, UK.

Coleman, D.C., Brussard, L., Beare, M.H., Hendrix, P.F., Hassink, J., Heijnen, C.E. and Marinissen, J.C.Y. (1989). *Microbial–Faunal Interactions as they Influence Soil Organic Matter dynamics.* pp.175–179. Kyoto, Japan.

Collins, N.M. (1983). Termite populations and their role in litter removal in Malaysian rainforests. In: *Tropical Rainforests: Ecology and Management* (Ed. by S.L. Sutton, T.C. Whitmore and L.O. Chadwick). Blackwell, Oxford, UK.

Cooke, R.C. and Godfrey, B.E.S. (1964). A key to the nematode-destroying fungi. *Trans. Bri. mycol. Soc.* **47**, 61–74.

Cortez, J., Hameed, R. and Bouché, M.B. (1989). C. and N. transfer in soil with or without earthworms fed with ^{14}C. and ^{15}N-labelled wheat straw. *Soil Biol. Biochem.* **21** (4), 491–497.

Coûteaux, M.M. (1985a). Relationships between testate amoebae and fungi in humus microcosms. *Soil Biol. Biochem.* **17** (3), 339–345.

Coûteaux, M.M. (1985b). Relation entre la densité apparente d'un humus et l'aptitude à la croissance de ses Ciliés. *Pedobiologia* **28**, 289–303.

Cowie, R.H., Logan, J.W.M. and Wood, T.G.(1991). Termite damage and control in tropical forestry. In *Advances in Management and Conservation of Soil Fauna* (Ed. by G.K. Veeresh *et al.*), pp.161–167. Oxford & IBH, New Dehli, India.

Cox, G.W., Gakahu, C.G. and Waithaka, J.M. (1989). The form and small stone content of large earth mounds constructed by mole rats and termites in Kenya. *Pedobiologia* **33**, 307–314.

Curl, E.A. (1988). The role of soil microfauna in plant-disease suppression. *CRC Crit. Rev. Plant Sci.* **7** (3), 175–196.

Curry, J.P. and Boyle, K.E. (1987). Growth rates, establishment and effect on herbage yield of introduced earthworms in grassland on reclaimed cutover peat. *Biol. Fertil. Soils* **3**, 95–98.

Darbyshire, J.F. and Greaves, M.P. (1973). Bacteria and Protozoa in the rhizosphere. *Pesticide Sci.* **4**, 349–360.

Darlington, J.P.E.C. (1982). The underground passages and storage pits used in foraging by a nest of the termite *Macrotermes michaelseni* in Kajiado, Kenya. *Kenya J. Zool.* **198**, 237–247.

Darwin, C. (1881). *The Formation of Vegetable Mould Through the Action of Worms with Observations on Their Habits.* Murray, London, UK.

Dash, M.C., Senapati, B.K. and Mishra, C.C. (1980). Nematode feeding by tropical earthworms. *Oikos* **34**, 322–329.

Deleporte, S. (1987). Rôle du Diptère Sciaridae *Bradysia confinis* (Winn., Frey) dans la dégradation d'une litière de feuillus. *Rev. d'Ecol. Biol. du Sol* **24**, 341–358.

Deleporte, S. and Rouland, C. (1991). Etude préliminaire de l'équipement digestif osidasique de *Bradysia confinis* (Diptera, Sciaridae): implications dans la dégradation de la matière organique. *Compte-Rend. l'Acad. Sci. Paris*, **312** (III), 165–170.

Derouard, L., Tondoh, J., Vilcosqui and Lavelle, P. (1996). Species-specific effects in the response of tropical annual crops to the inoculation of earthworms. Short-scale experiments at Lamto (Côte d'Ivoire). *Soil Biol. Biochem.*, (in press).

126   P. LAVELLE

Didden, W.A.M. (1990). Involvement of Enchytraeidae (*Oligochaeta*) in soil structure evolution in agricultural fields. *Biol. Fertil. Soils* **9**, 152–158.

Dokuchaev, V.V. (1889). The zones of Russia (in Russian). *Akad. Nauk Moscow*, **6**.

Elkins, N.Z., Sabol, G.Z., Ward, T.J., & G, W.W. (1986). The influence of subterranean termites on the hydrological characteristics of a Chihuahuan desert ecosystem. *Oecologia* **68**, 521–528.

Elliott, E.T., Anderson, R.V., Coleman, D.C. and Cole, C.V. (1980). Habitable pore space and microbial trophic interactions. *Oikos* **35**, 327–335.

Elliott, E.T., Hunt, H.W. and Walter, D.E. (1988). Detrital foodwebs interactions in north american grassland ecosystems. *Agric. Ecosyst. Environ.* **24**, 41–56.

Eschenbrenner, V. (1986). Contribution des termites à la macro-agrégation des sols tropicaux. *Cahiers ORSTOM, Sér. Pédolog.* **22** (4), 397–408.

Ferrière, G. and Bouché, M. (1985). Première mesure écophysiologique d'un débit d'éléments dans un animal endogé: le débit d'azote de Nicodrilus longus longus Ude (Lumbricidae, Ologochaeta) dans la prairie de Citeaux. *Comptes Rend. l'Acad. Sci. Paris* **301**, 789–794.

Freckman, D.W. (1988). Bacterivorous nematodes and organic-matter decomposition. *Agric. Ecosyst. Environ.* **24**, 195–217.

Galli, E., Tomati, U. and Di Lena, G. (1990). Effect of earthworm casts on protein synthesis in Agaricus bisporus. *Biol. Fertil. Soils* **9**, 290–291.

Garnier-Sillam, E., Villemin, G., Toutain, F. and Renoux, J. (1987). Contribution à l'étude du rôle des termites dans l'humification des sols forestiers tropicaux. In: *Micromorphologie des Sols – Soil Micromorphology* (Ed. by N. Fedoroff, L.M. Bresson and M.A. Courty), pp. 331–343. AFES, Paris, France.

Garnier-Sillam, E., Braudeau, E. and Tessier, D. (1991). Rôle des termites sur le spectre poral des sols forestiers tropicaux. Cas de *Thoracotermes macrothorax* Sjöstedt (Termitinae) et de Macrotermes mülleri (Sjöstedt) (Macrotermitinae). *Insect. Soc.* **38**, 397–412.

Gilot, C. (1994). *Effets de l'introduction du ver géophage tropical Millsonia anomala Omodeo en systèmes cultivés sur les caractéristiques des sols et la production végétale en moyenne Côte d'Ivoire.* PhD thesis, Paris VI/INAPG, France.

Gourbière, F. (1982). Pourriture blanche de la litière d'Abies alba Mill. I. – Evolution de la litière sous l'action des basidiomycètes du genre Collybia. *Rev. d'Ecol. Biol. du Sol* **19** (2), 163–175.

Grassé, P.P. (1984). *Termitologia.* Masson, Paris, France.

Griffiths, B.S., Wood, S. and Cheshire, M.V. (1989). Mineralisation of ¹⁴C-labelled plant material by *Porcellio scaber* (Crustacea, Isopoda). *Pedobiologia* **33**, 355–360.

Haimi, J., Huhta, V. and Boucelham, M. (1992). Growth increase of birch seedlings under the influence of earthworms. A laboratory study. *Soil Biol. Biochem.* **24** (12), 1525–1528.

Hanlon, R.D.G. (1981). Some factors influencing microbial growth on soil animal faeces. II. Bacterial and fungal growth on soil animal faeces. *Pedobiologia* **21**, 264–270.

Hanlon, R.D.G. and Anderson, J.M. (1980). Influence of Macroarthropod feeding activities on microflora in decomposing oak leaves. *Soil Biol. Biochem.* **12**, 255–261.

Harris, W.V. (1969). *Termites as Pests of Crops and Trees.* Commonwealth Institute of Entomology, London, UK.

Hassal, M. and Rushton, S.P. (1982). The role of coprophagy in the feeding strategies of terrestrial isopods. *Oecologia* **53**, 374–381.

Hassall, L., Turner, J.G. and Rands, M.R.W. (1987). Effects of terrestrial isopods on the decomposition of woodland leaf litter. *Oecologia* **72**, 597–604.

Hattori, T. (1988). Soil aggregates as microhabitats of microorganisms. *Rep. Inst. Agric. Res., Tohaten University*, **37**, 23–36.

Hendrix, P.F., Parmelee, R.W., Crossley, D.A. Jr., Coleman, D.C., Odum, E.P. and Groffman, P.M. (1986). Detritus food webs in conventional and non-tillage agroecosystems. *BioScience* **36** (6), 374–380.

Hoeksema, K.J. and Jongerius, A. (1959). On the influence of earthworms on the soil structure in mulched orchards. *Proc. Int. Symp. Soil Struct.* pp.188–194, Ghent.

Holt, J.A. and Coventry, R.J. (1982) Occurence of termites (isoptera) on cracking clay soils in northeastern Queensland. *J. Aust. Entomol. Soc.* **21**, 135–136.

Hoogerkamp, M., Rogaar, H. and Eijsackers, H.J.P. (1983). Effect of earthworms on grassland on recently reclaimed polder soils in the Netherlands. In: *Earthworm Ecology: from Darwin to vermiculture* (Ed. by J.E. Satchell), pp.85–105. Chapman and Hall, London, New York.

Hopp, H. (1973). *What every Gardener Should Know About Earthworms.* Garden Way Publishing Co., Vermont, USA.

Hunt, H.W., Coleman, D.C., Ingham, E.R., Ingham, R.E., Elliott, E.T., Moore, J.C., Rose, S.L., Reid, C.P.P. and Morley, C.R. (1987). The detrital food web in a shortgrass prairie. *Biol. Fertil. Soils* **3**, 57–68.

Ingham, R.E., Trofymow, J.A., Ingham, E.R. and Coleman, D.C. (1985). interactions of bacteria, fungi, and their nematode grazers: Effects on nutrient cycling and plant growth. *Ecolog. Monographs* **55**, 119–140.

Ingham, E.R., Trofymow, J.A., Ames, R.N., Hunt, H.W., Morley, C.R., Moore, J.C. and Coleman, D.C. (1986). Trophic interactions and nitrogen cycling in a semiarid grassland soil II. System responses to removal of different groups of soil microbes or fauna. *J. Appl. Ecol.* **23**, 615–630.

Ingham, E.R., Coleman, D.C. and Moore, J.C. (1989). An analysis of food web structure and function in a shortgrass prairie, a mountain meadow and a lodgepole pine forest. *Can. J. Soil Sci.* **66**, 261–272.

Jeanson, C. (1979). Structuration du sol par la faune terricole, incidences sur les concentrations organo-minérales. In: *Migrations Organo-minérales dans les Sols Tempérés*, pp.113–123. CNRS, Nancy, France.

Jenkinson, D.S. (1966). The Priming Action. *J. Appl. Rad. Isotopes Suppl.* pp.199–208.

Jenkinson, D.S. and Ladd, J.N. (1981). Microbial biomass in soil: measurement and turnover. In: *Soil Biochemistry* (Ed. by J.N. Ladd and E.A. Paul), pp.415–471. Dekker, New York, USA.

Jenkinson, D.S. and Rayner, J.H. (1977). The turnover of soil organic matter in some of the Kothamsted classical experiments. *Soil Sci.* **123**, 298–305.

Kladivko, E.J., Mackay, A.D. and Bradford, J.M. (1986). Earthworms as a factor in the reduction of soil crusting. *Soil Sci. Soc. Am. J.* **50** (1), 191–196.

Kretzschmar, A. (1982). Description des galeries de vers de terre et variations saisonnières des réseaux (observations en conditions naturelles). *Rev. d'Ecol. Biol. du Sol* **19**, 579–591.

Kretzschmar, A. (1987). Caractérisation microscopique de l'activité des lombriciens endogés. In: *Micromorphologie des Sols – Soil Micromorphology* (Ed. by N. Fedoroff, L.M. Bresson and M.A. Courty), pp.325–331. AFES, Paris, France.

Kretzschmar, A. (1989). Galeries de lombriciens en réseaux: structures fonctionnelles et signatures comportementales. PhD thesis, Paris XI, France.

Kuikman, P.J. and van Veen, J.A. (1989). The impact of protozoa on the availability of bacterial nitrogen to plants. *Biol. Fertil. Soils* **8**, 13–18.

Kumarasinghe, N.C. (1991). Studies on the damage by termites to seed setts and to postharvest stubbles of sugarcane plantations in the dry zone of Sri Lanka. In: *Advances in Management and Conservation of Soil Fauna*, (Ed. by G.K. Veeresh, D. Ranagopal and C.A. Viraktamath), pp.147–154. Oxford & IBH, New Dehli, India.

128 P. LAVELLE

Lal, R. (1988). Effects of macrofauna on soil properties in tropical ecosystems. *Agric. Ecosyst. Environ.* **24**, 101–116.

Lamparski, F., Kobel-Lamparski, A. and Kaffenberger, R. (1987). The burrows of *Lumbricus badensis* and *Lumbricus polyphemus*. In: *On earthworms* (Ed. by A.M. Bonvicini Pagliai and P. Omodeo), pp.131–140. Mucchi editore, Modena, Italy.

Lavelle, P. (1978). Les Vers de Terre de la savane de Lamto (Côte d'Ivoire): peuplements, populations et fonctions dans l'écosystème. Thèse d'Etat, Paris VI Publication du Laboratoire de Zoologie de l'ENS n°12.

Lavelle, P. (1984). The soil system in the humid tropics. *Biol. Int.* **9**, 2–15.

Lavelle, P. (1988). Earthworm activities and the soil system. *Biol. Fertil. Soil,* **6**, 237–251.

Lavelle, P. and Gilot, C. (1994). Priming effects of macroorganisms on microflora: A key process of soil function? In: *Beyond the Biomass,* (Ed. by K. Ritz, J. Dighton and K. Giller), pp. 176–181. Wiley–Sayce, Chichester, UK.

Lavelle, P., Barois, I., Cruz, C., Hernandez, A., Pineda, A. and Rangel, P. (1987). Adaptative strategies of *Pontoscolex corethrurus* (Glossoscolecidæ, Oligochæta), a peregrine geophagous earthworm of the humid tropics. *Biol. Fertil. Soils,* **5**, 188–194.

Lavelle, P., Alegre, J., Barois, I., Fragoso, C., Gilot, C., Gonzalez, C., Kanyonyo, ka Kajondo, Martin, A., Melendez, G., Moreno, A., Pashanasi, B., Patron, C. and Schaefer, R. (1992a). *Conservation of Soil Fertility in low Input Agricultural Systems of the Humid Tropics by Manipulating Earthworm Communities.* CCE–STD2 programme. Final report. ORSTOM, Bandy, France.

Lavelle, P., Melendez, G., Pashanasi, B., Szott, L. and Schaefer, R. (1992b) Nitrogen mineralization and reorganization in casts of the geophagous tropical earthworm *Pontoscolex corethurus (Glossoscolecidae). Biol. Fertil. Soil* **14**, 49–53.

Lavelle, P., Spain, A.V., Blanchart, E., Martin, A. and Martin, S. (1992c). The impact of soil fauna on the properties of soils in the humid tropics. In: *Myths and Science of Soils of the Tropics* (Ed. by P.A. Sanchez and R. Lal), pp. 157–185. SSSA Special Publication, Madison, Wisconsin, USA.

Lavelle, P., Blanchart, E., Martin, A., Martin, S., Barois, I., Toutain, F., Spain, A.and Schaefer, R. (1993). A hierarchical model for decomposition in terrestrial ecosystems. Application to soils in the humid tropics. *Biotropica* **25** (2), 130–150.

Lavelle, P., Lattaud, C., Trigo, D. and Barois, I. (1994b). Mutualism and Biodiversity in soils. *Pl. Soil* **170** (1), 23–33.

Lee, K.E. (1985). *Earthworms: Their Ecology and Relationships with Soils and Land use.* Academic Press, New York, USA.

Lee, K.E. and Wood, T.G. (1971). *Termites and Soils,* Academic Press, London, UK.

Lepage, M.G. (1974) *Les termites d'une savane sahélienne (Ferlo septentrional, sénégal): peuplement, populations, consommation, rôle dans l'écosystème.* PhD thesis, Dijon, France.

Lepage, M.G. (1979). *La récolte en strate herbacée de Macrotermes aff. subhyalinus* (Isoptera: Macrotermitinae) dans un écosystème semi-aride (Kajiado–Kenya). C.R. UIEIS section Française, Lausanne, 7–8 September 1979.

Lepage M.G. (1981). L'impact des populations récoltantes de *Macrotermes michaelseni* (Sjöstedt) (Isoptera, Macrotermitinae) dans un écosystème semi-aride (Kajiado–Kenya). I.–L'activité de récolte et son determinisme. *Insect. Soc.* **28** (3), 297–308.

Lewis, D.H. (1985). Symbiosis and mutualism: crisp concepts and soggy semantics. In: *Biology of Mutualism* (Ed. by D.H. Boucher), pp. 29–42. Croom Helm, Beckenham, UK.

Lopes–Assad, M.L (1987). *Contribution a l'étude de la macroporosité lombricienne de différents types de sols de France.* Université, USTL, Montpellier, France.

Lòpez–Hernàndez, D. and Febres, A. (1984). Changements chimiques et granulométriques produits dans des sols de Côte d'Ivoire par la présence de trois espèces de termites. *Rev. Ecol. Biol. du Sol* **21** (4), 477–489.

Lòpez–Hernandez, D., Fardeau, J.C. and Lavelle, P. (1993). Phosphorus transformations in two P-sorption contrasting tropical soils during transit through *Pontoscolex corethrurus* (Glossoscolecidae, Oligochaeta). *Soil Biol. Biochem.* **25** (6), 789–792.

MacKay, W.P. and Whitford, W.G. (1988). Spatial variability of termite gallery production in Chihuahuan desert plant communities. *Sociobiology* **14** (1), 281–289.

MacKay, W.P., Blizzard, J.H., Miller, J.J. and Whitford, W.G. (1985). Analysis of above-ground gallery construction by the subterranean termite *Gnathamitermes tubiformans (Isoptera: Termitidae). Environ. Entomol.* **14** (4), 470–474.

Marinissen, J.C.Y. and Van den Bosch, F. (1992). Colonization of new habitants by earthworms. *Oecologia* **91**, 371–376.

Martikainen, E. and Huhta, V. (1990). Interactions between nematodes and predatory mites in raw humus soil: a microcosm experiment. *Rev. d'Écol. Biol. du Sol* **27** (1), 13–20.

Martin, A. (1991). Short–term and long–term effect of the endoge ice earthworm *Millsonia anomala* (Omodeo) (Megascolecidae, Oligochaeta) of a tropical savanna, on soil organic matter. *Biol. Fertil. Soil.* **11**, 234–238.

Martin, A. and Lavelle, P. (1992). Effect of soil organic matter quality on its assimilation by Millsonia anomala, a tropical geophagous earthworm. *Soil Biol. Biochem.* **24** (12), 1535–1538.

Martin, A. and Marinissen, J.C.Y. (1993). Biological and physico–chemical processes in excrements of soil animals. *Geoderma* **56**, 331–347.

Martin, A., Cortez, J., Barois, I. and Lavelle, P. (1987). Les mucus intestinaux de Ver de Terre, moteur de leurs interactions avec la microflore. *Rev. d'Ecol. Biol. du Sol* **24** (4), 549–558.

Martin, A., Mariotti, A., Balesdent, J., Lavelle, P. and Vuattoux, R. (1990). Estimates of the organic matter turnover rate in a savanna soil by the ^{13}C natural abundance. *Soil Biol. Biochem.* **22** (4) 517–523.

Martin, A., Mariotti, A., Balesdent, J. and Lavelle, P. (1991). Soil organic matter assimilation of a geophagous tropical earthworm based on ^{13}C measurements. *Ecology* **73** 118–128.

Meyer, J.A. (1960). Résultats agronomiques d'un essai de nivellement des termitières réalisé dans la cuvette centrale Congolaise. *Bull. d'Agric. Congo Belge.* **51**, 1047–1059.

Minderman, G. (1968). Addition, decomposition and accumulation of organic matter in forests. *J. Ecol.* **56** 355–362.

Mitsui, Y. (1985). Distribution and ecology of nematode-trapping fungi in Japan. *JARQ* **18** (3), 182–193.

Moore, J.C. (1988). The influence of microarthropods on symbiotic and non-symbiotic mutualism in detrital-based below-ground food webs. *Agric. Ecosys. Environ.* **24**, 147–159.

Moore, J.C. and de Ruiter, P.C. (1991). Temporal and spatial heterogeneity of trophic interactions within below-ground foodwebs. *Agri. Ecosyst. Environ.* **34**, 371–397.

Moore, J.C., DeRuiter, P.C. and Hunt H.W. (1993). Soil invertebrate/micro-invertebrate interactions: disproportionate effects of species on food web structure and function. *Vet. Pathol.* **48**, 247–260.

Moorhead, D.L. and Reynolds, J.F. (1989). The contribution of abiotic processes to buried litter decomposition in the northern Chihuahuan Desert. *Oecologia* **79**, 133–135.

Muller, P.E. (1887). *Studien über die naturlichen Humusformen und deren Einwirkung auf vegetation und Boden.* Springer, Berlin, Germany.

Nye, P.H. (1955). Some soil-forming processes in the humid tropics. IV – The action of the soil fauna. *J. Soil. Sci.* **6** (1), 73–83.

Okello–Oloya, T. and Spain, A.V. (1986). Comparative growth of two pasture plants from northeastern Australia on the mound materials of grass and litter-feeding termites (Isoptera: Termitidae) and on their associated surface soils. *Rev. d'Écol. Biol. du Sol* **23** (4), 381–392.

Okwakol, M.J.N. (1987). Effects of *Cubitermes testaceus* (Williams) on some phy sical and chemical properties of soil in a grassland area of Uganda. *Afr. J. Ecol.* **25**, 147–153.

Pashanasi, B., Melendez, G., Szott, L. and Lavelle, P. (1992). Effect of inoculation with the endogeic earthworm *Pontoscolex corethrurus* (Glossoscolecidae) on N availability, soil microbial biomass and the growth of three tropical fruit tree seedlings in a pot experiment. *Soil Biol. Biochem.* **24** (12), 1655–1660.

Pashanasi, B., Lavelle, P. and Alegre, J. (1996). Effect of inoculation with the endogeic earthworm *Pontocolex corethrurus* on soil chemical characteristics and plant growth in a low-input agricultural system of Peruvian Amazonia. *Soil Biol. Biochem.* **28** (6), 801–810.

Pedro, G. (1989). Geochemistry, mineralogy and microfabric of soils. In: *Soils and their Management: a Sino-European Perspective* (Ed. by W.T. Maltby), pp. 59–90. Elsevier Applied Science, Bruxelles, Belgium.

Petersen, H. and Luxton, M. (1982). A comparative analysis of soil fauna populations and their role in decomposition processes. *Oikos* **39** (3), 287–388.

Petz, W., Foissner, W., Wirnsberger, E., Krautgartner, W.D. and Adam, H. (1986). Mycophagy, a new feeding strategy in Autochthonous soil ciliates. *Naturwissenschaften* **73**, 560–561.

Piearce, T.G. and Phillips, M.J. (1980). The fate of ciliates in the earthworm gut: an *in vitro* study. *Micro. Ecol.* **5**, 313–319.

Pop, V.V. and Postolache, T. (1987). Giant earthworms build up vermic mountain rendzinas, In: *On Earthworms* (Ed. by A. Bonvicini Pagliai and P. Omodeo), pp. 141–150, Mucchi, Italy.

Rajagopal, D., Sathyanarayana, T. and Veeresh, G.K. (1982). Physical and chemical properties of termite mound and surrounding soils of Karnataka. *J. Soil Biol. Ecol.* **2**, 1831.

Rajagopal, D., Veeresh, G.K. and Kumar, N.G. (1991). Assessment of losses of nutrients in dung and farmyard manure due to termite foraging In: *Advances in Management and conservation of soil fauna* (Ed. by G.K. Veeresh, D.R. Jagopal, C.A. Viraktamath. *et. al.*), pp. 155–160. Oxford and IBH, New Dehli, India.

Reyes, V.G. and Tiedje, J.M. (1976). Metabolism of 14C-labelled plant materials by woodlice (*Tracheoniscus rathkei* (brandt) and soil micro-organisms. *Soil Biol. Biochem.* **8**, 103–108.

Rose, C.J. and Wood, A.W. (1980). Some environmental factors affecting earthworms populations and sweet potato production in the Tari Basin, Papua New Guinea Highlands. *Papua N. G. Agric. J.* **31**, 1–10.

Rouelle, J. (1983). Introduction of amoebae and *Rhizobium japonicum* into the gut of *Eisenia foetida* (Sav.) and *Lumbricus terrestris* L. In: *Earthworm Ecology: from Darwin to Vermicuture* (Ed. by J.E. Satchell), pp. 375–381. Chapman and Hall, London.

Rouland, C., Civas, A., Renoux, J. and Petek, F. (1988). Synergistic activities of the enzymes involved in cellulose degradation, purified from *Macrotermes mülleri* and from its symbiotic, fungus *Termitomyces* sp. *Comp. Biochem. Physiol.* B **91** (3), 459–465.

Rouland, C., Brauman, A., Keleke, S., Labat, M., Mora, P. and Renoux, J. (1990). Endosymbiosis and exosymbiosis in the fungus-growing termites. In: *Microbiology in Poecilotherms* (Ed. by R. Lésel), pp. 79–82. Elsevier Science Publishers, B.V. (Biomedical Division), Amsterdam, The Netherlands

Rouland, C., Lenoir, F. and Lepage, M. (1991). The role of symbiotic fungus in the digestive metabolism of several species of fungus-growing termites. *Comp. Biochem. Physiol.* A **99** (4), 657–663.

Schaefer, M. and Schauermann, J. (1990). The soil fauna of beech forests: comparison between a mull and a moder soil. *Pedobiologia* **34**(5), 299–314.

Scharpenseel, H.W., Becker, Heidmann, P., Neue, H.U. and Tsutsuki, K. (1989). Bomb-carbon, ^{14}C-dating and ^{13}C-measurements as tracers of organic matter dynamics as well as of morphogenetic and turbation process. In *The Science of Total Environment*, pp. 99–110. Elsevier Science, Amsterdam, The Netherlands.

Seastedt, T.R., Todd, T.C. and James, S.W. (1987). Experimental manipulations of arthropod, nematode and earthworm communities in a north American tallgrass prairie. *Pedobiologia* **30**, 9–17.

Senapati, B.K., Pani, S.C. and Kabi, A. (1985). Effects of earthworm and green manuring on paddy. *Proc. Soil Biol. Symp,* Hisar, February 1985, pp. 71–75.

Setälä, H. and Huhta, V. (1991). Soil fauna increase *Betula pendula* growth: laboratory experiments with coniferous forest floor. *Ecology* **72** (2), 665–671.

Setälä, H., Martikainen, E., Tynismaa, M. and Huhta, V. (1990). Effects of soil fauna on leaching of nitrogen and phosphorus from experimental systems simulating coniferous forest floor. *Biol. Fertil. Soils* **10**, 170–177.

Setälä, H., Tyynismaa, M., Martikainen, E. and V, H. (1991). Mineralization of C, N and P in relation to decomposer community structure in coniferous forest soil. *Pedobiologia* **35** (5), 285–296.

Sharpley, A.N. and Syers, J.K. (1976). Potential role of earthworm casts for the phosphorus enrichment of run-off waters. *Soil Biol. Biochem.* **8**, 341–346.

Shaw, C. and Pawluk, S. (1986). The development of soil structure by *Octolasion tyrtaeum, Aporrectodea turgida* and *Lumbricus terrestris* in parent materials belonging to different textural classes. *Pedobiologia* **29** 327–339.

Sohlenius, B., Boström, S. and Sandor, A. (1988). Carbon and nitrogen budgets of nematodes in arable soil. *Biol. Fertil. Soils* **6**, 1–8.

Spain, A.V., John, R.D. and Okello-Oloya, T. (1983). Some pedological effects of selected termite species at three locations in north-eastern Australia. In: *New Trends in Soil Biology*, (Ed. by P.H. Lebrun *et al.,*), pp. 143–149. Dieu-Brichart, Ottignies-Louvain-la-Neuve, Louvain-la-Neuve, France.

Spain, A.V., Lavelle, P. and Mariotti, A. (1992). Stimulation of plant growth by tropical earthworms. *Soil Biol. Biochem.* **24** (12), 1629–1634.

Stockdill, S.M.J. (1959). Earthworms improve pasture growth. *N. Z. J. Agric.* **98**, 227–233.

Stockdill, S.M.J. (1982). Effects of introduced earthworms on the productivity of New Zealand pastures. *Pedobiologia* **24**, 29–35.

Stork, N.E. and Eggleton, P. (1992). Invertebrates as determinants and indicators of soil quality. *Am. J. Altern. Agric.* **7** (1) (2) 38–47.

Stout, J.D. and Heal, O.W. (1967). Protozoa. In: *Soil Biology* (Ed. by A Burges and F Raw), pp. 149–195. Academic Press, London, New York.

Swift, M.J. (1986). Tropical soil biology and fertility (TSBF): inter-regional research planning workshop. Report of the third Workshop of the decade of the Tropics/TSBF program. *Biology Int.* (Special Issue 13).

Swift, M.J., Heal, O.W. and Anderson, M.J. (1979). *Decomposition in Terrestrial Ecosystems.* Blackwell Scientific, Oxford, UK.

Swift, M.J., Frost, P.G.H., Campbell, B.M., Hatton, J.C. and Wilson, K.B. (1989). N cycling in farming systems derived from savanna: perspectives and challenges, In: *Ecology of Arable Lands* (Ed. by M. Clarholm and L. Bergström), pp. 63–76. Kluwer Academic Publishers, The Netherlands.

Syers, J.K., Sharpley, A.N. and Keeney, D.R. (1979). Cycling of nitrogen by surface-casting earthworms in a pasture ecosystem. *Soil Biol. Biochem.* **11**, 181–185.

Tomati, U., Grappeli, A. and Galli, E. (1988). The hormone-like effect of earth worm casts on plant growth. *Biol. Fertil. Soil* **5**, 288–294.

Toutain, F. (1987). Activité biologique des sols, modalités et lithodépendance. *Biol. Fertil. Soil* **3**, 31–38.

Toutain. F., Villemin, G., Albrecht, A. and Reisinger, O. (1982). Etude ultra structurale des processus de biodégradation II. Modèle Enchytraeides-litière de feuillus. *Pedobiologia* **23**, 145–156.

Trigo, D. and Lavelle, P. (1992). Changes in respiration rate and some physico-chemical properties of soil during gut transit through *Allolobophora molleri* (Lumbricidae). *Biol. Fertil. Soil* **15**, 185–188.

Trofymow, J.A. and Coleman, D.C. (1982). The role of bacterivorous and fungivorous nematodes in cellulose and chitin decomposition., In: *Nematodes in Soil Ecosystems* (Ed. by D.W. Freckman), pp. 117–138. University of Texas Press, Austin, USA.

Urbanek, J. and Dolezal, F. (1992). Review of some case studies on the abundance and on the hydraulic efficiency of earthworm channels in Czechoslovak soils, with reference to the subsurface pipe drainage. *Soil Biol. Biochem.* **24** (12), 1563–1571.

Van Noordwijk, M. and de Willigen, P. (1986). Quantitative root ecology as element of soil fertility theory. *Neth. J. Agric. Sci.* **34**, 273–281.

Vannier, G. (1985). Modes d'exploitation et partage des ressources alimentaires dans le système saprophage par les microarthropodes du sol. *Bull. d'Ecolog.* **16** (1), 19–34.

Verhoeff, H.A. and De Goede, R.G.M. (1985). Effects of Collembolan grazing on nitrogen dynamics in a coniferous forest. In: *Ecological Interactions in Soil; Plants, Microbes and Animals* (Ed. by D. Atkinson, A.H. Fitter, D.J. Read and M.B. Usher), pp. 367–376. Blackwell Scientific Publications, Oxford, UK.

Vinson, S.B. (ed) (1986). *Economic Impact and Control of Social Insects.* Praeger, New York, USA.

Whitford, W.G. (1989). Abiotic controls on the functional structure of soil food webs. *Biol. Fertil. Soil* **8**, 1–6.

Wielemaker, W.G. (1984). *Soil formation by termites, a study in the Kisii area, Kenya.* Doctorate thesis, Wageningen University, The Netherlands.

Williams, M.A.J. (1968). Termites and soil development near Brooks Creek, Northern Territory. *Aust. J. Soil Sci.* **31**, 153–154.

Wood, T.G. (1978). Food and feeding habits of termites. In: *Production Ecology of Ants and Termites,* (Ed. by M.V. Brian), pp. 55–80. Cambridge University Press, Cambridge, UK.

Wood, T.G. (1988). Termites and soil environment. *Biol. Fertil. Soils* **6**, 228–236.

Wolters, V. (1991). Soil invertebrates – Effets on nutrient turnover and soil structure – a review. *Z. Pflanzenernähr Bodenk* **154**, 389–402.

Yeates, G.W. (1981). Soil nematode populations depressed in the presence of earthworms. *Pedobiologia* **22**, 191–204.

Zhang, B.G., Rouland, C., Lattaud, C. and Lavelle, P. (1993). Origin and activity of enzymes found in the gut content of the tropical earthworm *Pontoscolex corethrurus* Müller. *Eur. J. Soil Biol.* **29** (1), 7–11.

Terrestrial Plant Ecology and ^{15}N Natural Abundance: The Present Limits to Interpretation for Uncultivated Systems with Original Data from a Scottish Old Field

L.L. HANDLEY AND C.M. SCRIMGEOUR

ADVANCES IN ECOLOGICAL RESEARCH VOL. 27
ISBN 0–12–013927–8

I. SUMMARY

Progress in understanding $\delta^{15}N$ patterns in terrestrial systems is severely constrained by the lack of mechanistic models for interpreting $\delta^{15}N$. The present literature suffers from the uncritical proliferation of unverified interpretations, many of which are based on the false assumption that $\delta^{15}N$ can be used as a reliable tracer of source nitrogen, and many that were meant only as suggestions for further testing when first published. In contrast to enriched ^{15}N, $^{15/14}N$ at natural abundance level ($\delta^{15}N$) is not a tracer of nitrogen from source to sink. It undergoes large fractionations relative to sample values, and one cannot infer that isotopic signatures in an ecosystem are related to each other. Most of the $\delta^{15}N$ literature attempts to use $\delta^{15}N$ as a tracer of nitrogen fluxes from source to sink. The literature is also rife with papers in which the authors attempt to use the isotope to 'prove' something of keen interest to them, rather than exploring data and using hypothesis testing. The widely published two-source method for estimating N₂-fixation assumes (implicitly) that soil nitrogen sources can be traced by $\delta^{15}N$, and much of the literature on this subject is an ever more detailed attempt to 'demonstrate' that $\delta^{15}N$ can be forced to estimate N₂-fixation. Recently, a number of papers have argued that $\delta^{15}N$ foliar values are sufficient, in themselves, to demonstrate that plants use different sources of soil nitrogen. This over-interpretation of foliar $\delta^{15}N$ hides at least one untested assumption (that soil nitrogen sources have different $\delta^{15}N$ values) and one false assumption (that $\delta^{15}N$ can be used as a reliable tracer of nitrogen from soil to plant leaves). This source–sink approach also suffers from the problem that many investigators want to know the source(s) of plant nitrogen, and they have drafted $\delta^{15}N$ into the search without much serious thought as to whether it is reasonable to expect foliar $\delta^{15}N$ to report on its nitrogen sources in a uniform and unbiased way.

Another serious problem is that interpretations are being transferred across unrelated vegetation types and climates and across different functional plant types without adequate recognition of the variety of plant strategies in nature or the potential for genetic variation within one taxon, e.g. one author uncritically transferred interpretations of $\delta^{15}N$ from herbaceous soybean cultivars growing in agricultural fields in the mid-Western U.S. to a California desert and leguminous, phreatophytic trees. Some authors quote relationships for plant $\delta^{15}N$ derived for herbaceous, agricultural plants and use them as substantiation for arguments about woody perennials. Functional relationships are frequently misquoted. A recent study of two species of mycorrhizal fungi, given only inorganic nitrogen, was cited by another author as evidence that 'mycorrhizas', in general, do not fractionate organic nitrogen very greatly.

Too few terrestrial systems and and too few plant taxa have been studied in sufficient detail to validate generalisations about δ^{15}N between sites or between plant taxa within a single site.

Ecologists now have automated mass spectrometers which make large numbers of δ^{15}N measurements, but lack trustworthy interpretations for the data they generate, and in many cases appear to be 'rummaging through the dusty attics' of previous authors' desperation and mistakes rather than furthering an understanding of the real mechanisms.

On the positive side, however, automated mass spectrometers have enabled generation of the large data sets required for biological and ecological research. Detailed δ^{15}N measurements of plants and soils frequently produce very strong patterns which are also strongly reproducible and may even appear to be similar across sites. Such strong reproducibility, now observed at many sites from many climates, must contain much valuable information on ecosystem nitrogen cycling. However, because of the multiplicity of processes which lead to similar δ^{15}N values and patterns, their correct interpretation cannot be achieved without manipulative experimentation and hypothesis testing.

The use of δ^{15}N is a powerful tool for obtaining insights through δ^{15}N pattern analysis, and for deriving new questions to be tested, but it is not a reliable tracer of nitrogen fluxes in soils or in plants growing in soils. To interpret δ^{15}N in terms of sources and sinks, one must know the source δ^{15}N value(s), be able to quantify the fractionation(s) occurring between source and sink, and frequently be able to estimate nitrogen fluxes. Given the widespread absence of such knowledge, we question whether much of the published literature on δ^{15}N for terrestrial systems can have been interpreted correctly.

II. INTRODUCTION

Here we give the first report of ^{15}N natural abundance (δ^{15}N) patterns in a transitional northern British vegetation type (broom-hawthorn-dominated old field). Attempting to interpret our own ecosystem patterns of δ^{15}N provided the impetus for this literature review.

With the exception of a small amount of radioactive material, all matter is composed of stable isotopes, which differ from each other only in their mass-to-charge ratios. Nitrogen (N) has two stable isotopes, masses 14 and 15. About 78% of air is N_2 gas, and, globally, only 0.3663% of N is ^{15}N. Hence, shifts in the ratio of ^{15}N to ^{14}N are small compared with, e.g. carbon, where 1.1% of global carbon occurs as the heavy isotope.

Natural abundance levels of stable isotopes are expressed as delta (δ) in parts per thousand (‰). For N,

$$\delta^{15}N = [(^{15/14}R_{sample} - {}^{15/14}R_{standard})/^{15/14}R_{standard}] \times 10^3,$$

where R is the isotopic ratio and the standard is N_2 of air, and atmospheric N_2 has a δ value of 0‰.

An isotopic ratio is the ratio of heavy to light isotopes (e.g. $^{15}N/^{14}N$); a fractionation is a change in that ratio and, when the difference between source and sink ratios can be calculated, that difference is called a discrimination, sometimes designated by Δ. A system for isotopic work consists of the source and the sink (the analysed sample) together with any fractionations which occur between. The source is not always intuitively obvious or steady in isotopic value, and the fractionations can cancel each other out in the net sample.

Both physico-chemical and biochemical processes can fractionate $^{15/14}N$. One such physico-chemical reaction for N is the gaseous loss of NH_3-N, in which the lighter isotope tends to be lost more readily than the heavier one. One biochemical process which fractionates nitrogen isotopes is denitrification, in which the first step may consist of microbial reduction of NO_3^--N to NO_2^--N and is thought to have a potential discrimination of about −30‰, relative to source $\delta^{15}N$ (Handley and Raven, 1992).

Different patterns of fractionations occur, depending on whether the reaction causing the change is an equilibrium reaction or a kinetic reaction, and whether the kinetic reaction occurs in a closed system or in an open system. An equilibrium reaction for N is the pH-dependent one of NH_4^+-N with NH_3-N; this equilibrium can result in a $\delta^{15}N$-difference between the two forms of ≤ 20‰ (Yoneyama, 1995). A kinetic reaction is one in which substrate A combines with substrate B to form product C. All enzyme reactions are considered to be kinetic. One such reaction for N is the transformation of NH_4-N to glutamine by the enzyme, glutamine synthetase (GS), with a potential $\delta^{15}N$ fractionation of ≤ -17‰ (Yoneyama et al., 1993) for the product (glutamine) relative to the substrate (NH_3-N).

In nature there are no completely closed systems. However, some natural systems behave isotopically, as if they were closed, because the replenishment rate of the substrate is slow relative to the reaction. One example is the carbon discrimination accompanying photosynthesis (Farquhar and Richards, 1984; O'Leary, 1988). The source CO_2 for fixation is in the stomatal cavity, inside the leaf; if the stomatal aperture is partially closed, then replenishment by external CO_2 is slow relative to carbon fixation, and the isotopic signature of the source CO_2 rises because the carboxylating enzyme, RuBISCO preferentially uses ^{12}C; this ultimately causes a ^{13}C-enriched plant signature. In the general case of closed systems, the product is initially more depleted in the heavy isotope than the reactants; as the reaction proceeds, more of the residual heavy isotope is converted to the product. In a closed system both the source and the product continually fractionate during the reaction. If the reaction goes to completion, then no net fractionation is expressed because all of the isotopes, both heavy and light, have moved from substrate to product. All isotopic changes conform to the law of conservation of matter. For a net fractionation to occur, there must be a physical loss or gain of either heavy or light isotopes.

In an open system, typical of most natural ones, there is effectively an infinite supply of substrate relative to the demands of the reaction. For instance, in the process of biological N_2-fixation, the supply of atmospheric N_2 is infinite compared with the demands of the N_2-fixing enzyme, nitrogenase. The fixed N is thought to be fractionated by $\leq 2\%o$ relative to atmospheric N_2, but the $\delta^{15}N$ of the large pool of atmospheric N_2 is unaffected isotopically, so that the source $\delta^{15}N$ value remains steady, while the sink $\delta^{15}N$ is fractionated.

Many enzymes have large potential fractionations which are not fully expressed at the level of the analysed sample because: (1) substrate is limited or (2) the remaining isotope (that portion not transformed by the enzyme) is not lost from the system analysed. An example of the first case is seen in GS, which has a potential fractionation of almost $-17\%o$ in the product (Yoneyama et al., 1993), but plant $\delta^{15}N$ values below $-7\%o$ are seldom reported, even for plants known to be heavily dependent on NH_4^+-N and, hence, on GS assimilation. The second case can be exemplified by the conversion of NO_3^--N to NH_4^+-N inside the plant. The enzyme, nitrate reductase, fractionates the first reduced product, NO_2^--N, relative to both the external NO_3^--N and relative to the remaining, unreduced internal NO_3^--N (Yoneyama and Kaneko, 1989); however, the expression of this fractionation is not measurable on a whole plant basis or on a whole tissue basis, because all of the NO_3^--N atoms which entered the plant are still present. (For illustrative purposes, we ignore NO_3^--N efflux.).

The reader will encounter in the literature an α-value, which is the relative reactivities of $^{14}N/^{15}N$. A value of $\alpha > 1$ indicates that the product is ^{15}N-depleted relative to the substrate; for example, an α value of 1.010 is equivalent to a discrimination of $-10\%o$ in the product relative to the substrate, where $\delta^{15}N = (\alpha - 1) \times 10^3$.

It is worth emphasising the contrast between applications of $\delta^{15}N$ and $\delta^{13}C$. Approximately fixed points exist for atmospheric CO_2, for marine dissolved inorganic carbon, for the products of C_3 and C_4 photosynthesis and fossil carbon; these are easily measured, and have errors which are small relative to the large natural differences. Useful mathematical models exist relating changes in $\delta^{13}C$ to physiological processes such as photosynthesis and water–use efficiency. In food webs and plant–soil systems, $\delta^{13}C$ can be used directly as a tracer of source carbon (DeNiro and Epstein, 1981; Wada et al., 1995). For $\delta^{15}N$, no parallels exist for any of the above. The increase in $\delta^{15}N$ with trophic levels (Wada et al., 1995) is one of the few general rules that apply to $\delta^{15}N$ over a range of natural systems, and even this can be complicated by surprising $\delta^{15}N$ values (Scrimgeour et al., 1995) which violate the general rule of about 0‰ to 3.5‰ ^{15}N-enrichment per trophic level.

A final and basic concept in natural abundance studies is the isotopic mass balance, which relates the isotopic signature of the entire system to its parts. For whole plant $\delta^{15}N$ this is

$$\delta^{15}N_{(whole\ plant)} = [(\delta^{15}N_{(shoots)} \times mg\ N\ in\ shoots)$$

$$+ (\delta^{15}N_{(roots)} \times mg\ N\ in\ roots)]/(mg\ N\ in\ whole\ plant).$$

III. MATERIALS AND METHODS FOR ORIGINAL DATA

A. Site Description

The study site is on the north bank of the Tay Estuary (c. 56° N, 3° W) in eastern Scotland. It is a typically small pocket (c. 2.5 ha) of abandoned agricultural land, last cultivated about 1960 (E. Gauldie, historian, personal communication, 1993).

The soil (D. Hopkins, Dundee University, personal communication) is Carey Series of the Carpow Association, a sandy clay loam. The soil layer is shallow, averaging 40 cm depth, and overlying Old Red Sandstone boulders. Tests at SCRI show that it has a poor water-holding capacity (20% dry weight) and is freely draining. Soil pH is approximately 5.4 in the top 40 cm.

The climate (MacKerron, 1994) is maritime. Long-term mean monthly air temperatures vary annually between about 0°C to 18°C; rainfall is evenly distributed throughout the year, averaging about 15 mm per week. Strong soil moisture deficits occur in May through September. Soil temperatures (to 40 cm) average between 2°C and 15°C seasonally, with the major rise in soil temperature occurring in late April/early June.

The vegetation consists of a mosaic of N_2-fixing broom bushes (Cytisus scoparius L. Link), and two rosaceous species, hawthorn (Crataegus monogyna Jacq.), and wild rose (mainly Rosa canina L.), underlain by a continuous sward of chiefly perennial grasses and dicotyledonous herbs. The herbaceous plants are not listed in detail because the main comparisons are between herbaceous and woody and among woody vegetation. Individual herbaceous species were not analysed or studied separately.

Rooting depths of plants were examined by excavation. Rose, hawthorn and broom bushes all shared the same rooting depth, to the bottom of the soil (about 40 cm); the main root mass for the grass–herb sward was in the top 10 cm of soil – the same depth to which earthworms were found.

In Scottish broom, some nitrogenase activity continues through most of the year with a maximum in late spring and early summer (Wheeler et al., 1979). In 1994, nodules were excavated from three broom bushes used in the study and judged to be of typical size and condition. Acetylene reduction assays

(Turner and Gibson, 1980) were performed on these by Dr C.T. Wheeler, Glasgow University.

B. Experimental Design

Field sampling in 1991–92 was done over the course of an entire growing season, and the same bushes of hawthorn and broom were repeatedly sampled. Whole branches of broom bushes (phyllodes and reduced leaves) were harvested and analysed for %N and δ^{15}N. All leaves from whole branches of hawthorn were stripped and pooled for the same analyses. Hawthorn bushes were chosen at varying distances from the nearest N_2-fixing broom (0 to 11 m). In the grass-herb sward, 10 cm^2 quadrats were harvested for shoots at 50-cm intervals, starting from a broom bush and extending along an 11-m transect through the shrubless sward.

In 1992–93 new broom, rose, and hawthorn bushes (13 of each) were chosen for sampling; a new transect was established in the grass–herb sward and sampled as described above with the addition of taking soil samples to 10 cm, roses and hawthorn were sampled repeatedly through the growing season on a total of five occasions. Broom was sampled on nine occasions between February and December. Plant parts from shrubs were analysed by age: youngest leaves, mature leaves, oldest leaves and similarly for three ages of stems and for three ages of fruits.

Soil was taken on each sampling date and location (to approximately 10 or 20 cm, as appropriate).

Much of the apparent variability in the data arises from sampling different individuals or soil locations at a given time; where taxa or sites are combined, variation also arises from combining data which contain large, but systematic, seasonal changes. The numbers of samples and the length of sampling vary among taxa because of their different growth periods. The same data have been deliberately plotted in different ways (e.g., spatially, temporally, by taxa or plant parts) in order to illustrate how δ^{15}N patterns may change with frame of reference and how this can influence δ^{15}N interpretation.

C. Sample Preparation, Isotope and Statistical Analyses

Both %N and δ^{15}N were measured by continuous-flow isotope ratio mass spectrometry (CF-IRMS). This approach allows a large sample throughput, and offers sufficient precision to allow statistically significant sampling for ecological studies. Plant samples were prepared and measured as described by Handley et al. (1993), using 100 μg N contained in 1–5 mg amounts of sample for each analysis. Soil samples contained too little N for this method to be applied directly, as rapid ash accumulation and incomplete combustion were likely with the large ($\geq$25 mg) samples which would have been required. The use of a delayed oxygen pulse (Preston and McMillan, 1988; Owens and Rees, 1989) allowed samples with only 25 μg N to be analysed satisfactorily using samples of less than 10 mg total weight. The precision possible with this method is not as high as with plant material, but provided that the soils do not have too high a C:N ratio and sufficient N for samples is contained in less than 10 mg of sample, a precision of 1.5‰ or better is achieved.

Statistical analyses were carried out using Statistica® software (StatSoft, Tulsa, Oklahoma, USA). Regression analysis, ANOVA, Pearson's Product-Moment Correlation and Students' t-test were used as appropriate.

IV. SCOTTISH OLD FIELD RESULTS

A. Soil N: Total and Mineral N

Total soil N was 0.2% to 0.4% on a dry weight basis and was chiefly organic N. Mineral N (Figure 1) was consistently less than organic soil N by a factor of 10^3, and NO_3^--N was consistently less than NH_4^+-N. The concentrations of soil NO_3^--N and NH_4^+-N were largely unchanging through the year and across the site, except during a dry period in late August, when soil NH_4^+-N dropped to 50% of its early spring value.

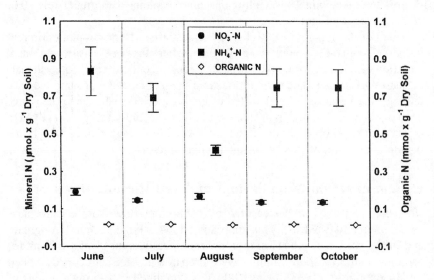

Fig. 1. Scottish old field. Original data. Concentrations of mineral and organic N in soil versus time of growing season, means and standard errors. Note that organic N was 10^3 times more abundant than mineral N ($n = 69$).

Concentrations of NO_3^--N, NH_4^+-N and organic N were statistically the same under all major plant taxa (Figure 2), excepting under roses where amounts of soil NH_4^+-N were more variable.

B. Total soil δ^{15}N

Total soil δ^{15}N underlying shrubs (Figure 3) was statistically the same for soil under rose, hawthorn and broom (averaging $\approx$ 4.0‰ to 4.2‰). Total soil δ^{15}N under the grass–herb sward was significantly ($p < 0.05$) more enriched than under shrubs.

The δ^{15}N of total soil N became more enriched as the growing season progressed (Figure 4). The parallel ^{15}N enrichment of above-ground biomass of the grass–herb sward (Figure 5) appears to eliminate N remobilisation and below-ground storage as the sole mechanism for soil organic matter enrichment at the end of growing season. By isotopic mass balance, mobilised shoot N appeared to be ^{15}N-depleted, and would at least, depress the end-of-season δ^{15}N of total soil N. Hence, other processes, which do not necessarily exclude root storage of N, must be operative.

C. Plants: Analyses Among Seasonal Means and Taxa

Seasonal averages of δ^{15}N for all leaves of all plants were more depleted than for total soil N (Figure 3). Hawthorn was more depleted than grass ($p < 0.01$);

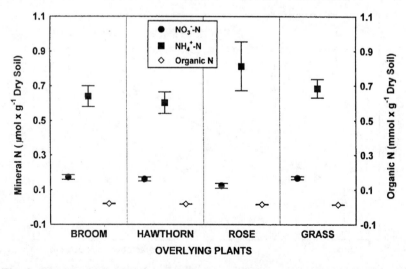

Fig. 2. Scottish old field. Original data. Concentrations of mineral N and organic N in soil underlying different plant types, seasonal means and standard errors. Note that the standard error of NH_4^+-N was large under roses ($n = 69$).

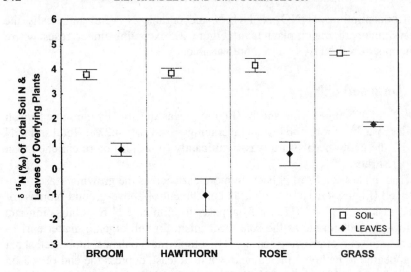

Fig. 3. Scottish old field. Original data. $\delta^{15}N$ of photosynthetic parts of all major plants and total soil N, annual means and standard errors. Much of the variation is due to temporal shifts in sampled $\delta^{15}N$. For hawthorn, there is additional scatter due to proximity of a N_2-fixing broom bush (See Figure 7). Rose and broom were indistinguishable for $\delta^{15}N$; total soil N and all other taxa were statistically different (ANOVA) at $p < 0.001$. (n for soil = 69, hawthorn = 112, broom = 312, rose = 72, grass = 233).

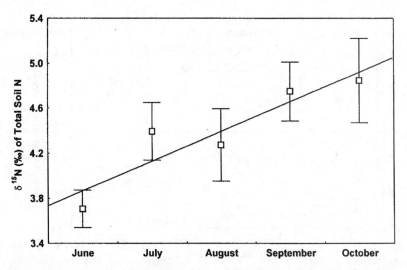

Fig. 4. Scottish old field. Original data. $\delta^{15}N$ of total soil N (to 10 cm) versus time of growing season, means and standard errors, which include soil from all overlying plant types. The regression ($\delta^{15}N = 0.26$ date $+ 3.6$) is significant at $p < 0.01$ ($n = 69$).

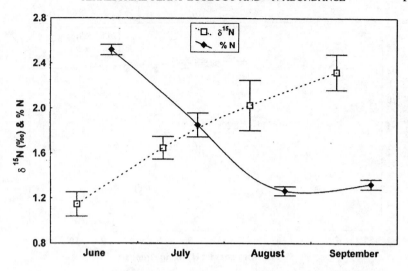

Fig. 5. Scottish old field. Original data. δ^{15}N and %N of grass–herb sward, shoots versus sampling date. Note that as %N declines with growing season, δ^{15}N increases ($n = 233$). δ^{15}N of soil also increased (see Figure 4).

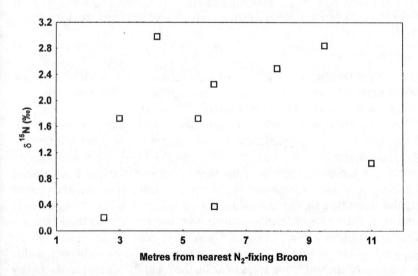

Fig. 6. Scottish old field. Original data. δ^{15}N of all leaves from single branches of rose bushes, taken on a single date, June 1993; nine individual bushes varying in distance from the nearest broom bush. All bushes chosen for similar size and appearance. There was no significant regression relating δ^{15}N of rose leaves and distance from the nearest N_2-fixing broom bush.

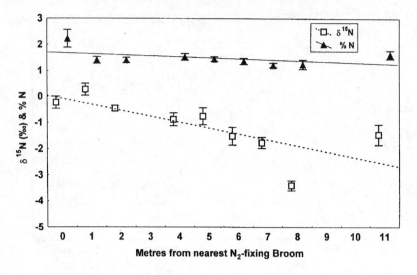

Fig. 7. Scottish old field. Original data. $\delta^{15}N$ of all leaves on single branches of hawthorn bushes growing at varying distances from the nearest N_2-fixing broom bush ($n = 112$), means and standard errors. The same individual bushes were sampled on each occasion. The regression ($\delta^{15}N = 0.17-0.233$ m) was significant at $p < 0.001$; bushes at 0–3 m were statistically different at $p < 0.001$ from bushes at 4–11 m (Students' t-test). Much of the variation at each distance was due to seasonal trends in $\delta^{15}N$.

and all plants (except rose and broom) were different from each other at $p < 0.001$. Statistically ($p < 0.6$), there was no significant difference in the seasonal mean $\delta^{15}N$ values of broom and rose (Figure 3).

Hawthorn and rose are both Rosaceous shrubs and occupy the same habitat, but they had distinctly different $\delta^{15}N$ signatures and responded differently (Figures 6 and 7) to the proximity of a N_2-fixer. N_2-fixing broom showed a seasonally variable pattern in whole stems (Figure 8) varying from around +2.5‰ to −1‰. New leaves and phyllodes emerged throughout the year at irregular intervals and, the low values shown in Figure 8 (sampling dates 2, 7 and 9 in the figure) corresponded to new flushes of growth. Acetylene reductions (C.T. Wheeler, Glasgow University) done on one occasion on excavated broom nodules demonstrated active N_2-fixation in broom at this site with acetylene reduction rates of 2.28, 0.69 and 0.73 μmol C_2H_4 g^{-1} dry weight of nodules h^{-1}.

The %N of broom, after a small early decline (sampling date 4, Figure 8), remained constant throughout the growing season, showing no correlation between %N and $\delta^{15}N$.

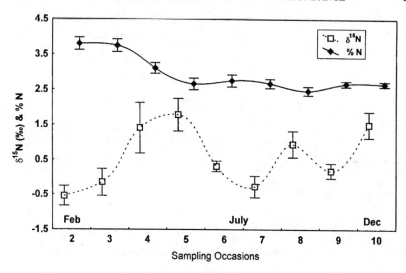

Fig. 8. Scottish old field. Original data. δ^{15}N and %N (dry weight basis) of entire broom shoots (reduced leaves and phyllodes). Means and standard errors for different bushes sampled on each occasion ($n = 312$).

D. Plant δ^{15}N and Proximity to a N_2-Fixing Symbiosis

The quadrats of whole shoots sampled from the 11-m herb–grass sward transect (Figure 9) and ending in a broom bush showed only minor and random variations in both %N and δ^{15}N which were unrelated to the distance from the nearest broom. Likewise, sampling of foliage from nine rose bushes on a single date revealed no correlation of either δ^{15}N or %N with distance from a broom (Figure 6). However, repeated sampling of leaves from the same hawthorn bushes showed a δ^{15}N response to distance from a broom, but in the opposite direction to that usually reported (Figure 7). With increasing distance from a broom, hawthorn leaves became isotopically more negative (regression of δ^{15}N versus distance at $p < 0.001$). The hawthorn bushes at 0 to 3 m from broom were indistinguishable from broom on the basis of foliar δ^{15}N, while those at 4–11 m were a statistically different group ($p < 0.001$). Per cent leaf N did not change with distance from a broom. It may be relevant that rose and broom have arbuscular mycorrhizal associations and hawthorn is ecto-mycorrhizal (Grime *et al.*, 1988).

The Scottish old field data suggest that there is something fundamentally different about plant N acquisition near a N_2-fixing broom, that not all plants responded to the proximity of a broom in the same way (Figures 6, 7 and 9), and that mechanisms for these differences in N acquisition cannot be determined from this type of field sampling, alone. The δ^{15}N of hawthorn in the old

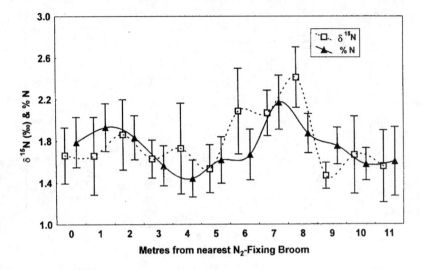

Fig. 9. Scottish old field. Original data. $\delta^{15}N$ and %N (dry weight basis) of shoots of grass–herb sward versus distance from nearest N_2-fixing broom bush. Seasonal means and standard errors of samples taken monthly every 50 cm along an 11-m transect ($n = 233$). Much of the variation is due to a directional seasonal change (see Figure 5). Similar plots for individual sampling dates were statistically indistinguishable from this composite plot of all sampling dates.

field crossed the range of $\delta^{15}N$ attributable to N_2-fixation, but appeared to be due to negative soil source(s); the $\delta^{15}N$ of rose was highly variable but contained some negative values.

E. Maturational Variability of $\delta^{15}N$

Each type of plant showed its own age-related patterns of $\delta^{15}N$. For the major N_2-fixer, broom, whole-shoot $\delta^{15}N$ changed cyclically (Figure 8) through the year, varying from +2.5‰ to –1‰. Because the low values coincided with flushes of new growth, the temporal pattern may be related to internal remobilizations. The most positive values (+1‰ to +2‰; Figure 8) occurred during the period in which N_2-fixation is most active in Scottish broom; and the values nearest to those of air N_2 (0‰ to –1‰; Figure 8) occurred during seasonal periods known to coincide with low nitrogenase activity in Scottish broom (Wheeler *et al.*, 1979, 1987).

No 'snapshot' sample on any one date gave a 'true' $\delta^{15}N$ ranking among taxa. The rankings of $\delta^{15}N$ for mature leaves (Figure 10) of each plant type varied between sampling dates.

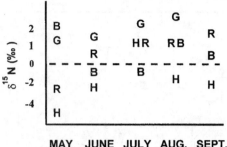

Fig. 10. Means of foliar δ^{15}N for broom (B), grass–herb sward (G), rose (R) and hawthorn (H) on each sampling occasion throughout the main growing season. Note that the rankings for δ^{15}N are different on each occasion.

Figure 11a, b and c summarises δ^{15}N of youngest, mature, and third youngest plant parts (1, 2 and 3, respectively) for the shrubs. Leaves in hawthorn and rose became ^{15}N-depleted by around 2‰ with age, and leaves of broom became enriched by about +2.5‰ to 3‰. In stems, temporal changes were generally less pronounced and not significant in hawthorn. Broom stems changed by about 2‰ (^{15}N-depletion) from youngest to third youngest stems. Rose showed an age-dependent δ^{15}N reversal, with initial depletion between youngest and second youngest stems, then an enrichment in third youngest stems.

Fruits of broom became slightly enriched with increasing age. For hawthorn fruits, δ^{15}N did not change with age. Rose fruits showed the most dramatic maturational changes in δ^{15}N (around 3‰ ^{15}N-depletion with age). δ^{15}N variability was greatest for fruits of hawthorn and rose.

While some patterns emerged for %N and δ^{15}N on the basis of plotted means (Figure 11a, b and c), no interpretable statistical patterns related %N to δ^{15}N in different parts and ages of plant parts or hinted at explanations for N mobilisations in the plants. There were no significant correlations of %N and δ^{15}N for hawthorn or rose leaves, stems, or fruits or among various ages of these parts. In broom %N and δ^{15}N were correlated (Pearson's Product Moment Correlation = +0.42 at $p < 0.05$) in leaves and phyllodes but not in fruits.

Results of one-way ANOVAs followed by Scheffe's *post-hoc* test ($p < 0.05$) showed no significant differences among parts or ages of parts for hawthorn, consistent with the lack of significant correlations. In rose, these tests showed significant differences ($p = 0.02$) between youngest leaves and youngest stems for %N but not for δ^{15}N; this %N distribution was not reflected in the correlations discussed above. For broom, significant differences ($p < 0.05$) were found for δ^{15}N between leaves aged 2 and 3 versus stems aged 2 and 3 and between fruits of all ages versus stems aged 2 and 3; however, for %N,

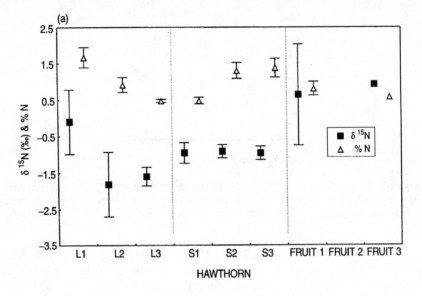

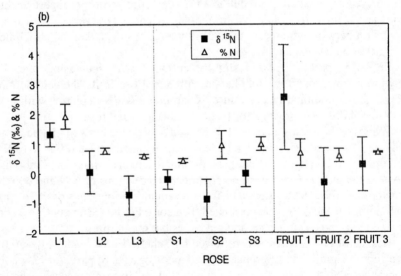

Fig. 11. (caption opposite)

almost all parts were significantly different ($p < 0.05$) from each other. Hence, even the good overall correlation of %N and $\delta^{15}N$ for broom was not validated as a reliable indicator of N mobilisations when tested in detail by ANOVA.

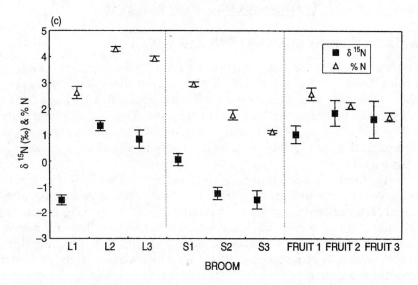

Fig. 11. δ^{15}N and %N for major plant parts from emergence to senescence for hawthorn (a), rose (b) and broom (c). L3 is the senescent stage of leaves; S3 is several ages younger than the main stems of any of the shrubs, but the third youngest stems; and F3 is fully ripe fruit.

F. The Two-Source Natural Abundance Model for Estimating N₂-Fixation Could Not Be Used

Almost none of the stipulated assumptions (see Section VP of this article and Shearer and Kohl, 1986) were satisfied in this very common type of vegetation. Rose was isotopically indistinguishable from broom; the grass–herb sward was statistically distinguishable from broom but had a shallower rooting depth for the majority of root biomass and was different in life form and phenology from rose and hawthorn; some hawthorns (Figure 7) were isotopically indistinguishable from broom and some were isotopically different from broom. Hence, no suitable non-N₂-fixing reference plant was available. The δ^{15}N of broom, itself, varied cyclically and foliage δ^{15}N nearest to that of air N₂ (0‰) occurred at a time of year known to be a lull in nitrogenase activity. No one sampling date or age of plant parts emerged as representative of a whole growing season and the inter-taxa rankings among these (for mature leaf δ^{15}N) changed with time (Figure 10). The data (Figure 3) also suggested (but did not demonstrate without controlled experiments) that several types of N-sources may have been used by the plants. Where more than two N sources exist, a two-source mixing model cannot be calculated.

V. DISCUSSION AND REVIEW

A. Heterogeneity of Soil N: δ^{15}N and %N

Soil δ^{15}N is not easily measured. There are no methods compatible with high sample throughput for analysing the δ^{15}N of separate forms of N in soils, e.g., NH_4^+-N, NO_3^--N, various amino acids and other organic forms. Hence, it was not possible to measure soil δ^{15}N in the same detail that we measured plant δ^{15}N. Chemically reducing digestions and distillations of NH_4^+-N have the drawback that statistically meaningful numbers of samples cannot be analysed routinely.

It is unlikely that much further progress will be made until novel approaches are developed which are appropriate both to the scale of the heterogeneity in soil N sources and to the low concentrations found in unmanaged systems. Existing preparation methods cannot be improved sufficiently to fill present and future needs, and new compound-specific techniques are required which can measure δ^{15}N of ng amounts of individual compounds directly isolated from the soil matrix.

In the old field most of the soil N was bound in soil organic matter. This is consistent with other reports for temperate forest soils (see review by Nadelhoffer and Fry, 1994, and references therein). Vitousek and Matson (1984) found that organic N in the microbial biomass was the most important pool for retaining N in these systems. In various systems reports of spatial variability of soil N amounts appear to coalesce on two major factors: (1) scale of sampling, with variation appearing to be random over short distances and systematic over longer ones, as shown by Goovaerts and Chiang (1993) and Sutherland *et al.* (1993) and (2) variations in soil water relations (Selles *et al.*, 1986; Groffman *et al.*, 1993; Kim and Craig, 1993; Sutherland *et al.*, 1993) as controlled *inter alia* by topographic relief, soil textural differences and (e.g., Kim and Craig, 1993) climate.

It is also is important to realise that a large proportion of N entering an ecosystem may be derived from point sources and is not evenly distributed. The fate of N derived from point sources is very poorly known – fixed atmospheric N_2 being one conspicuous example. Other point sources of singular δ^{15}N values might be animal excreta, even decaying bodies of animals, or fungal fruiting bodies. The spatial fates of point-source-derived plant N can be related, qualitatively, to both climate and plant community architecture through controls on N mineralisation and rooting depth of plants.

Total soil N may not be directly associated with proximity to a N_2-fixing plant. We found no correlation between %N of soil and proximity of a broom bush in the old field. Working in a vegetation type in Kenya, which was architecturally similar (tropical savannah) to the Scottish old field, Belsky *et al.* (1989) found that soil N concentrations were not dependent on the proximity of a woody N_2-fixing plant. Unlike our data set, however, they found that soil N concentrations were correlated with the presence/absence of trees. Lajtha and Schlesinger (1986) observed similar patterns for North American desert plants. In the Sonora desert, Shearer *et al.* (1983), Virginia and Jarrell (1983) and Johnson and Mayeux (1990) found that total N concentrations were greater within the stands of the N_2-fixing tree, *Prosopis*, than between. Wheeler and Dickson (1990) found greater soil N concentrations under N_2-fixing *Spartocytisus* spp. shrubs than between them on the volcanic island of Tenerife.

Some of the reported spatial variation of soil N may be due to the below-ground distribution of N-rich plant parts. Johnson and Mayeux (1990) reported the variation in nodule distributions of N_2-fixing trees as occurring both spatially (close to and distant from the main stem) and shallow as well as deep. In the context of heterogeneity of soil N sources, nodule N may comprise an abnormally large point source of ^{15}N-enrichment which is different from other immediate sources. Nodules of some N_2-fixing plants are isotopically enriched relative to whole plant δ^{15}N (Bergersen *et al.*, 1988). Data for broom (Wheeler *et al.*, 1979, 1987), and for *Casuarina equisetifolia* (Srivastava and Ambasht, 1995) suggest that nodule activity is greatly variable seasonally, and that nodule turnover may be high.

In excavating plants of *Hippophae* rhamnoides (an actinorrhizal N_2-fixing shrub) on the grounds of SCRI in Scotland, we found all of the nodules clustered underneath the main stem of the tree and near the soil surface (unpublished data); only one or two very small nodules were found on lateral roots. These trees varied in age from 5 to 10 years and many of the nodules were greater than 1 cm in cross-section. For the actinorrhizal N_2-fixer, *C. equisetifolia*, Srivastava and Ambasht (1995) found in the field, that root nodule decomposition was rapid and contributed an average of >100 mg m^{-2} of newly available N to the soil annually by this means. This is not a large amount on the reported m^2 basis, but is a substantial resource when considered as the point sources that they actually are. Roskoski (1981) demonstrated similar rates of N released from the root nodules of the rhizobially N_2-fixing tree, *Inga jinicuil*. Hence, these aggregated soil N-patches may be the general rule where N_2-fixing vascular plants exist.

Roots of non-N_2-fixing species, containing low concentrations of N may constitute an important, spatially heterogeneous source, given that roots may contribute up to 85% of total plant productivity (Fitter, 1987, and references therein).

Above- and below-ground contributions of previously fixed N may be dependent on a number of factors, including taxon-specific plant architecture, phenology, rate of decomposition and distribution by wind, animals or other factors, which would also be site-dependent. Goovaerts and Chiang (1993) found, however (in agricultural fields) that once landscape-scale patterns of soil N concentrations were established, they persisted over the two years of their study, changing in the winter but re-establishing again in the spring. Whether this finding applies to longer time periods and to natural systems is unknown.

In the old field, we found %N no more variable in soils than total soil $\delta^{15}N$, and both were uniform across the site (Figures 1–3). Selles et al. (1986) found soil N concentrations more variable than $\delta^{15}N$. In examining the $\delta^{15}N$ of soil organic matter crop residues, van Kessel et al. (1994a) found $\delta^{15}N$ randomly distributed. Ledgard et al. (1984) found only c. 5‰ variation in $\delta^{15}N$ of total soil N across 400 km^2 of Australian pasture and no correlation with %N. Shearer et al. (1983) found little variation in soil $\delta^{15}N$ across a Sonoran Desert site, but did find systematic variations in soil %N. At another USA desert site, where a cryptobiotic crust of N_2-fixing organisms supplied most of the soil N, Evans and Ehleringer (1994) demonstrated a close correlation between $\delta^{15}N$ and %N of the soils. Andreux et al. (1990) reported no correlation between soil $\delta^{15}N$ and %N for a tropical system. Broadbent et al. (1980) reported for both native and cultivated USA sites, that the coefficient of variation for total soil N was greater than for $\delta^{15}N$. Since the residual $\delta^{15}N$ of soil is related to both the signature and amount of inputs and outputs, and also to the relative preponderance of contributing microbial processes (Yoshida, 1988; Delwiche and Steyn, 1970; Kim and Craig, 1993), a direct correlation between $\delta^{15}N$ and remaining %N is not expected to be a consistent rule.

B. Total soil $\delta^{15}N$

Variable relationships between soil $\delta^{15}N$ and $\delta^{15}N$ of overlying vegetation are reported from different sites. In a desert system, Shearer et al. (1983) found no significant difference for total soil $\delta^{15}N$ under the canopies of N_2-fixing trees and between them, although N concentrations varied within and between canopies. The nodules of these trees were located deep in the soil, and in the exposed, windy desert environment, above-ground litter may have been widely dispersed (Coûteaux et al., 1995).

We found that the shallower soil samples under the old field grass–herb sward were more enriched than the deeper soil samplings. Schimel et al. (1989) reported for a California grassland, that more than half of the soil N mineralisation occurred in the upper 5 cm. Since mineralisation may lead to loss of ^{15}N-depleted products (Blackmer and Bremner, 1977), it is not surprising that the remaining soil is isotopically enriched. Farquhar et al. (1983) suggested that volatilisation of NH_4^+-N may play a role in the enrichment found at the top of a soil profile. However, the magnitude of loss would be pH dependent, with more NH_4^+-N being lost from sites with high soil pH.

C. δ¹⁵N of Soil Mineral N

Although it is widely published that mineralisation leads to ^{15}N-enriched N pools in soils, this is only a half-truth. A more accurate statement would be that the soil N pools reflect the net ^{15}N-enrichments and -depletions of N mineralisation processes. Nömmik et al. (1994) found total hydrolysable NH_4^+-N more ^{15}N-depleted than total soil N by 3‰ to 8‰ in an unfertilised Swedish pine forest, and in mature forest soils of temperate climates (Nadelhoffer and Fry, 1994) mineral N is frequently ^{15}N-depleted relative to air.

This situation is consistent with environments in which: (1) gaseous losses through denitrification and losses from leaching of ^{15}N-depleted NO_3^--N are slow, so that the expected sequence of relative δ¹⁵N values is organic N > NH_4^+-N > NO_3^--N or (2) ^{15}N-depleted mineral N pools are balanced by ^{15}N-enrichments in another part of the system. It is not possible to determine which is the case at a specific site with only δ¹⁵N of one or two components of the system. The first scenario is also consistent with that recently suggested for soil N in Alaska (Nadelhoffer et al., 1995) where plants shown to be mainly using soil NO_3^--N had low δ¹⁵N foliar values relative to other taxa showing less dependence on NO_3^--N. In South Africa, however, Stock et al. (1995) found NO_3^--N-using trees to be the most ^{15}N-enriched.

Nadelhoffer and Fry (1994) suggested that the major source of ^{15}N-depleted mineral N in forest soils was heterotrophic assimilation followed by excretion (internal isotopic partitioning) rather than slow losses from the ecosystem.

Where temperatures vary during the growing season, and strong nitrification is expected (Garcia-Mendez et al., 1991), there are reports of ^{15}N-enriched mineral N (relative to total soil N). For a southern hemisphere jarrah forest, Hansen and Pate (1987) found total soil δ¹⁵N ranging from –1‰ to +6‰, and positive values for total soil mineral δ¹⁵N (0–11‰). However, in an Australian pasture Ledgard et al. (1984) documented that mineral soil N was consistently ^{15}N-depleted relative to total N. Turner et al. (1987) found in Australia that the δ¹⁵N of mineral N in the top 10 cm of soil changed seasonally, becoming more enriched as the growing season progressed under oats and more ^{15}N-depleted in cultivated fallow over the same time period.

In a Tennessee forest, Garten (1993) found that ridge and valley bottom soils varied little in their total surface soil δ¹⁵N (3.8‰ and 3.2‰ for ridges versus valley bottoms, respectively) but that signatures of NO_3^--N varied from slightly negative to slightly positive (–4.4‰ to +2.3‰).

Even in mature northern forests the mineral N fractionations can be different from any preconceived general patterns. Binkley et al. (1985) found a complex site-dependent picture for δ¹⁵N of total soil N, NH_4^+-N, NO_3^--N and plant-harvested N at four forested sites in British Columbia. At all sites, total soil N was positive, varying among sites from about +1‰ to almost +6‰. At three sites, soil NO_3^--N and NH_4^+-N were greatly depleted in ^{15}N (δ¹⁵N of

$NO_3^--N \leq -10\%o$). At a fourth site, NH_4^+-N under mixed conifers was slightly positive (c. +1‰) as was NO_3^--N (c. +0.5‰), while under conifer–alder mixed forest soil NH_4^+-N, only, was negative (c. −1.5‰). Alder was present at other sites, so the negative signature of NH_4^+-N cannot be due solely to N_2-fixation by alder. This is substantiated by the extremely low values which exceed those reported for N_2-fixation. Plant-harvested N in all trees was negative except at one site, where both soil and plant harvested N were positive, while both NO_3^+-N and NH_4^+-N were negative in signature.

For agricultural soils, Black and Waring (1977) found that NO_3^--N formed from fertiliser additions to soil was up to 12‰ more enriched than the applied fertiliser. Cheng et al. (1964) found mineralised N and fixed NH_4^+-N of several soils more depleted (by 0–6‰) than their presumed precursors, but still positive in sign. Black and Waring (1977) found that NO_3^--N obtained during laboratory incubations of field soils was 2–4‰ depleted in ^{15}N relative to soil total N, while NO_3^--N taken from boreholes and a stream at the same locales were more enriched by 1–2‰ than total soil N. In studying a transect from inland agriculture across a salt marsh to the sea, Page (1995) used extreme variations of NO_3^--N $\delta^{15}N$ (6.5–54.6‰) to strengthen the argument for extensive denitrification in the narrow inland border of the marsh. NH_4^+-N $\delta^{15}N$ values over the same transect only ranged from 15.6‰ to 7.6‰, becoming more depleted in the direction of the transect over which NO_3^--N $\delta^{15}N$ became more enriched.

Lengthy soil incubations were used in the past in an attempt to determine the $\delta^{15}N$ of potentially mineralisable N. Feigin et al. (1974a) incubated agricultural soils in a closed system at 35°C for 42 weeks and concluded that the final $\delta^{15}N$ of the NO_3^--N produced was representative of field values. Their time versus $\delta^{15}N$ curves (Figure 12) showed large variations in $\delta^{15}N$; however, if the incubation were run to completion, the $\delta^{15}N$ values of this soybean soil should have begun to approach the starting value. Complete conversion of substrate to product yields no *net* fractionation (see also section II.B), and this is unlikely in nature. Their observed ^{15}N-depletion of NO_3^--N relative to NH_4^+-N could be interpreted (Focht, 1973) in terms of the fractionation factor of NH_4^+-N being converted to NO_3^--N and never completely nitrifying all of the organic N pool in the field.

The plant values reported by Feigin et al. (1974a) need not be related to the incubation-induced $\delta^{15}N$ of NO_3^--N and could have arisen, e.g. as use of both NH_4^+-N and NO_3^--N with internal signature distributions differing between root and shoot. This is a point which needs continual emphasis: the same net $\delta^{15}N$ signatures can occur via many unrelated processes. No unique $\delta^{15}N$ values are known.

Herman and Rundel's (1989) 40-day incubation (Figure 13) of recently burned chaparral soil yielded a curve of the same general shape as the fertiliser addition curve reported by Feigin et al. (1974b), i.e., a sharp divergence

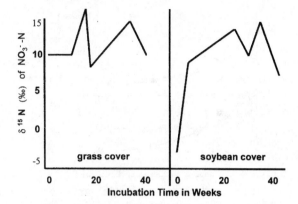

Fig. 12. Adapted and simplified from Feigin *et al.* (1974a). Soil incubation curves showing the variability of δ^{15}N in soil NO_3^--N. Had the incubations gone to completion, the values for soil NO_3^--N under soybean would have returned to its starting value as it did under grass.

of δ^{15}N for NH_4^+-N versus NO_3^--N followed by what may have been a trend toward reconvergence with time (Figure 14) had the experiment been carried on longer.

The data of Cheng *et al.* (1964) for a 2-week incubation at 30°C may be a closer approximation of many field conditions, where nitrification is likely to proceed at relatively slow rates punctuated by bursts of high activity (due, *inter alia*, to episodic drying and wetting) and never completely converting all of the available organic N.

It is important to note that when Herman and Rundel (1989) incubated chaparral soils from burned and unburned areas, the beginning and final δ^{15}N for NH_4^+-N was *c.* 0‰, the same value attributed to dinitrogen fixation. NO_3^--N also began with δ^{15}N (*c.* 0‰) and became very negative (*c.* −12‰ and −18‰) as the incubation approached completion. Feigin *et al.* (1974b) provided a curve (Figure 14) showing that fertiliser N added to field soil resulted in near-zero δ^{15}N values of soil mineral N for both NO_3^--N and NH_4^+-N only for a brief period after application. This suggests that plant δ^{15}N values attributed to fertiliser N instead originate from other sources, instead.

Herman and Rundel (1989) proposed a model of ^{15}N-depleted NO_3^--N leaching and near-surface absorption of relatively enriched NH_4^+-N. Their qualitative model is consistent with the frequently observed relative ^{15}N-enrichment in shallow-rooted annuals versus the depletion of deeply rooted perennials. Where perennials use both deep and shallow nutrient sources seasonally, dilution by the shallow soil NO_3^--N or NH_4^+-N (here *c.* 0‰) could not be distinguished isotopically from dinitrogen fixation as a source of plant N. Examination of the literature suggests that it is a very common phenomenon for the δ^{15}N of plant available N from many sources to approximate 0‰.

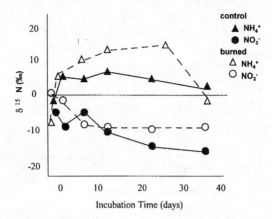

Fig. 13. Adapted and simplified from Herman and Rundel (1989). $\delta^{15}N$ of NO_3^--N and NH_4^+-N from recently burned, solid figures, and unburned, open figures, chapparal soils. Note that initial values, as well as final NH_4^+-N values are centred on 0‰.

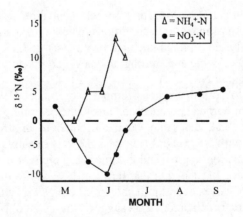

Fig. 14. Adapted and simplified from Feigin *et al.* (1974b). Changes in $\delta^{15}N$ of NO_3^--N and NH_4^+-N in top 30 cm of profile following addition of anhydrous ammonia to field soil. Note that in this early study, where mass spectrometry was manual on a single-inlet instrument with wet chemical sample preparation, samples were pooled and were not true field replicates. In the original publication, error bars were drawn for means of three pooled samples. The pooled samples each comprised 12–16 individual field samples. Much information would have been lost by this method.

From the data available, it appears that the isotopic change of ^{15}N during mineralization is more variable in nature than laboratory-derived incubations would predict. Traditional laboratory soil incubations are likely to reveal little about natural abundance level fractionations of 'potentially' mineralisable soil N. Some of the conflicts between field and laboratory conditions are that: (1) no new inputs replenish the substrate of a laboratory soil incubation, and, as a closed system, it tends toward quantitative conversion of substrate to product; and (2) in the field, mineralisation does not proceed under uniform conditions. Changing soil moisture, microbial population compositions, and temperatures, and also seasonal plant demands for N, continually alter the isotopic mass balance of soil mineral N both temporally and spatially.

As a final comment on soil mineral δ^{15}N, Lajtha and Marshall (1994) restated a commonly held view (attributed by them to Shearer and Kohl, 1989) that only the δ^{15}N variations of mineralised soil N were related to isotopic variations in plants. This description of soil N sources is too simplistic for the study of many, if not most, ecosystems, because plants also acquire organic N with and without microbial assistance (see sections 1.2 and M.1).

D. Exchange Resins

We attempted to recover soil mineral N from the Scottish old field for δ^{15}N analysis using exchange resins, but the results were unreliable. Exchange resins have been successfully used (Binkley and Matson 1983; Hübner et al., 1991) for estimating quantities of mineral N in soil solution. Pate et al. (1993) used mixed bed exchange resins to recover NO_3^--N from Australian soils and analysed the recovered NO_3^--N for δ^{15}N. In these highly nitrifying soils, where presumably NH_4^+-N was in low supply, they were not successful in extracting NH_4^+-N from resins (G.R. Stewart, personal communication, 1994), and found the results highly variable, as we did in Scotland.

Fractionation of N recovered from the resins is not a serious problem *per se*, provided that initial concentrations of either NO_3^--N or NH_4^+-N are high enough for good recovery. Delwiche and Steyn (1970) showed that single-pass depletions of NH_4^+-N exposed to cation exchange resin and to clay were less than 0.1‰; for NO_3^--N the equivalent fractionation was about +0.2‰, with NH_4^+-N in solution becoming depleted and NO_3^--N in solution becoming enriched. These discriminations are negligible in the context of the ecological variability of δ^{15}N in soils or sediments (Owens, 1987). Because NH_4^+-N is unlikely to move very far in most soils, it can be inferred from these data that the so-called 'soil chromatographic fractionations' (postulated cation and anion exchanges acting on soil solution N so that it becomes ^{15}N-fractionated as it moves down a soil profile) of NH_4^+-N are generally small. NO_3^--N, however, may move large distances in soil solution and in downward-percolating waters; however, its soil chromatographic enrichment may also be small, because the anion exchange capacity of most soils is relatively low.

There is, however, one publication of data which convincingly suggests that the existence of various levels and kinds of 'soil chromatography' may be complex and variable. Herbel and Spalding (1993) examined soil NO_3^--N, which was extracted using distilled water (NO_3^--N in solution) and using 2 molar KCl (all of the NO_3^--N, both in solution and adsorbed to any exchange sites). They found large differences in the $\delta^{15}N$ values of the two types of extracts; the KCl-extracted NO_3^--N was more depleted in ^{15}N than the distilled water-extracted NO_3^--N. In one soil core the relative depletion was correlated with the occurrence of clay minerals; in the second core there was no such correlation. We are not aware of any follow-up studies.

E. Adequate Sampling and Appropriate Comparisons

Agricultural systems are managed in various ways to provide, in so far as possible, a uniform, nutrient-rich and pathogen-free environment. In uncultivated systems, variability will remain a greater challenge and, high sample throughput is a prerequisite for describing variability. However, for each system the appropriate scale and amount of sampling must be determined. This has seldom been done for soil $\delta^{15}N$, with Selles et al. (1986), van Kessel et al. (1994a) and Piccolo et al. (1994), being notable exceptions.

Two methods which appear to yield favourable results are the grid quartile approach (Sutherland et al., 1993) and the geostatistical approach (Goovaerts and Chiang, 1993; Robertson and Gross, 1994). The geostatistical approach, as explained by Robertson and Gross (1994), employs semivariances and kriging (a type of interpolation) to quantify the autocorrelation of samples and, therefore, the best scale of sampling for a given parameter at a particular site. This technique can be used also to quantify the 'patchiness' of a resource in the soil.

Some samples are impossible to measure for $\delta^{15}N$ and, therefore, interpretations must be conservative. Soil $\delta^{15}N$ values are difficult to measure for NH_4^+-N and for NO_3^--N in natural systems, where mineral N is in low concentration. There are presently no methods for obtaining $\delta^{15}N$ for soil organic N fractions which might be direct sources for fungi or higher plants. Gebauer and Dietrich (1993) interpreted the $\delta^{15}N$ of fungal fruiting bodies (Figure 15) as evidence of an organic N source for potentially mycorrhizal fungi without being able to measure the source signatures of N for the fungus. Gebauer et al. (1991, 1994) tried to relate the effects of atmospheric pollution to the $\delta^{15}N$ values of tree foliage in declining and healthy trees not on the basis of trees and their N sources, but on the basis of comparing declining and healthy stands at different sites, on different soils and without measuring the N source values or taking into account differences in the δ^{15} of soil mineral N at the two sites; in addition, the 'declining site' contained both healthy and declining trees.

The importance of correct comparisons is highlighted by two recent papers (Androsoff et al., 1994 and Stevenson et al., 1995) which overturned much

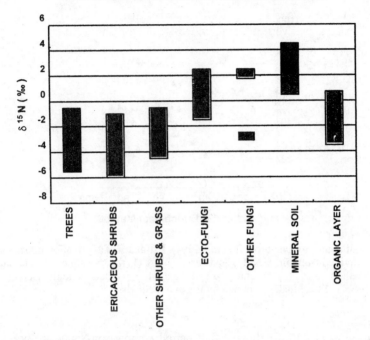

Fig. 15. Adapted and simplified from Gebauer and Dietrich (1993). δ^{15}N values for components of a mixed conifer and deciduous forest in Germany. Values for vascular plants are for leaves or needles only. The δ^{15}N values of ectomycorrhizal fungi overlap the ranges of values for all components of the system.

published literature on comparing methods for estimation of N_2-fixation. This was done solely by making a rigorous application of statistical experimental design. Androsoff et al. (1994) and Stevenson *et al.,* (1995) did field experiments on a large scale and made pairwise comparisons between the N_2-fixation estimates obtained by δ^{15}N and ^{15}N isotope dilution methods. They found, in agreement with many previous studies (Domenach *et al.,* 1989; Bremer and van Kessel, 1990) that the mean estimates of N_2 fixed by pea plants were in fair agreement. However, in these new studies the individual estimates were not correlated (Figure 16). The authors interpreted this lack of correlation as evidence that the two methods measured different aspects of soil/plant N relations, i.e. mechanistically the two methods were different. This productively opens a new area of investigation into the detailed differences between the two methods, and thus offers some hope of resolving whether either method is reliable in the field.

Shearer and Kohl (1986), on the other hand, wrote of the δ^{15}N estimation method that in natural ecosystems, '... the effect of any existing variability can usually be reduced by collecting N_2-fixing and reference plants in pairs'. Experience by investigators working in both agricultural and non-agricultural systems has not substantiated this statement. Androsoff *et al.* (1994) and

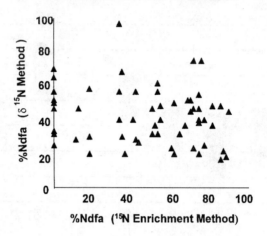

Fig. 16. Adapted from Androsoff *et al.* (1994). Estimates of N_2-fixation in pea (% of plant N derived from air, %Ndfa) as determined by (1) the two-source natural abundance method versus (2) ^{15}N-enriched isotope dilution method. Note the lack of correlation between estimates obtained by the two methods.

Stevenson *et al.* (1995) showed that paired comparisons may not be valid, even in a relatively uniform agricultural setting (Figure 16), and hence, even more care should be taken not to over-interpret data in natural systems (e.g. Lajtha and Marshall (1994) on the statistically inappropriate use of means in e.g. Shearer *et al.* (1983))).

On a temporal scale, snapshots of δ^{15}N only reveal the current relationships among plants and between plants and soil. Our old field data (Figure 10) showed that rankings of δ^{15}N among taxa or types of plants change through the growing season. Hence, the δ^{15}N patterns obtained for this ecosystem were entirely dependent on the time of sampling.

F. Soil Water and Soil or Plant-harvested δ^{15}N

One of the processes related to soil water status is the evolution and gaseous loss of N_2O. By isotopic mass balance (Section II.B), the residual soil δ^{15}N may be affected by this fractionating loss. Kim and Craig (1993) found, for Hawaiian systems (either consistently wet or alternately wet and dry), that the δ^{15}N of evolved N_2O gas was related to the periodicity of soil wetting and also to soil moisture content. This dependence on water relations is probably a combination of chemical effects (e.g., pH, salinity, aeration) and the relative balance of specific fractionations attributable to the microbial community composition (Delwiche and Steyn, 1970; Karamanos *et al.*, 1981; Garten, 1993; Kim and Craig, 1993). Handley *et al.* (1994b) found a correspondence between plant-harvested δ^{15}N and water availability at various sites sampled in Kenya. Karamanos and Rennie (1980), Selles *et al.* (1986) and Sutherland *et al.* (1993) reported correlations between soil moisture (as a function of topographic relief) and total soil δ^{15}N.

Soil δ^{15}N in Tennessee (Garten, 1993) was related to topographic relief. The δ^{15}N of soil surface extractable NH_4^+-N was negatively correlated with elevation, and with average soil water content, ridges and valleys showing means of +2.4‰ and +9.4‰, respectively.

Plant δ^{15}N can be either depleted or enriched relative to total soil δ^{15}N. It is common for foliage of woody perennials in moist temperate ecosystems to be ^{15}N-depleted relative to total soil δ^{15}N (Domenach *et al.*, 1989; Nadelhoffer and Fry, 1994, and references therein), and this depletion is commonly thought to be associated with major dependence on NH_4^+-N and/or ectomycorrhizas, although this cannot be taken as an absolute rule. Nadelhoffer *et al.* (1995) found that plants mainly relying on NO_3^--N in Alaska were more depleted than taxa at the same site known to be heavily dependent on NH_4^+-N. Yoneyama *et al.* (1991b) found for non-nodulating plants in upland, drained and in paddy fields that plant-harvested δ^{15}N was ^{15}N-depleted relative to total soil N in paddy conditions. In the old field (Figure 3) all plants were, on average, ^{15}N-depleted relative to total soil δ^{15}N. Bergersen *et al.* (1990) found that crop plants growing in well drained conditions in South-East Asia and Australia were ^{15}N-depleted relative to soil mineral N. In a California salt marsh, Page (1995) found a close correspondence between the δ^{15}N of pore water NH_4^+-N and that of *Salicornia* plants, which extended to the *Cuscuta* parasitising *Salicornia*.

G. δ^{15}N Patterns Attributed to Land Use

Karamanos *et al.* (1981) found no difference in total soil δ^{15}N between cultivated sites in Canada and native grasslands. Shearer *et al.* (1978) reported surface soil δ^{15}N from a large number of sites in North America (both cultivated and non-cultivated) to vary from about +1‰ to +13‰, the mode occurring between +7‰ and +9‰. Because their reported between-site means were different by about only 1‰, the major variations must have been within sites. Piccolo *et al.* (1994) found relatively high values for soil samples taken to 1 m depth in tropical native forests (+9.8‰ to +13.6‰), in contrast to the generally negative values (Nadelhoffer and Fry, 1994) reported from temperate forest systems. Piccolo *et al.* (1994) noted that conversion from native forest to pasture or cultivation resulted in a depletion of total soil δ^{15}N, in contrast to the enrichment reported from temperate systems for the same change of land use (Mariotti *et al.*, 1980b). Riga *et al.* (1971) found soil more ^{15}N-depleted in temperate forest soils than in cultivated ones. Broadbent *et al.* (1980) reported that soil N concentrations were less variable in cultivated soil than in virgin soils. Biggar (1978) found the opposite.

Herman and Rundel (1989) reported for soil mineral N in the fire-prone chaparral of California that the amount of mineral N was doubled by the burning; the amount of NH_4^+-N was unchanged, but the amount of NO_3^--N was increased. The opposite was true for $\delta^{15}N$: there was a significant change in the $\delta^{15}N$ of NH_4^+-N and no change in the $\delta^{15}N$ of NO_3^--N. Over a 37-day incubation of burned and unburned soils, both NO_3^--N and NH_4^+-N assumed 0‰ or near-0‰ values at some time.

Stewart et al. (1993) studied eight sites in Western Australia of varying ages since the last burning. Both NO_3^--N and NH_4^+-N were present in soil solution throughout the chronosequence, but NO_3^--N predominated in recently burned sites. For $\delta^{15}N$ at the same sites (Pate et al., 1993), only herbaceous plants responded to previous burns with changes in signatures and only on sites burned less than five years previously.

For an agricultural system, Doughton et al. (1991), showed that the $\delta^{15}N$ of barley (a strongly NO_3^--N-dependent crop; Pearson and Stewart, 1993) was related to previous soil history, which included fallow and sorghum cultivation followed by varying managements of the sorghum stubble (incorporation into soil, removal or left as surface debris). Feigin et al. (1974b) reported no ^{15}N-depletion of soil mineral N due to previously cropped N_2-fixing soybean. Ledgard et al. (1984) also found no correlation between the period under *Trifolium subterraneum* and total soil $\delta^{15}N$ of Australian pastures. Yoneyama (personal communication, 1994) however, found that sorghum $\delta^{15}N$ was more negative in proportion to the number of previous crops of N_2-fixing plants on agricultural soils in India.

For fynbos and strandveldt in South Africa, Stock et al. (1995) used $\delta^{15}N$ of soil and plants to show that invasion by exotic N_2-fixing acacias substantially altered ecosystem N-cycling.

For all components of soil, reported $\delta^{15}N$ patterns are complex and almost as numerous as the number of systems investigated. How much of the complexity is historical inaccuracy and how much is real, remains to be seen.

H. Total Soil $\delta^{15}N$ and Depth

Total soil $\delta^{15}N$ may become enriched with depth in the profile (Shearer et al., 1978; Karamanos and Rennie, 1980; Mariotti et al., 1980b; Guo-Qing et al., 1991; Nadelhoffer and Fry, 1994; Nadelhoffer et al., 1996), but the pattern is not consistent across sites (Shearer et al., 1978; Karamanos and Rennie, 1980; Andreux et al., 1990). Gebauer and Schulze (1991) found that total soil $\delta^{15}N$ increased with depth in a temperate coniferous forest, from $-3‰$ at the surface mineral horizon to $+4‰$. Nömmik et al. (1994) found a similar trend under unfertilised *Pinus sylvestris*. In Canadian agricultural soils Karamanos and Rennie (1980) found that the $\delta^{15}N$ of total soil N became ^{15}N-depleted with depth of profile in well-drained topographic depressions; on upper slopes, total soil δ^{15}N maintained the same signature throughout the depth profile. Broadbent *et al.* (1980) examined three virgin soils in California and demonstrated that several depth patterns are possible for δ^{15}N. At a high altitude wet site (mixed conifers, alder and willow) total soil N increased in δ^{15}N with depth in the profile (+2.6‰ to +10‰). In an oak savannah, the total soil δ^{15}N values were positive near the surface (+3.8‰) and decreased to −2.2‰ at 60 cm depth, then increased again to +3.1‰ at 100 cm, then became more negative at 120 cm (−5.4‰). At a third site in the San Joaquin Valley (vegetation not noted) surface soil δ^{15}N was near 0‰, increased to a maximum of +3‰ at a depth of 60–90 cm, then dropped again to near 0‰. Such depth-dependent reversals of ^{15}N enrichment are mentioned by Nadelhoffer and Fry (1994).

The often observed increase in δ^{15}N with soil depth has been interpreted as age-dependent mineralisation (Marrioti *et al.*, 1980b; Nadelhoffer and Fry, 1994) plus soil chromatographic fractionations (Delwiche and Steyn, 1970; Karamanos and Rennie, 1978), with younger organic N less mineralised and less enriched. The apparent correlation of enrichment with age of organic matter is secondary and indirect; the underlying relationship being ^{15}N-enrichment driven by amount of mineralisation or selective retention of N isotopes. The dominance of mineralisation in producing particular soil N isotopic signatures and reversals of those signatures in a soil depth profile was most convincingly developed by Schmidt and Voerkelius (1989) using a dual isotope approach (δ^{18}O and δ^{15}N) in which the signature of NO_3-O identified the mineralisation process (denitrification or nitrification) responsible for producing the NO_3-N (Figure 17) sampled at each depth in the soil profile.

Deep soil layers may be more consistently moist than surface soils, and soil moisture content is often correlated with enriched δ^{15}N by enhancing mineralisation rates and hence losses of ^{15}N-depleted N. Ledgard *et al.* (1984) examined depth profiles in Australian pasture soils. They found that the ^{15}N-enrichment which was correlated with depth and with the amount of heavily mineralised (clay-sized) organic matter. Less mineralised fractions (sand- and silt-sized) were more depleted than the clay-sized fraction. Denitrification, as a major fractionating mineralisation step, is not confined to surface layers, as documented for groundwater denitrification (Vogel *et al.*, 1981; Amberger and Schmidt, 1987; Mariotti *et al.*, 1988; Böttcher *et al.*, 1990).

Wedin *et al.* (1995) showed isotopic enrichment of both δ^{13}C and δ^{15}N in soil litter with increasing amounts of decay. In contrast to the above arguments, they suggested that the enrichments observed in litter δ^{13}C were not due to selective loss of ^{13}C through microbial respiration (Melillo *et al.*, 1989; Berg *et al.*, 1993a; Gleixner *et al.*, 1993) but were caused, instead, by immigration of carbon to the decaying organic matter, possibly *via* fungal hyphae and other microbes. This new information leads us to wonder how much of the observed ^{15}N-enrichment in decaying organic matter and soil profiles may

164   L.L. HANDLEY AND C.M. SCRIMGEOUR

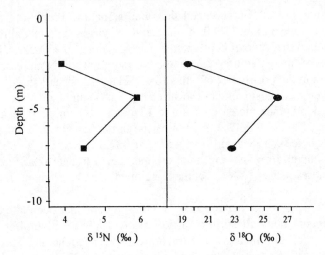

Fig. 17. Adapted and simplified from Schmidt and Voerkelius (1989). Soil depth profiles showing $\delta^{15}N$ and $\delta^{18}O$ values for residual nitrate at different depths. With dual isotope information it is possible to interpret the source of the nitrate (industrial production, nitrification or atmospheric processes) and whether denitrification has occurred. In this graph, the $\delta^{15}N$ (low values) and $\delta^{18}O$ (high) values show that nitrate in the top 3.5 m arose from application of industrially produced chemical fertilisers. The increase of both values at 4.5–5 m depth indicates that a partial denitrification has taken place.

be due to mineralisation processes, *per se*, and how much may be caused by a similar pattern of immigrating external N.

Much has been made of so-called soil chromatography (section V.D) to explain observed $\delta^{15}N$ changes with soil depth. It is our opinion that these effects are probably slight, even over time, compared with documented mineralisation effects, and as shown by the low enrichment factors for clay and exchange resins given in Delwiche and Steyn (1970). Karamanos and Rennie (1978) examined the limits of soil chromatographic $\delta^{15}N$ fractionations of NH_4^+-N in the extreme case, where clays were totally saturated with either Ca^{2+}, K^+, or NH_4^+-N, as well as the case of mixtures of cations. They found that the greatest fractionation was expressed when clays were saturated with Ca^{2+}. However, all fractionations done on clays saturated with cation mixtures (the expected natural situation) were on the order of only –1‰ to –2‰ at equilibrium. These slight fractionations stand in stark contrast to the relatively large fractionations documented for microbial processes (Delwiche and Steyn, 1970; Mariotti *et al.*, 1981; Macko and Estep, 1984; Yoshida, 1988) and ranging from –60‰ to +22‰ for N mineralisation and heterotrophic assimilation. Ledgard *et al.* (1984) provided soil depth data consistent with assuming a small fractionation due to soil chromatography. They found ^{15}N-enrichment correlated with depth for total N and the most heavily mineralised organic fraction, but mineral N had the same δ^{15}N value at all depths. The data of Karamanos and Rennie (1980) and Garten (1993) also support the presumption of a major role for soil water relations and consequent microbial processes in observed changes of δ^{15}N related to profile depth. Riga *et al.* (1971) found ^{15}N-enrichment followed by reversal at depth in both cultivated and forest soils in Belgium. Focht (1973) modelled the enrichment of soil NO_3-N with depth as microbial processes. It is worth speculating, that the frequently observed depth reversals of δ^{15}N are related to the relative amounts of NO_3-N retained at depths where both sufficient water and organic matter are present for extensive denitrification versus the amount which moves to depths where there is insufficient organic matter to provide energy for denitrification.

The most convincing counter-argument in favour of 'soil chromatography' comes from data presented by Herbel and Spalding (1993) where strong 'chromatography' of $^{15/14}$N was demonstrated and shown sometimes to be related to the presence of clays and sometimes not. In our view this work requires further investigation and deserves following-up.

I. Litter and Soil Organic Matter Inputs

Soil organic matter δ^{15}N is influenced by the type of inputs from above-ground litter, root turnover, and microbial biomass. (In some systems, exogenous wind-derived organic matter doubtless plays a role, as shown in the geological record for aeolian transport.) Contributions of both above- and below-ground organic matter to soil δ^{15}N are sketchy. For above-ground litter of northern European pine forests Berg *et al.* (1993b) showed that evapotranspiration was the chief constraint on decay of litter with other climate factors contributing. Coûteaux *et al.* (1995) concluded that there was a progressive shift in the dominant processes from biotic in temperate areas to abiotic in arid regions. In very arid areas, litter was concentrated discontinuously, hindering the development of a surface decomposer community, and processes such as physical fragmentation and photochemical degradation assumed greater importance. In wet tropical systems, they reported that litter quality, rather than climate factors, determined rates of litter decomposition. These types of variations in the recycling of litter help to illustrate why the temporal and spatial variations of soil and plant δ^{15}N do not conform to a simple model.

For temperate forest soils, Nadelhoffer and Fry (1994) presented a plausible qualitative model which presupposed ^{15}N-depleted plant organic matter inputs and subsequent soil organic matter ^{15}N enrichment, causing an increase with soil profile depth. This model was also used by Gebauer and Schulze (1991), Gebauer and Dietrich (1993) and Nömmik et al. (1994) to explain soil δ^{15}N increases with depth. Litter inputs generally shift soil δ^{15}N toward the value of the litter by isotope dilution, while decomposition of the same litter may cause enrichment relative to source litter (Nadelhoffer and Fry, 1988). Alternatively, at least some of the observed ^{15}N-enrichment (Wedin et al., 1995) may be due to immigration of microbial biomass containing new and ^{15}N-enriched N (section H).

Högberg et al. (1995) concluded that ^{15}N-depletion of a Swedish pine forest (which part of the 'forest' was not defined) was not due to inputs of depleted litter (Shearer and Kohl, 1986; Nadelhoffer and Fry, 1988), but was due, instead, to tight cycling of N in the forest ecosystem. This conclusion is unwarranted because the data set contained no information on losses from the forest ecosystem, and an isotopic mass balance was not done and probably would have been impossible. It also poses a false antithesis because there is no conflict to be resolved between the isotopic patterns within the system (foliage was analysed) and those external to it (N inputs and losses for the whole forest).

Foliage from plants which are known to rely primarily on NH_4^+-N are commonly reported to be ^{15}N-depleted (Domenach et al., 1989; Guo-Qing et al., 1991; Pate et al., 1993; Stewart et al., 1993; Nadelhoffer and Fry, 1994; Schulze et al., 1994). The negative values of the foliage of NH_4^+-N-assimilating plants can be explained by the signature of source N plus assimilatory fractionations (Yoneyama et al., 1991b) and internal plant partitioning. At least in theory, NH_4^+-N assimilation, coupled with the return of ^{15}N-depleted litter to soils, could maintain a low δ^{15}N for the entire system, despite large losses to atmosphere and/or groundwater.

With regard to a broader spectrum of ecosystems, contributions to soils by plant organic matter (leaves, roots, leaves and nodules of N_2-fixers) can be more enriched than the inputs from temperate forest plants. In our Scottish old field, senescent leaves and other parts of broom ranged from about –2‰ to +1‰; ripe fruits of various shrubs ranged from +1.5‰ to +4.9‰; senescent shoots of the grass–herb sward averaged +2.4‰, while total soil δ^{15}N was an average of c. 4‰. Shearer et al. (1983) found the non-N_2-fixing plants in their Sonoran Desert site to be positive in ^{15}N signature as did Handley et al. (1994b) for African savannas.

Roots can be more or less ^{15}N-enriched than other plant parts (Shearer and Kohl, 1986; Bergersen et al., 1988; Gebauer and Schulze, 1991; Gebauer and Dietrich, 1993), and nodules of N_2-fixing plants may be both N-rich and highly ^{15}N-enriched relative to other plant parts (Bergersen et al., 1986, 1988; Shearer and Kohl, 1986). Yoneyama et al. (1990b) found that nodules of wild-growing leguminous trees had δ^{15}N ranging from +2.1‰ to +11.6‰. For Frankia-associated N_2-fixers, Yoneyama and Sasakawa (1991) found that the occurrence of nodule enrichment was variable among plant taxa and among associated Frankia strains. These ranged from –1.6‰ to +5.5‰. Hence, the organic matter contribution of below-ground plant parts can be initially ^{15}N-enriched rather than depleted.

1. Microbial Degradation of Organic Matter and Litter Quality

For dead organic matter the fractionations attendant on microbial N assimilation of the released N can be large. While some measurements have been made of fractionations due to dissimilatory microbial N oxidation and reduction reactions such as nitrification, denitrification and physical processes such as volatile NH_3 loss in air (Focht, 1973; Handley and Raven, 1992, and references therein), we are unaware of any studies of soil microorganisms for N assimilatory fractionations. There has been, however, considerable work on δ^{15}N consequences to N assimilation in aquatic microorganisms. It is speculative to extrapolate from aquatic studies to soil organisms, but because we lack any similar studies of soil organisms, the marine studies will be discussed in order to suggest the potential magnitude of the δ^{15}N effects caused by soil microbial processes.

However, before we discuss the relevant marine literature it seems appropriate to mention one possible source of terminological confusion which may arise for a terrestrial ecologist when reading this literature. Some marine scientists (Hoch et al., 1994; Goericke et al., 1994) working on δ^{15}N of single-celled marine organisms appear to use the term 'uptake' to comprise both physical transport into the cell and enzymatic assimilation. This usage is also occasionally found in the terrestrial plant literature. Most marine plants lack the complication of a vascular system. In using δ^{15}N with terrestrial vascular plants, it is important to make a clear distinction between uptake (a physical process which apparently causes no fractionation) and assimilation (a biochemical process which may cause fractionation).

When marine heterotrophic bacteria were fed different N-source substrates Macko et al. (1986, 1987), whole-organism fractionations varied, depending on substrate, from $-12.9‰$ to $+22.3‰$. Velinsky et al. (1991) reported substrate NH_4^+-N enrichment up to $+21‰$ and instantaneous enrichment factors of $5–30‰$ due to bacterial assimilation of NH_4^+-N in marine anoxic basins.

Net assimilatory discrimination is not expressed in the whole organism (as opposed to the various parts in the case of vascular plants) unless there are parallel losses of N with a different isotopic signature to that remaining in the organisms, i.e. an isotopic mass balance is maintained. Various mechanisms have been proposed to explain the observed discriminations, relative to source, found for microbes: (1) parallel influx and efflux of inorganic N accompanied by assimilation (Hoch et al., 1992, 1994); (2) a postulated 'membrane fractionation' (Fogel and Cifuentes, 1993) which leaves a fractionated portion of the substrate on the exterior of the cell membrane; we add here a third possible mechanism, as yet unexamined for δ^{15}N – the loss of organic N (Bronk et al., 1994) having an isotopic signature substantially different from that of the whole organism.

One of the potentially fractionating losses is NO_3^--N efflux. This has been said to be energetically costly and an active transport process (attributed by

Goericke et al., 1994, to Wheeler, 1983) in plants. It has been amply demonstrated and reported elsewhere (Lee and Stewart 1978; Handley and Raven, 1992; Trebaçz et al., 1994) that in photolithotrophs, NO_3^--N efflux occurs passively (i.e., as slippage down an electrochemical potential gradient via exchange of internal for external NO_3^--N through the $2H^+ + NO_3^-$-N symporter, or as leakage via an anion uniport channel). While the loss of NO_3^--N costs the organism energy in terms of net NO_3^--N influx, the process of loss itself is passive and does not directly consume energy. Although NO_3^--N influx, efflux and assimilation were modelled for wheat using enriched ^{15}N label (Devienne et al., 1994a, b), the effect of these processes on plant $\delta^{15}N$ is unknown. The Shearer et al. (1991) $\delta^{15}N$ model for NO_3^--N efflux versus influx in Synechococcus was experimentally flawed. The CO_2 concentration used (5%) altered both the carbon metabolism and N^- use efficiency of this alga by impairment of the CO_2 accumulating mechanism (Badger and Price, 1994, and earlier references therein) so that the results were merely artifacts of the experimental design.

The concept of 'membrane fractionation' for either NO_3^--N or NH_4^+-N appears to be apocryphal as we can find no data substantiating its existence, nor can we imagine, a priori, why the presence of a semi-permeable membrane would directly fractionate either NH_4^+-N or NO_3^--N. All of the existing evidence for higher plants (Mariotti et al., 1982; Yoneyama and Kaneko, 1989; Yoneyama et al., 1991a) indicates that fractionation of exogenous N takes place at the first assimilatory enzyme (first irreversible step) and not upon simple physical uptake or transport.

A sequence of complex ideas are being developed in the marine literature to explain puzzling fractionations in marine bacteria under varying external concentrations of NH_4^+-N (Hoch et al., 1992, 1994), and these arguments require the presence of a biological membrane in order to maintain separation of pools of N with different $\delta^{15}N$ values. However, the reasoning of these authors does not imply that the membrane, itself, acts as the kind of fractionating isotopic sieve which some later authors have called upon to explain their results (Høgh-Jensen and Schjoerring, 1994).

Remobilisation of combined N within the plant requires the action of enzymes which do fractionate N isotopes, so that fractionations by transport (per se) is not a required postulate for explaining the observed fractionation correlated with N remobilisation. Nor should diffusion limitation (per se) at the cell boundary layer cause fractionation relative to source. On the contrary, theory dictates that it should lessen fractionation by decreasing substrate supply relative to enzyme demand. It seems likely to us, on balance, that the observed whole-organism discriminations of ^{15}N for microbes (if we extrapolate from the aquatic experience) are due to losses of ^{15}N-enriched or ^{15}N-depleted organic N products in addition to the demonstrated efflux of inorganic N. Although we are unaware of any papers estimating these effects for

δ^{15}N, Bronk *et al.* (1994) recently demonstrated up to 41% loss of organic N by phytoplankton using ^{15}N-enriched tracers. This was observed in healthy populations from oceanic, coastal and estuarine waters, and these findings are consistent with earlier work, e.g. Hellebust (1974) reported that cyanobacteria lost 30% of fixed N_2 as organic N. The isotopic signature of the lost organic N will reflect its source and its enzymatic history (Macko *et al.*, 1986, 1987). However, if the lost organic N is very different from whole organism δ^{15}N, then the use of δ^{15}N, alone, will not explain the fractionations of δ^{15}N, observed for microorganisms because the total number of potentially fractionating processes (unknowns) exceeds the number of measurable isotope signatures.

Näsholm (1994) reported high concentrations of arginine in senescent pine needles of a highly fertilised conifer plantation. If the fractionation factors determined by Macko and Estep (1984) hold for microbial decay of soil organic matter, then degradation of these needles might lead to pools of enriched microbial biomass (+4.6‰ to +6.5‰ more positive than the arginine substrate). Hence, one of the isotopic consequences of conifer plantation fertilisation could be ^{15}N enrichment of soil organic matter. Berg (1988) found, however, in an ^{15}N-enrichment experiment for decomposition of pine needles, that the ^{15}N content of the residual pine needles was slightly depleted (relative to the beginning value) at the end of the experiment.

In addition to the loss of molecular species of N, the δ^{15}N of whole vascular plants can be changed by loss of selectively ^{15}N-enriched or ^{15}N-depleted organs (e.g., leaves, twigs, fruits and flowers; Handley and Raven, 1992). We suggest that the literature, so far, may have over emphasised direct efflux of inorganic N at the expense of underestimating the effects of organic N losses in the whole organism δ^{15}N.

The quality of above-ground litter influences its final δ^{15}N through the rates and types of degradation it undergoes (Palm and Sanchez, 1991). Northup *et al.* (1995) found that the polyphenol concentration of decomposing pine litter controlled the proportion of N released in dissolved organic forms relative to mineral forms and thereby regulated the extent to which mineralization was short-circuited by mycorrhizal associations. Isotopic fractionations were not addressed.

2. Fungal Influences on δ^{15}N

If we are to understand plant and soil δ^{15}N, it is important to understand how mycorrhizas affect these signatures and how they mobilise organic N sources for reuse, because mycorrhizas can affect N cycling in several potentially fractionating ways. Fungi can assimilate large amounts of N (up to 1560 g m^{-2} yr^{-1};

Fogel and Hunt 1979); they can transport N from soil to plants, from plant to plant (van Kessel *et al.*, 1985; Arnebrant *et al.*, 1993); they can breakdown and use organic N; they enhance N_2-fixation in N_2-fixing plants thereby bringing more N with $\delta^{15}N = 0‰$ into the system; and all mycorrhizas (arbuscular, ecto, ericoid) can readily transport NH_4^+-N from soil to plant (Allen, 1991, and references therein) accessing small amounts of NH_4^+-N as mineralization gradually makes it available in N limited systems; they have also been shown to especially proliferate into pockets of soil NH_4^+-N which may have atypical signatures, such as subterranean termite or ant nests (Allen, 1991) or discrete pockets of plant organic matter (Read, 1991; Hutchings and de Kroon, 1994).

In controlled studies, Handley *et al.* (1993) found that AM infection influenced whole-plant $\delta^{15}N$ of *Ricinus communis* and the partitioning of $\delta^{15}N$ within the plant by slightly less than 2‰. These findings were consistent with the field results of Pate *et al.* (1993), in which the N of non-mycorrhizal nitrate-using plants was more ^{15}N-enriched (*c.* +2‰) than mycorrhizal ones. In South Africa, Stock *et al.* (1995) also found non-mycorrhizal *Protea* spp. to be ^{15}N-enriched relative to other nearby plants. In the experiment by Handley *et al.* (1993) plant $\delta^{15}N$ dilution by initial seed N was only indirectly addressed by using seeds of the same weight, size and origin. This does not affect the overall treatment differences between infected and non-infected plants, but it does leave unresolved whether treatment differences were due to fungal assimilation of N or to growth effects on the plants caused by the infection treatments.

It is known that AM infection can increase plant N accumulation and assimilation (Frey and Schüepp 1993). Infection may also alter the relative uses of chemical types of N. Cliquet and Stewart (1993) found enhanced N uptake in AM-infected maize and increased (nearly double) activity of both glutamine synthetase and nitrate-reductase; NH_4^+-N use was particularly increased over non-infected controls. Barea *et al.* (1991) used ^{15}N-enrichment to show that AM infection enhanced total plant N assimilation and suggested that this possibly represented a N-source shift. A major shift of N source has the potential to change foliar $\delta^{15}N$ in at least three ways: (1) a shift in source $\delta^{15}N$ and (2) a change in assimilatory $\delta^{15}N$ fractionation incurred on assimilation by different enzymes located in different parts of the plant and (3) increasing assimilatory discrimination by making the external N supply larger, relative to enzyme demand.

The interaction of ectomycorrhizal associations and plant $\delta^{15}N$ is probably complex, and just a few of the considerations are that: organic N sources may be more or less ^{15}N-enriched than soil NH_4^+-N or NO_3^--N and variations in the N assimilation pathways (spatially and chemical) are dependent on the particular combination of host plant and fungus (Martin and Botton 1993) as well as on the age of the host and N source (Martin *et al.*, 1994). Read (1991) described ecto-mycorrhizal fungi on temperate forest trees as functionally and taxonomically diverse assemblages, some of which rely primarily on N mineralised by other organisms (chiefly NH_4^+-N) and proteolytic fungi capable of accessing organic N sources, pioneer and climax forms of fungi, which grow outward from the roots '... into a complex mosaic of organic substrates made up of qualitatively distinctive resource types'. Hence, the fungi on any one tree may be functionally diverse, in N assimilating capacities and in using different major N sources with accompanying differences of δ^{15}N among sources. Hence, it cannot be assumed, in the field, that 'the mycorrhiza' has access to a particular N source with a single definable δ^{15}N.

There may be variation in fungal loss of ^{15}N-depleted NH_3 during growth and senescence of fruiting bodies (and hence, in the isotopic signatures of the soil-residual fungal biomass and in N transferred to a host plant). Ingelög and Nohrstedt (1993), measured up to 20 mg N lost as NH_4^+-N per gram of dry weight of fungal biomass. They also found that the NH_4^+-N absorbed into the forest soil from the deliquescing fungus substantially raised the soil pH. This new and higher pH may have further effects, locally, on NH_4^+-N loss to the atmosphere and local soil δ^{15}N. The N for this process was presumably moved from a large soil volume and concentrated by the fungus into a point (the mushroom); if Ingelög and Nohrstedt's (1993) data are typical of NH_4^+-N fluxes through hyphae to fruiting bodies of forest fungi, then the role of fungi in redistributing NH_4^+-N in the forest floor may be at least as important as their direct symbiotic role in accessing various forms of N for the host plant. Because roots selectively exploit concentrated pockets of resources, the maintenance of N-rich patches, could be an important stabilising mechanism in some systems. To put this into proportion, we cite Bååth and Söderström (1979) who found that 20% of total soil N in a Swedish conifer forest was bound in fungal hyphae.

Michelsen et al. (1996) concluded that patterns of foliar δ^{15}N at a sub-Arctic site were explained by the source values of organic N presumably accessed by mycorrhizas. The patterns were consistent with this hypothesis, but were insufficient, in themselves, to support this conclusion.

In the 1993 controlled study, Handley et al. found no δ^{15}N effect of ecto-mycorrhizal infection on Eucalyptus seedlings when only organic N was supplied. Högberg et al. (1994) reported a pot experiment in which pine seedlings were infected, or not, with ectomycorrhizal fungi. However, they applied enriched NH_4^+-N (1.2342 atom % = 2369‰ as δ^{15}N), thereby masking the expected treatment effects of 1‰ to 2‰, and concluded from this that there were no effects at the natural abundance level. While the absence of treatment effects is consistent with the findings of Handley et al. (1993) for ectomycorrhizas and tree seedlings, the experimental design precluded interpretation at the natural abundance level. There were also random enrichments in the potting mixture, which were too large for the authors' suggested explanation of soil cation exchange (Delwiche and Steyn, 1970; sections D and H).

We are aware of only one other mycorrhizal infection versus non-infection $\delta^{15}N$ experiment. Michelsen and Sprent (1994) carried out $\delta^{15}N$ analyses on material grown in an outdoor nursery in Ethiopia where many of the tree seedlings grew roots out of the containers onto plastic sheeting, potentially accessing unknown and unmeasured N sources. Other confounding effects included use of river and rain water, both containing unknown amounts of N of unmeasured $\delta^{15}N$, application of dung (a highly variable and enriched ^{15}N source not measured for $\delta^{15}N$) in the potting mixture, and variable amounts of root pruning. Only shoots were analysed for $\delta^{15}N$. The data were then used in a calculation which is appropriate only for whole plants (e.g., Bergersen et al., 1988); no soil $\delta^{15}N$ values were analysed. The anticipated treatment differences (1–3‰) were small relative to the possibly conflicting signatures from other sources. For these reasons, the data cannot be used to assess the impact of mycorrhizas on plant $\delta^{15}N$. Field studies by Högberg (1990) in the dry forests of Tanzania, East Africa revealed patterns in which tree foliar $\delta^{15}N$ was correlated with type and/or occurrence of root symbiotic associations. Later research in humid forests in West Africa (Högberg and Alexander, 1995) showed that root symbioses (ecto-, arbuscular or N_2-fixing) could not be separated reliably on the basis of foliar $\delta^{15}N$. However, a general pattern may yet emerge if cases are restricted to woody taxa in semi-arid or arid environments. In Western Australia, Pate et al. (1993) also could not distinguish between N_2-fixing trees and mycorrhizal plants on the basis of foliage $\delta^{15}N$. However, foliage from nitrate-using non-mycorrhizal trees was +2‰ more ^{15}N-enriched than foliage from mycorrhizal or N_2-fixing ones. In another semi-arid environment, the South African fynbos, Stock et al. (1995) found that non-mycorrhizal members of the Proteaceae with little or no nitrate reductase activity were the most ^{15}N-enriched of a community of plant taxa.

With regard to the fungi, themselves, Gebauer and Dietrich (1993) found fungi in a German forest that were ^{15}N-enriched and ^{15}N-depleted relative to nearby plant foliage, plant litter and soil (Figure 15). We collected fruiting bodies of fungi from Lamington National Forest, a high-altitude rain forest in Queensland, Australia (Handley et al., 1996). Analyses showed that fruiting bodies of both potentially mycorrhizal fungi and fungi found on dead and live wood had $\delta^{15}N$ values significantly enriched over total soil N, plant litter N and leaves of nearby living plants (fungi ranged from +10.2‰ to +4.2‰; litter was +4.1‰ to +0.2‰, and tree leaves were +3.5‰ to +1.2‰). The same patterns were seen for forest fungi from Argyll on the west coast of Scotland, (Handley et al., 1996). Fungal fruiting bodies varied from +11‰ to −3‰; fungal substrates of all kinds ranged from +2.4‰ to 0‰, and tree foliage ranged from +2.8‰ to +1.2‰. In eastern lowland Scotland (grounds of SCRI) fifty individuals of the fruiting bodies of *Agrocybe* sp. were collected from a wood-mulched decorative garden. Specimens of this single flush of a single species, collected simultaneously from an area of about 4 m^2, showed δ^{15}N values ranging from +2.4‰ to +4.5‰; caps were significantly different from stipes and relatively ^{15}N-depleted. Thus, the total δ^{15}N of the fungus depended to some extent on the amount of stipe recovered. Neither visual inspection of size and condition nor measurement of height or diameter accurately predicted the subsequent δ^{15}N-value class.

Hence, field measurements show that in many widely separated places having different vegetations and climates, fungi can be highly ^{15}N-enriched relative to other nearby organisms, soil or other substrates. A naive analysis of these data would suggest that the fungi assimilate soil N, retain the ^{15}N-enriched portion and pass the ^{15}N-depleted portion to the plant with which it is associated. This is especially appealing as an hypothesis for potentially mycorrhizal fungi, because ecto-mycorrhizal trees may have very negative δ^{15}N foliar signatures. The situation is almost certainly not this simple, and the full explanation of these now widely observed patterns awaits controlled experimentation.

Bearing in mind Ingelög and Nohrstedt's (1993) data on NH_4^+-N loss from fungi, the ^{15}N-enrichments observed in field-collected fungal samples could be largely explained by variable amounts of organic N gained from the hyphae during growth and NH_4^+-N lost during subsequent growth and senescence of variously aged specimens. It is also not known how variable these factors might be among different taxa of fungi. For the present, we view the interpretation of δ^{15}N values of field-collected fungal specimens with caution.

J. Chronic N Eutrophication of Soils

Chronic N eutrophication of soils can arise from, *inter alia*: (1) an overabundance of N$_2$-fixing organisms, (2) anthropogenic fertilisation, or (3) atmospheric deposition (pollution) of both NH_4^+-N and oxidised forms of N. Consequences for soil can include changes in soil pH, loss of soil cations, changes in the plant and microbial communities and also dysfunctional physiological responses in existing vegetation (Pearson and Stewart, 1993).

1. N$_2$-Fixation

Sandhu *et al.* (1990) recorded large N losses from the litter, stems and roots of the N$_2$-fixing shrub, *Leucaena leucocephala*. This shrub was seeded from the air to revegetate battle-devastated Guam after World War II. Guamanian ground waters are now several times richer in NO_3^--N than potable limits, and this NO_3^--N pollution is thought to derive from release and mineralisation of N fixed by *L. leucocephala* (Marsh, Agana, Guam, personal communication). High concentrations of NO_3^--N in stream waters of *L. leucocephala*-covered watersheds on the island of Oahu, Hawaiian Islands, were documented in water quality surveys by Water Resources Research Center, University of Hawaii (Lau et al., 1975; Young et al., 1976). Evans and Ehleringer (1994) found substantial N enrichment and $\delta^{15}N$ depletion of soils where cryptobiotic soil crusts were undisturbed. Using $\delta^{15}N$ and hydrological techniques (Vogel et al., 1981), Heaton et al. (1983) were able to model past concentrations of NO_3^--N and measure present concentrations of up to 40 mg l^{-1} NO_3^--N in groundwater under the Kalahari Desert (twice the potable limit) and to show through use of $\delta^{15}N$ that these high NO_3^--N concentrations were due to N_2-fixation by native acacia trees and possibly also to N_2-fixing desert cryptogamic crusts (Ryckert et al., 1978). Several reports now exist (Riga et al., 1971; Tobita et al., 1994; van Kessel et al., 1994b) demonstrating that legumes leave substantial N in soil organic matter for subsequent mineralisation and loss, sometimes with devastating effects to local water supplies.

2. Deliberate Chemical Fertilisation

In some infrequently cultivated systems, e.g. plantation forestry, large amounts of fertiliser are applied. To estimate the effects of chronic anthropogenic fertilisation, most (but not all) of the values for changes in soil $\delta^{15}N$ have been measured as plant-harvested $\delta^{15}N$. This ignores the fact that some of the observed effects for $\delta^{15}N$ could be due to changes in plant assimilation and N sequestration at high levels of available N rather than due entirely to direct changes in the $\delta^{15}N$ of plant-available soil N (plants are not reliable samplers of soil $\delta^{15}N$). Therefore, plant-harvested $\delta^{15}N$ effects of chronic eutrophication are actually effects of eutrophication plus the internal plant fractionations in response to the level of N inputs; if mycorrhizal status of the sampled plants is affected by eutrophication, then the internal distributions of plant $\delta^{15}N$ may also be changed, regardless of whether total plant $\delta^{15}N$ is greatly affected (Handley et al., 1993; Martin and Botton, 1993). Freyer and Aly (1974) found the relationship between fertiliser N levels and soil mineral $\delta^{15}N$ complex inconclusive and argued that biochemical fractionations of $\delta^{15}N$ prevent its general use as a tracer.

Hauck et al.'s (1972) argument that $\delta^{15}N$ fractionations preclude its use as a tracer in soils applies also to plant foliage. Unusual instances have arisen where $\delta^{15}N$ could be used as a direct tracer or could be used with appropriate ancillary data and modelling techniques which account for fractionations between source and sink. In the Minnesota sand-plain, Komor and Anderson (1993) traced aquifer NH_4^+-N to its sources of animal feedlots, cultivation or sewage because percolation was rapid and little microbial fractionation occurred en route to the groundwater. Heaton (1985) modelled the past contribution of native N_2-fixing vegetation to groundwater NO_3^--N concentrations, taking into account relevant fractionations. In food web studies, $\delta^{15}N$

can be used as a tracer when the fractionations among trophic levels can be determined (DeNiro and Epstein, 1981; Wada *et al.*, 1991, 1995; Scrimgeour *et al.*, 1995).

Näsholm's (1994) study of fertilised conifer plantations provides a good case for discussion. High fertilisation levels resulted in more positive pine needle δ^{15}N, but also in sequestration of arginine in senescing needles. There are several possible reasons for the δ^{15}N enrichments observed, some of which are argued at different levels of organisation:

(1) The accumulation of a terminal metabolic N product, not recycled into other plant parts, can lead to singular ^{15}N signatures for those plant parts in which the terminal product is sequestered. This is an extreme case of the generalisation that intra-plant distributions of different metabolic products are the chief reason for the variations of δ^{15}N observed within plants.

(2) The Näsholm (1994) work raises additional questions (as pointed out by the author) of the role of gaseous N loss from plants as a mechanism for ^{15}N-enrichment of plant parts. The author suggested that the loss of NH_4^+-N (Farquhar *et al.*, 1983) might be responsible for elevated δ^{15}N at needle senescence, but also noted that the amounts documented in the literature were insufficient to explain the observed ^{15}N enrichment.

(3) An additional gaseous loss for consideration is NO. In animals, NO is produced by the conversion of arginine to citrulline and is a gaseous product which is as likely to be ^{15}N-depleted as is NH_4^+-N. It is not known whether arginine conversion to citrulline in plants produces NO as a by-product (Raven *et al.*, 1993). However, work by Farquhar *et al.* (1983) confirmed the loss of small amounts of oxidised forms of N from plants, and theoretical reasons exist (Raven *et al.*, 1993) for supposing that plants possess the capacity to produce oxidised forms of N.

(4) Without evidence of gaseous losses, the most intuitively appealing explanation for the observed ^{15}N-enrichments lies in the observed foliar arginine amounts. In all treatments, some arginine disappeared and was perhaps retranslocated to other plant parts. Medina *et al.* (1982) measured the instantaneous enrichment factor associated with the enzyme, arginase, and found it to be relatively large ($\alpha = 1.0104 = 10.4\%$). If one takes approximate amounts from graphs of Näsholm (1994) (combined by us into one graph; Figure 18), it can be calculated that a partial expression of this enzyme would account for all of the enrichment observed in the older pine needles. However, this only suggests a line of investigation.

Recently Högberg *et al.* (1995) summarised several years' work on Swedish pine plantations. Their findings were consistent with the generally acknowledged observation that increasing amounts of N also increase rates

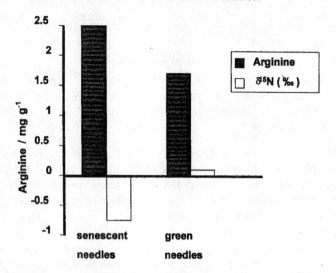

Fig. 18. Adapted and simplified from Näsholm, 1994. Arginine concentrations (mg g⁻¹ dry weight) and $\delta^{15}N$(‰) of current and senescent pine needles in Sweden. The discrimination due to expression of arginase (10.5‰) could account for the discrimination of $^{15/14}N$ between current and senescing needles. Using approximations from the published graph to calculate an isotopic mass balance: $[(2.5 \text{ mg}) \times (-0.75) - (0.8 \text{ mg}) \times (X \text{‰}) = (1.7 \text{ mg}) \times (0.1\text{‰})]$, where X is the predicted needle signature. Here $X = 2.6$‰; graphically shown needle enrichments were 1.25‰.

of N mineralisation. However, $\delta^{15}N$ alone has little quantitative predictive power and has long since proved disappointing in predicting saturation amounts of fertilisation. In 1975, Meints *et al.* found that agricultural soil $\delta^{15}N$ was not correlated with fertilisation levels applied over a period of 20 years. They concluded that $\delta^{15}N$ should not be used as an indicator of fertiliser levels and requirements. Yoneyama *et al.* (1990a) found depletion of plant-harvested (not soil) $\delta^{15}N$ due to chronic chemical fertiliser additions to upland crops and paddy rice. Riga *et al.* (1971) reported the total N of fertilised Belgian soils to be depleted in ^{15}N relative to unfertilised controls. Fertilisation appears to be as site-dependent in its effects on the $\delta^{15}N$ of soil and plant N pools as are many other empirical correlations for $\delta^{15}N$. A more generally useful approach, in our opinion, is the simultaneous use of dual isotopes (i.e., $\delta^{15}N$ and $\delta^{18}O$), as used by, for example, Durka *et al.* (1994), Hedin (1994) and Böttcher *et al.* (1990), to suggest, the extent of N cycling versus saturation and/or flow-through by the unassimilated N inputs. Widespread use of this approach, however, awaits technical developments in automated pyrolysis.

Johannisson and Högberg (1994) measured the δ^{15}N of both soils and understory grass along a fertilisation gradient in pine forests and concluded that chronic fertilisation changed the plant values more than it changed the δ^{15}N of total soil N. This is consistent with the common assumption that plants preferentially use the newest N added to a system. Nömmik et al. (1994) examined δ^{15}N effects of high fertilisation rates on pines and also concluded that pine needles were ^{15}N-enriched at high fertilisation levels. They also found a correlation between amount of fertilisation and δ^{15}N of the litter and mineral soil, with urea fertilisation producing the greatest enrichment. Black and Waring (1977) measured δ^{15}N of NO_3^--N formed from fertiliser additions to soil cultures and found that the NO_3^--N was enriched, relative to fertiliser inputs, from -4.1% to 1.9%.

Högberg and Johannisson (1993) surmised, from data collected in a Swedish conifer plantation, that fertilisation acts in the short-term on plant available pools of δ^{15}N, while total soil δ^{15}N reflects longer term processes. In any soil–plant system where mineral N is a smaller pool than organic N, herbaceous plant δ^{15}N will more closely reflect the δ^{15}N of the mineral N pool than the organic N pool. This is also consistent with published work on the fates of added ^{15}N (Schimel et al., 1989; Jackson et al., 1989; Zak et al., 1990; Groffman et al., 1993).

Data published by Black and Waring (1977) also serve to illustrate that fertiliser δ^{15}N can be very different from 0‰. Macko and Ostrom (1994) provided data on the distribution of fertiliser δ^{15}N showing that the mode was near 0‰, but that common values ranged from $-8‰$ to $+7‰$. We have measured fertiliser NO_3^--N with δ^{15}N as low as $-14‰$, and Freyer and Aly (1974) reported sodium nitrate with a δ^{15}N of $+22‰$. Högberg et al. (1995) found that the δ^{15}N of NH_4NO_3 applied to Swedish pine plantations was $-3.0‰$, and the urea was $-10.8‰$. It should not be assumed that different batches of fertiliser from the same manufacturer have similar δ^{15}N. Medina et al. (1982) reported that the instantaneous fractionation due to urease could be as much as 8‰ of ^{15}N-depletion in the NH_4^+-N product. Subsequent volatile NH_4^+-N loss may provoke a large fractionation, leaving an enriched NH_4^+-N pool in the soil, but only at neutral to basic pH. Plants may also assimilate varying and unknown proportions of both the partially remaining, and fractionated urea (directly) and the produced and fractionated NH_4^+-N. It has been suggested (Nömmik et al., 1994) that canopy foliage can intercept (and perhaps assimilate) at least some of the ^{15}N depleted gaseous NH_4^+-N lost from the soil. The general trend toward plant enrichment by fertilisation is a qualitative observation, verified on more than one occasion for Swedish pine forests. However, we doubt that this observation can be used quantitatively because of the many complicated and possible fractionating fates for applied N, as mentioned above. Lightly ^{15}N-enriched fertilisers of known distinctive signature could help trace the fate of N in forest systems (Fry et al., 1995).

3. Atmospheric Deposition

We did not measure atmospheric combined N at our Scottish site. However, Pearson and Stewart (1993) reported an average range of 25–56 kg ha^{-1} yr^{-1} atmospheric deposition of all forms of N for Scotland as a whole – a substantial input to a non-agricultural system. We cannot rationalise differences in the foliar δ^{15}N signatures in our old field by calling on atmospheric N deposition. The woody plants had negative signatures, the grass–herb sward was positive (Figure 3), and atmospheric N, both oxidised and reduced, presumably falls equally on all parts of this small, windy site, although interception and use may vary.

There is a misconception in the literature (e.g., Abbadie et al., 1992) that all atmospheric N (Figure 19) deposition is negative in δ^{15}N signature; however, the δ^{15}N values of atmospheric N cover a wide range globally, from about −20‰ to +20‰, depending on source and relative amounts of NO$_4^+$-N and NO$_3^-$-N, as well as on the relative proportions of dry versus dissolved or particulate N. At any one site, presently sparse data suggest a typical range of about 15‰. The δ^{15}N values also vary seasonally by a range of about 13‰ (Heaton, 1986), and seasonal variation in amount must also be taken into consideration. Overlaying the seasonal variation of amounts are diurnal variations in amounts (Schulze, 1989). Whether δ^{15}N values vary diurnally is probably site-specific and may be due to both source and diurnal effects of temperature and relative humidity on interaction with other chemicals in the atmosphere; e.g. McLeod et al. (1995) documented the enhancement of atmospheric N deposition by co-deposition of sulphur.

Geographic remoteness is no guarantee of non-anthropogenic N sources, and one cannot assume 'natural' ranges of isotopic values in the absence of both isotopic and mass measurements as well as knowledge of prevailing wind patterns.

Garten (1992) reported negative to slightly positive δ^{15}N for NH$_4^+$-N and NO$_3^-$-N in rainfall and cloud vapour for valley and mountain locations in Tennessee and southwestern Virginia (−5.5‰ to +2.3‰). This range of values is consistent with the information provided by F.R. Fosberg (personal communication, 1981) that much of the air pollution in the southeastern USA was derived from the industrial heartlands located hundreds of miles to the north (principally Chicago, Illinois and Gary, Indiana) and is moved southward by prevailing weather patterns. These values from a rural area are not greatly different from those recorded in Tokyo (−9.7‰ for NH$_4^+$-N to +4‰ for NO$_3^-$-N; Wada et al., 1975). Values recently recorded (Heaton et al., 1996) for rainfall NH$_4^+$-N and NO$_3^-$-N in Yorkshire (Witton Fell, Wensleydale) were also broadly within the same range as those from the southeastern USA and Tokyo (−8.6‰ to −7.0‰ for NH$_4^+$-N and −1.5‰ to +0.9‰ for NO$_3^-$-N) and may indicate anthropogenic influences in this area about 50 miles from the city of

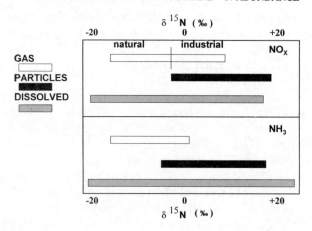

Fig. 19. Adapted from a more detailed figure in Heaton (1986). Specific data gathered prior to 1986 are given or cited in Heaton (1986); other references, whose data still fall within the ranges given above, are Garten (1992), Wada et al. (1975), Heaton (1987), and Heaton et al. (1996). The last found for rural Yorkshire, 1993–94 in which rainfall NO_3^--N ranged from −1.5‰ to +0.9‰, and rainfall NH_4^+-N was −8.6‰ to −7.0‰.

Leeds. The extent of the spread of industrially sourced atmospheric N was recently highlighted by Cornell et al., (1995) who reported, '... analyses of dissolved organic nitrogen in rain and snow which show that it is a ubiquitous and significant component of precipitation, even in remote marine areas'. The δ^{15}N of these marine samples ranged from about −7‰ to about +7‰. Consistent with the findings of Stewart et al. (1994), however, the lowest values were associated with the most geographically remote areas.

Heaton and Crossley (1995) used artificial misting of N compounds onto Sitka spruce and found that C assimilation rates were depressed by 10% (as calculated from δ^{13}C measurements). This effect on C assimilation should also have a depressive effect on N assimilatory capacity of the trees. Further simulated deposition studies of lightly ^{15}N-enriched 'atmospheric N' might help to resolve the fates of atmospheric N.

Vegetation rooted in soil is a poor bio-sampler of atmospheric N because of the influence of soil N sources. Epiphytes, however, obtain almost all of their nutrition directly from the air, with the possible exception of some nutrition derived from stem flow or throughfall and a few epiphytes which have N_2-fixing symbioses or symbioses with animals (Treseder et al., 1995). Stewart et al. (1994) correlated the δ^{15}N of non-symbiotic epiphytes with the influence of potentially large sources of atmospheric N from urban areas and found no correlation between the morphology/strategy of the epiphyte and its δ^{15}N. The δ^{15}N of epiphytes collected near urban centres was enriched relative to the signatures of those from more remote areas that were not downwind of urban centres. Vitousek *et al.* (1989) used the $\delta^{15}N$ of a non-N_2-fixing lichen as a bioindicator of the $\delta^{15}N$ of atmospheric N in Hawaii. J.A. Raven (unpublished data), however, found coastal lichens in Scotland to be highly variable in $\delta^{15}N$ and, therefore, an unreliable tracer of N-source.

Plants assimilate oxidised and reduced forms of atmospheric N, and both wet and dry deposited forms are bound and/or assimilated to some extent by tree foliage (Anderson and Mansfield, 1979; Rogers *et al.*, 1979a, b). In forests, related processes include: (1) deposition on leaves or soil, (2) in leaves, either cuticular absorption, mesophyll storage, or assimilation into organic N (Pearson and Stewart, 1993), (3) surface-deposited N carried by throughfall from foliage to soil, and (4) at the soil surface, deposited N can be assimilated by plants, processed through the microbial biomass (assimilation, dissimilation), or be lost as dissolved N in percolate water (Durka *et al.*, 1994) or as gas. If soil N is lost gaseously, it may be partially trapped by canopy foliage (Nömmik, et al., 1994). These processes are discussed in some detail by Raven (1988), Pearson and Stewart (1993), Raven *et al.* (1992a,b, 1993) and Sutton *et al.* (1993).

Research by Garten (1992) suggested that at least for HNO_3 deposition, broadleaf trees catch (but not necessarily assimilate) much more of the deposition than conifers, as would be expected from the relative amounts of horizontal leaf area presented to the atmosphere and the wettability of leaves. If assimilation plus cuticular absorption are small in amount relative to total plant N, then the change in plant $\delta^{15}N$ will be small and possibly unmeasureable.

If foliage $\delta^{15}N$ changes can be correlated with atmospheric N inputs, then the relationship may be indirect, through the effects of added N on soil ecology and plant biochemical ion balance (e.g., through nutritional deficits, cation leaching and pH changes) (Raven, 1988; Gebauer and Schulze, 1991; Pearson and Stewart, 1993) rather than via direct absorption and accumulation of a given $\delta^{15}N$ signal into the plant.

Two papers suggested atmospheric N as a possible explanation for foliar signatures of plants found in remote locations on severely N-deficient soils. Kohls *et al.* (1994) found negative $\delta^{15}N$ for plants growing on N-poor glacial till in Canada with leaf values ranging from -6.4% to -3.3%. They noted that these values were consistent with isotopically depleted atmospheric deposition as a main source, but also discussed the possible consequences of slow growth in a cold location on plant $\delta^{15}N$. We found similar ranges of negative leaf values in the Scottish old field where soil NO_3^--N and NH_4^+-N concentrations were in the same range as measured by Kohls *et al.* (1994), i.e. $< 1 \mu g \ g^{-1}$ soil. If plant growth is very slow, then, $1-2 \mu g \ g^{-1}$ soil may be a sufficiently large external concentration, relative to enzyme demand, to induce assimilatory fractionations; or it may simply reflect the $\delta^{15}N$ of the source N or a combination of both.

Vitousek *et al.* (1989) found negative δ^{15}N for foliage of a pioneer evergreen tree (*Metrosideros collina* var. *polymorpha*) in Hawaii, and the δ^{15}N was positively correlated with the age of soils. The age of soils was also correlated with both total soil δ^{15}N and N content. Young soils (197 years old) had δ^{15}N of $-2‰$, while older soil (*c.* 67 000 years old) was found to have δ^{15}N of $-3.5‰$. Hence, foliage δ^{15}N was correlated with both soil age and with soil N concentration. The authors suggested that the negative signatures in *Metrosideros* foliage were due to either assimilatory discriminations or to the prevalence of a depleted N source such as atmospheric N. The question will remain unresolved until controlled experiments are done examining assimilatory fractionations of plants at low N concentrations with slow growth.

Gebauer and Schulze (1991) and Gebauer *et al.* (1994) attempted to relate levels of atmospheric N deposition (of unmeasured δ^{15}N) with forest decline and δ^{15}N of conifer foliage and twigs. This work produced original data on the δ^{15}N of various compartments of continental European forests, subjected to presumed atmospheric N inputs of unmeasured δ^{15}N. However, their usefulness for δ^{15}N interpretations is limited by at least three underlying problems. (1) Insufficient background information is given for the reader to follow the arguments, and consequently it is often difficult to see any relationship between data and conclusions; (2) the experimental design is not consistent with assessing the interaction of atmospheric N inputs and δ^{15}N of tree foliage, because it includes many confounding site factors and excludes measurement of the important N sources (Section V.E); (3) especially in the second paper (Gebauer *et al.*, 1994) some of the reasoning appears to us to be circular, and depends on propositions which are unsubstantiated in the paper (e.g., the authors concluded that the trees were 'using atmospheric N inputs for enhanced growth' but did not measure any aspect of growth or mention any directly in the paper) because they did not find either large accumulations of N or hypothesized (but unmeasured) atmospheric δ^{15}N signatures in the older needles, which they expected, if the needles accumulated atmospheric N).

K. Plants: Analyses Among Seasonal Means and Taxa

We are unaware of any detailed seasonal δ^{15}N studies with which to compare the patterns from the Scottish old field. Domenach *et al.* (1989) documented for foliage of *Alnus* spp. that δ^{15}N was constant over a mature period of growth. Shearer *et al.* (1983) traced δ^{15}N of *Prosopis juliflora* through the growing season in the California desert, and showed through manipulative experiments that those changes were probably related to amounts of N_2-fixation. However, most field studies (Schulze *et al.*, 1994; Nadelhoffer *et al.*, 1996) have assumed or directly argued that a grab sample of mature leaves (before senescence) provides a 'representative' δ^{15}N value of plant foliage for the purposes of obtaining the average growing season δ^{15}N as well as the relative $\delta^{15}N$ values among plants. One recent study (Högberg and Alexander, 1995) explicitly used foliar %N as indication that foliar $\delta^{15}N$ was stable. Our seasonal old field data (Figure 11a, b, c) for several taxa growing at the same site show that %N is not necessarily correlated with $\delta^{15}N$ or with leaf age and that patterns of $\delta^{15}N$ through the growing season, among plant parts and between taxa are not predictable. $\delta^{15}N$ may change where there is too little change in %N to be measured, and %N may change greatly with no corresponding change in $\delta^{15}N$ value. More worrying is the observation in the old field (Figure 10) that the rankings among plant taxa for foliar $\delta^{15}N$, of mature, non-senescent foliage, changed through the growing season, i.e., the $\delta^{15}N$ relationships among plant taxa growing at one site are not necessarily stable through time. This has serious implications for accessibility of sampling sites, choice of sampling regimes and the extent to which data from 'snapshot' sampling can be interpreted.

L. Plant $\delta^{15}N$ and Proximity to a N_2-Fixing Symbiosis

No one knows how previously fixed N is transferred from a N_2-fixer to a non-fixer or whether the transferred signal ($c.$ 0‰), often presumed to be due to N_2-fixation, is due to N_2-fixation or to some other subsequent discriminating mineralization step occurring after initial N_2-fixation. Certainly the hawthorn, rose and grass–herb sward in the Scottish old field (Figures 6, 7 and 9) responded differently to the proximity of a N_2-fixing broom, suggesting that selective mechanisms other than direct transfer were in operation. Pate *et al.* (1994) also found in Western Australian pasture that some non-fixing plants were influenced isotopically by proximity to a N_2-fixing plant and some were not.

Handley *et al.* (1994b) could find no relationship between plant $\delta^{15}N$ and distance to the nearest N_2-fixing tree in Kenya. In that study, soil and plant $\delta^{15}N$ were correlated with water availability and history of soil disturbance (presumably indirect measures of mineralisation) and unrelated to the proximity of a putative N_2-fixer. In the North American desert, Lajtha and Schlesinger (1986) found no correlation between $\delta^{15}N$ of non-N_2-fixing *Larrea* shrubs and proximity of a N_2-fixing *Prosopis* tree.

Kohls *et al.* (1994) found leaf $\delta^{15}N$ values closer to that of air N_2 for non-fixing plants occurring within 1 m of *Dryas* (actinorrhizal N_2-fixers) which was shown to be actively fixing N_2. There was no isotopic change in non-fixers growing within 1 m of non-nodulated *Dryas*. In Hawaii, van Kessel *et al.* (1994b) reported that over five years the $\delta^{15}N$ of understory grasses approached that of air in a stand of the leguminous shrub, *Leucaena leucocephala*. On the basis of existing lines of evidence, the influence of a N_2-fixing plant on $\delta^{15}N$ of a non-fixer appears to be, *inter alia*, a function of the spatial distribution of dead organic matter from the N_2-fixer and the rates and timing of organic matter mineralisation and the ability of the non-fixer to assimilate the predominant type of N produced.

Although it appears that direct transfer of recently fixed N is trivial in agricultural systems (Tobita et al., 1994), we think that there is scope for further investigation of same-season transfer to microbial biomass and non-fixing plants in uncultivated systems, because the possibilities for root damage, xylem leakage (Hatch and Murray, 1994), enhanced root turnover and mycorrhizal transfer may be greater than in agriculture.

There is some evidence that invertebrates grazing on roots and root nodules of N_2-fixing plants cause leakage of xylem fluids in sufficient amounts to transfer large amounts of previously fixed N (or large amounts of xylem N from any original source) (Murray and Hatch, 1994; Hatch and Murray, 1994). Other work showed that the grazing of soil invertebrates on roots and nodules increased the rates of N mineralization, making more mineral N available to plants (De Ruiter et al., 1993; Bouwman et al., 1994; Stephens et al., 1994). Nitrogen excreted from these grazing animals would be 0‰ to 5‰ more ^{15}N-depleted than the original plant N consumed (Scrimgeour et al., 1995 and references therein). Conversely, the whole bodies of these animals contribute an organic matter δ^{15}N to the soil which is 0‰–5‰ more ^{15}N-enriched than the original plant N, thereby creating lightly enriched patches of relatively high N concentration. On average, these soil invertebrates are 8‰–10% N, whereas plant roots tend to be less than 1% N by dry weight.

There is also some evidence that the presence of earthworms (most of which eat some fine roots, even if only inadvertently) enhances nodulation and N_2-fixation in clover, *Trifolium subterraneum* (Doube et al., 1994), thereby increasing the amount of N with an initial signature of 0‰ which enters the system and possibly increasing overall mineralization in the system through a N 'priming effect'.

M. Plants Use Various Soil N-Sources

1. Chemical Sources

Plants may directly use many chemical forms of combined N (NH_4^+-N, NO_3^--N, urea, purines, amino acids, amines) without symbiotic assistance (Raven et al., 1993) and, globally, NH_4^+-N is the major form of combined N used by photolithotrophs. Stewart et al. (1993) found clear evidence in a fire-prone woody ecosystem in Australia that plants sharing the same sites used at least four chemically different N sources. Stewart et al. (1988) also showed varying reliance on NO_3^--N and NH_4^+-N in Australian rain forest plants. The results of both of these studies are consistent with those of Rice and Pancholy (1972), Ellenberg (1977), Smirnoff and Stewart (1985), Stewart et al. (1993) and those reported by Raven et al. (1992a, b, 1993). It has long been known that ecto-mycorrhizal associations assist with acquisition of organic N (Read, 1991). Chapin et al. (1993) found that Arctic sedges can assimilate large amounts of organic N directly without mycorrhizal assistance. Kielland

(1994) determined that Arctic plants can acquire between 10% and 82% of their N directly as amino acids, although mainly in conjunction with mycorrhizal infection. He concluded that, 'The differences among species in capacities to absorb different forms of N provide ample basis for niche differentiation of what was previously considered a single resource'. The consequences to plant δ^{15}N values are presently unknown and cannot be inferred solely from patterns of foliar δ^{15}N.

Stewart *et al.* (1993) documented large differences in the ability of woody plants to use NO_3^--N, and this was related to their life strategies as pioneers on a recently burned substrate or as woody resprouters which persisted as geophytes on the same site. Pioneer (opportunistic) species, in general, appear to rely greatly on NO_3^--N (Smirnoff and Stewart, 1985; Pearson and Stewart, 1993) and to contain more of the chloroplastic form of glutamine synthetase (GS_2), which is thought (Pearson and Stewart, 1993) to play a large role in NO_3^--N assimilation and reassimilation of respired NO_3^--N. Longer-lived plants (persisters) contained more GS_1 (the cytoplasmic form of glutamine synthetase), which is associated with retranslocation and storage of previously assimilated N (Pearson and Stewart, 1993). In a survey of Chinese forest soils and plants, Guo-Qing *et al.*, (1991) found negative foliage δ^{15}N for all trees expected to be NH_4^+-N using, long-lived taxa (both conifers and oaks); positive foliage δ^{15}N was found only in birch, which is known to be a fast-growing NO_3^--N using taxon. Nothing is known about the possible effects of GS_1 versus GS_2 on plant δ^{15}N.

Schulze *et al.* (1994) presented data for δ^{15}N of plant foliage in Alaska, which suggested plant use of several N pools at shared sites. While their circumstantial evidence (foliage δ^{15}N only) is intriguing, their argument for plant use of different chemical N sources was unsupported by any measurements of δ^{15}N for external N sources or internal plant δ^{15}N variations; nor did they provide any other supporting analyses such as nitrate reductase or GS activities. Because only foliage was sampled, it remains unresolved whether the observed variations were differences among species in their internal δ^{15}N patterns (leaves versus stems versus roots, etc.), isotopic source shifts or varying assimilatory discriminations due to a variety of causes, e.g., stress, genetic differences (Handley *et al.*, 1996), changes in amount of external N. The same interpretative problems apply to Michelsen *et al.* (1996).

Nadelhoffer *et al.* (1996) also examined δ^{15}N signatures of various plant taxa sharing the same sites in Alaska. They also suggested that different taxa partitioned N as more than one resource. These authors added nitrate reductase assays of plant leaves (but not roots) to their data set and suggested that their hypothesis about N partitioning needed corroboration and interpretation via ^{15}N-enriched controlled experiments. This is also the first publication to point out an abnormally large range of δ^{15}N values among plants (*c.* 10‰) in the context of previously published values (typical published values for a large range would be 5–6‰). Since they pointed this out, we have noted similarly large ranges in some of our data sets. In the Scottish old field study spanning three years, mean δ^{15}N differences among plant taxa were not only at the large end of the previously reported range of values (c. 6‰), but on any one sampling occasion a range of 10–12‰ among all samples of all taxa was not rare. Since then, a study was completed of the native conifer, *Juniperus communis*, at a site in the Scottish highlands, where foliar δ^{15}N had a maximum range of slightly more than 11‰ (Hill et al., 1996).

For ash and oak grown in pots, Stadler et al. (1993) found both taxa equally capable of assimilating NH_4^+-N or NO_3^--N with total N uptake higher when both forms were supplied. More NO_3^--N was found in the xylem sap of ash *Fraxinus excelsior*, although both showed strong ability to assimilate NO_3^--N in the roots. These authors further suggested, on the basis of previously published field data, that the relative importance of root versus shoot NO_3^--N reduction varies with age and with growing conditions. In nature, competition and soil factors also influence relative dependence on the two N sources and also introduce the possibility of using organic N. Garten and van Miegroet (1994) showed diverse N sources for trees and ferns across an elevational and N-supply gradient in the USA. In mixed forest stands composed of many woody species, only two species (a conifer and a maple) had foliar δ^{15}N which could be correlated with N-mineralisation potentials. Except at a N-poor site, the δ^{15}N of maple foliage was correlated with potential nitrification. The studies mentioned above are clear indications that different plants 'sample' different N pools, both chemically and isotopically.

Several authors (Schulze et al., 1991b; Handley et al., 1994b; Högberg et al., 1995) have examined likely plant N sources (NO_3^--N, NH_4^+-N or N_2) versus δ^{13}C of foliage in field studies or suggested that there was some fundamental relationship between the two. The idea that δ^{15}N and δ^{13}C are mechanistically related so that δ^{13}C of field plants will reflect the chemical type of N source may have arisen from misunderstandings of studies by Allen et al. (1988), Allen and Raven (1982) and Raven et al. (1984) combined with confusion about the Farquhar and Richards' (1984) model linking water use efficiency (of some plants under some circumstances) and foliar δ^{13}C (water-use efficiency and δ^{13}C are not the same thing; they are linked by a mutual dependence on the ratio of CO_2 partial pressures inside and outside the leaf). Raven and Farquhar (1990) hypothesised that there were theoretical reasons to expect a linkage, on the biochemical level, between type of N-source and organic δ^{13}C. That hypothesized correlation cannot be measured in the whole organism. This is also made explicit by Raven et al. (1992b) whose discussion concerns the effects of N-source on the directly measured ratio of water-lost-to-carbon-gained, not on its sometime surrogate, δ^{13}C.

For both δ^{15}N and δ^{13}C we must distinguish between whole-organism effects and effects internal to the organism. The isotopic signature of the whole organism is determined by the exogenous source signature, as modified by assimilation and losses. The distribution of isotopic signatures within the organism is a function of the biochemistry of that organism and its relative investments in different chemical products carrying different $\delta^{15}N$ values. Non-RuBISCO carboxylations and the balance of organic acids may affect internal distributions of $\delta^{13}C$ within a plant, and this may be partially due to whether the N-source is primarily NH_4^+-N or NO_3^--N. However, the whole organism $\delta^{13}C$ (or $\delta^{15}N$) value is still determined by isotopic inputs minus isotopic losses.

In Allen *et al.* (1988), Allen and Raven (1987) and Raven *et al.* (1984), differences in whole-plant water-use efficiency (as measured by a water budget) were much too small to be identified in field data. In field surveys, Handley *et al.* (1994b) used a data set for plants at 12 sites in Kenya to test whether there were any correlations between $\delta^{15}N$ and $\delta^{13}C$. These sites contained a variety of plant forms (N_2-fixing trees, shrubs and herbs, non-N_2-fixing trees, shrubs and herbs, dicots and monocots) over a variety of climate types from humid tropical sea level to tropical high altitude and arid zones. No such correlation was found at any of these sites. They concluded that both $\delta^{15}N$ and $\delta^{13}C$ in field data from mixed vegetation reflected chiefly water supply: $\delta^{15}N$ being dependent on amounts of mineralisation, which were, in turn, dependent on amount and periodicity of soil wetting and disturbance, and foliar $\delta^{13}C$ being related to soil water supply via Farquhar and Richards' (1984) model.

Changes in real water-use efficiency (carbon accumulation versus water loss rather than $\delta^{13}C$) reported by Allen and Raven (1987), Allen *et al.* (1988) and Raven *et al.* (1984) were also related to root respiration, as also pointed out by Handley and Raven (1992). Root respiration is not reflected in foliar or whole plant $\delta^{13}C$, nor is it explicitly accounted for in Farquhar and Richards' (1984) model relating $\delta^{13}C$ to water-use efficiency. Knight *et al.* (1993) directly compared $\delta^{13}C$ of legumes by using varying amounts of atmospheric N_2-fixation versus those relying on mineral N sources and found no significant correlation with $\delta^{13}C$, even in this controlled experimental situation.

In the field, the primary effect of N on $\delta^{13}C$ is from varying amounts of N, but only when N is in such low quantity that it is limiting to growth (O'Leary, 1988) and then as a direct effect on RuBISCO activity, or on RuBISCO activity via leaf chlorophyll content (Madhavan *et al.*, 1995; Hill *et al.*, 1996). RuBISCO activity is explicitly part of Farquhar and Richards' (1984) whole-leaf $\delta^{13}C$ model. In the field, $\delta^{13}C$ overwhelmingly reflects plant water supply. Within populations sharing a common water supply, $\delta^{13}C$ can be correlated with potential water-use efficiency or used as an index to the amount of genetic variation within a population (Handley and Raven, 1992; Ehleringer, 1993; Ehleringer *et al.*, 1993; Handley *et al.*, 1994a,b). However, there is no evidence from field studies or from controlled studies that chemical type of N-source is directly linked to plant $\delta^{13}C$.

2. Spatial Distribution of N-sources

N occurs heterogeneously in the natural environment, and plants respond actively to heterogeneity by foraging and exploiting rich caches of N and other nutrients (Fitter, 1987; Caldwell and Pearcy, 1994). This makes simple tracer experiments difficult in natural systems, but carefully designed experiments using heterogeneous labelling might be a way forward. Hutchings and de Kroon (1994) reviewed the adaptations of plants to a heterogeneous resource environment. They discussed the non-random placement of plant 'resource acquiring structures', and pointed out that access to different concentrations of N can change internal plant growth patterns, N allocations within plants and that the responses to encountering N caches vary among taxa. They, and also Crabtree and Bernston (1994), argued that the plant response to a nutrient cache was independent of the chemical composition of the patch tested (NO_3^--N, NH_4^+-N, P, Ca) and that heterogeneity of supply was the factor determining changes in plant architecture. It was also evident from these papers that plant responses were not integrated, but local and highly specific in exploitation of aggregated nutrients. Where internal allocations are altered, internal variations of δ^{15}N may also be altered.

The reports of both root proliferation and mycorrhizal fungal proliferation into nutrient-rich nutrient patches (Allen, 1991; Read, 1991; Hutchings and de Kroon, 1994; Robinson et al., 1994) suggest that plants use the most concentrated source available. This is not new information, being first reported in 1862 (Fitter, 1987, and references therein). This being the case, plant δ^{15}N should closely resemble some function of the δ^{15}N of the most concentrated N source, which may not be representative of the surrounding soil N.

Schulze et al. (1994) argued that δ^{15}N of foliage from different taxa in Alaska could be partially a function of rooting depth (with attendant soil N source changes of δ^{15}N with depth in the soil profile), but they did not measure δ^{15}N of soil profiles or excavate tree rooting depths. In Germany, Gebauer and Dietrich (1993) found no differences in foliage δ^{15}N among three taxa of trees (coniferous and broad-leaved) with different rooting depths; although δ^{15}N for the soil organic layer and mineral soil were given, no depth profile versus δ^{15}N was shown in this study.

Heterogeneity can change both amount and type of available N. We have already discussed that fungi and plant roots can concentrate N in the soil as well as selectively access concentrations of soil nutrients. Högberg and Alexander (1995) found foliar δ^{15}N associated with varying tree density and litter accumulation in West Africa. This density variation presumably altered local patterns of N-cycling. Animals also create N heterogeneity, faeces, urine patches, carcasses, waste chambers in underground burrows and earthworm castes, inter alia, create locally rich N sources of relatively enriched $^{15/14}$N (DeNiro and Epstein, 1981; Minagawa and Wada, 1984; Wada et al., 1991, 1995; Scrimgeour et al., 1995).

Högberg and Alexander (1995) found foliar ^{15}N enrichment of trees growing near termite mounds in West Africa. This is consistent with the ^{15}N-enrichment expected from the NO_3^--N behaviour for Australian termite mounds reported by Barnes et al. (1992): NO_3^--N was 'wicked' to the outside of the mound, where it was subsequently washed into and through the soil with rainfall percolate water. Such NO_3^--N leaching would be expected to leave ^{15}N-enrichment of residual soil N. Soil disturbance (e.g., moles or pig digging) may create locally enhanced nitrification which leads to patches of plants with a different species composition (Hutchings and de Kroon, 1994) and thereby an altered $\delta^{15}N$ from the newly formed soil NO_3^--N source.

Treseder et al. (1995) made a study of an ant-epiphyte symbiosis in which $\delta^{13}C$ and $\delta^{15}N$ were used to demonstrate that the epiphyte received from the ants (a very localised source) about 39% of their fixed carbon and 29% of their N. Schulze et al. (1991a) used $\delta^{15}N$ to estimate the percentage of carnivorous plant N derived from insects as a food source; the key problem was insufficient knowledge of insects as N sources and the assumption that $\delta^{15}N$ could be used as a tracer isotope. Treseder et al. (1995) measured most of the parts of the system under study; little was assumed, much was measured (CO_2 from ant respiration, ant faeces, plant $\delta^{15}N$ with and without symbiotic ants and/or ant faeces). Then, crucially, these data were combined with prior knowledge about the habits and ecology of the ant–epiphyte relationship.

3. Temporal availability and phenological demand

The temporal data for the $\delta^{15}N$ of broom in the Scottish old field are in direct contradiction to the commonly held assumption that active N_2-fixation is reflected in the fixer host by foliage $\delta^{15}N \leq 0‰$ and that this is ^{15}N-depleted relative to co-occurring non-N_2-fixing plants or to other periods of growth during which fixation is not active. They are, however, consistent with phenological information reported by Hansen and Pate (1987) that the main period of N_2-fixation was separate from the main period of growth for some Australian legumes.

Plants use both internal and external sources of N. Understanding the phenology of plant N sources is essential to understanding plant and soil $\delta^{15}N$. The discussion below represents only a small sample of the information published on temporal N demand, and relevant, therefore, to a discussion of plant $\delta^{15}N$.

Soil N availability varies through the growing season (e.g., Jackson et al., 1989; Zak et al., 1990; Groffman et al., 1993), and plant phenologies also impose temporally varying patterns of N demand. In many ecosystems, most of the N is found in organic matter, both living and dead, and the effect of this partitioning on the $\delta^{15}N$ of residual plant-available soil N has never been studied. (To our knowledge, no one has yet attempted a complete isotopic mass balance of an ecosystem). In both herbaceous plants and woody perennials, patterns of temporal N-source use vary greatly (Millard, 1988). In some woody perennials much of the N required for early growth is taken from stores in the wood and remobilised in the spring; soil N is used directly in later growth. In some taxa the stored N originates in senescing leaves, and in some soil N is previously assimilated into roots and/or stems and only translocated to the leaves in the following spring (Millard and Proe, 1992; Okano et al., 1994). Some plants immediately begin using soil mineral N in the spring and rely little on stores or use both stores and soil N simultaneously. The woody N_2-fixing shrub, gorse (Ulex europaeus L.) relies on N_2-fixation, internal stores and soil N simultaneously (Thornton et al., 1995). In many plants the N requirements of new leaves can be met from leaves senescing during the same season (Wendler et al., 1995).

In alder (Côte et al., 1989), non-nodulated plants had no net re-absorption (storage) of leaf N in autumn, while nodulated alder showed a decline of leaf N in autumn, concurrent with cessation of nitrogenase activity. Thus, in this widespread, actinorhizal N_2-fixing tree, nodulated and non-nodulated treatments cannot be compared directly because internal translocations, stores and N losses concurrent with leaf litter are intrinsically different with and without symbioses.

There is little literature concerning the δ^{15}N phenology of plants. Bremer et al. (1993) and Unkovich et al. (1994) worked on this aspect of δ^{15}N with a view to estimating N_2-fixation. Bergersen et al. (1988) detailed the δ^{15}N signatures of all plant parts of soybean during growth under controlled conditions. There are also a few papers relating age of conifer needles with δ^{15}N (Gebauer and Schulze, 1991; Näsholm, 1994). Shearer et al. (1983) measured the seasonal δ^{15}N of Prosopis foliage.

N. The Plant N-Niche

We propose the term, 'plant N-niche', to describe the differentiation of the single resource, N, into several resources. Plant taxa occurring in natural mixed stands have N acquisition strategies that differ chemically, spatially, and temporally. Qualitatively, plant taxa vary in their abilities to acquire NO_3^--N, NH_4^+-N, and organic N, both with and without the assistance of microbial associates. (Symbiotic N_2-fixation is one extreme example of a specialist N-niche.) Spatial strategies may comprise root architecture, and the ability to exploit spatially heterogeneous sources, the ability to store and remobilise previously assimilated N and the location and relative activity of assimilating enzymes. Temporal strategies are related to phenological demand for both exogenous N and internal stores.

The older concept, largely derived from agriculture, that plants use only mineral N (and mostly NO_3^--N) (Shearer and Kohl, 1989; also reported by Lajtha and Marshall, 1994), is now known to be wrong for uncultivated systems. It should be noted here that, unlike the diverse of assemblages of plants found in nature, most of our agricultural plants are derived from ruderals *sensu* Grime *et al.* (1988), opportunistic species adapted to using predominantly, but not solely, NO_3^--N (Pearson and Stewart, 1993). Hence, our most intensively studied model system (agriculture) has provided an unrealistically simple view of the natural world, which has rather curiously persisted, despite repeated refutations that began decades ago (Gainey, 1917; Allison, 1931; Lee and Stewart, 1978).

For source–sink studies of plant–soil communities, a more accurate description of the situation may be that plant $\delta^{15}N$ depends on the relative proportions of NO_3^--N, NH_4^+-N and organic N in available N pools, including atmospheric sources, the differences in their signatures at the time of acquisition, the competition-weighted preferences of the plants for one form over the others, and the attendant fractionations due to assimilation, remobilisation, and preferential losses. The frequently small range of net $\delta^{15}N$ values for plants sharing the same site could be masking large differences of detailed processes.

O. $\delta^{15}N$ and Plant N Metabolism

In vascular land plants, fractionations are associated with initial assimilation of exogenous N from external sources, with internal allocations and remobilisations of N and with losses from the plant. In the case of both internal and whole-plant fractionations, the fractionations are incurred by enzymes, not by physical transport processes. There is no evidence, to date, that movement across a living membrane (transport or diffusion, active or passive) fractionates either NO_3^--N or NH_4^+-N. Other causes of fractionations (secondary to enzymatic fractionations and dependent upon them) are losses of fractionated plant parts (e.g. leaves or fruits) and gaseous losses from leaves (Farquhar *et al.*, 1983). Most of the known $^{15/14}N$ fractionation factors for plant enzymes are given in Handley and Raven (1992). From the little that is presently known about vascular plant $\delta^{15}N$, it appears that NO_3^--N assimilation may produce no significant whole-plant fractionation relative to external-source N, but may produce large internal fractionations among plant parts because of incomplete (and fractionating) assimilation. Conversely, NH_4^+-N assimilation may produce relatively large depletions in the whole plant relative to external N source, but little or no internal variation of $\delta^{15}N$ among plant parts. The amount of discrimination produced in both NO_3^--N and NH_4^+-N assimilation may be dependent on the concentration of external N.

Yoneyama (1995) has written an excellent review (complete with analytical methods) of the present state of knowledge on internal and biochemical distributions of $\delta^{15}N$ for plants. He noted that the $\delta^{15}N$ of individual amino acids can be highly variable, and individual N positions on molecules can vary by as much as 40‰. He also noted the need for both improved analytical methods and more research into plant metabolism utilising δ^{15}N.

1. NO_3^--N Assimilation (An External Source)

Mariotti *et al.* (1980a, 1982) and Kohl and Shearer (1980) reported whole-plant depletions in young herbaceous plants (but not older ones) relative to the δ^{15}N of the external NO_3^--N supplied. They further found that fractionation relative to source N increased with increasing concentration of the external NO_3^--N solution. No mechanism immediately springs to mind, which would explain both the fractionations in young plants and the lack of fractionations in older plants (relative to source). If these data are not artefactual, then we must postulate, as did Mariotti *et al.* (1982), that a profound (and yet undescribed) change in N metabolism and/or leakiness to NO_3^--N occurs during herbaceous plant development.

Yoneyama and Kaneko (1989) found no whole-plant depletion relative to source NO_3^--N in mature (29-day-old) individuals of a short-lived *Brassica* species (salad rocket), even at 12 mM external NO_3^--N concentration. Instead, they found differences among the internal plant N fractions, with δ^{15}N of unassimilated NO_3^--N being more ^{15}N-enriched than external NO_3^--N, and all organic N more depleted that source NO_3^--N.

Using a dual isotope approach (δ^{15}N and δ^{18}O) T. Olleros and H.-L. Schmidt (personal communication of previously unpublished data, 1995) made in *in vitro* measurements of the kinetic isotope effects of nitrate reductase in *Zea mays* and *Chlorella vulgaris*. They found that the kinetic nitrogen isotope effect (α) in both organisms was 1.030 ± 0.005 at 20°C, pH 7.6. This isotope effect was found to be independent of temperature, but varied with NADH-concentration. The total kinetic oxygen isotope effect in *Chlorella vulgaris* was 1.015 ± 0.008, and the intramolecular oxygen isotope effect was 1.055 ± 0.010, (both done at 25°C, pH 7.6: ^{15}N-analysis was done via reduction to NO by $TiCl_3$, direct measurement of NO or production of N_2 from NO by Tesla-discharge; ^{18}O-analysis of labelled NO_3^- through reduction to NO_2^- and H_2O and equilibration with CO_2).

This kinetic isotope effect for nitrogen (1.030) is somewhat greater than the ones reported for *in vivo* analyses, which generally lie between 1.003 and 1.02, although there was one report of an *in vivo* value of 1.03 for *Phaeodactylum tricornutum* (Table 3 in Handley and Raven, 1992).

When comparing chloroplastic and synthesised nitrate assimilatory systems (1:5 nitrate reductase to nitrite reductase), Ledgard *et al.* (1985) concluded that the fractionating step in NO_3^--N assimilation was NO_3^--N reduction to NO_2^--N. Mariotti *et al.* (1982) found that NO_3^--N fed plants became less depleted relative to source N as nitrate reductase activity increased.

2. NH_4^+-N Assimilation (External and Internal Sources)

In vascular plants GS is the primary assimilating enzyme for both exogenous NH_4N and for NH_3^--N generated internally from photorespiration. GS may be found in both root and shoot. This enzyme occurs in two forms (GS_1 and GS_2) the first of which is cytoplasmic and the second of which is chloroplastic. GS_2 is believed to assimilate NH_3-N from both photorespiration and from the NH_3-N generated by nitrate reductase. GS_1 is thought to be an important enzyme for mobilisation of N to translocation and storage (Pearson and Ji, 1994) and may occur in some plants only after the onset of senescence. The relative isotopic enrichment factors for these two forms of the enzyme are unknown. It is commonly known that the $\delta^{15}N$ of plants may change with age (e.g., Figure 11a,b and c), and the relative occurrence and expression of these two forms of GS may be implicated in the change. This is, however, only speculation.

Although N enters the plant as both NH_4^+-N and NH_3-N the substrate for glutamine is the NH_3-N form (Stewart et al., 1980). The enzyme is thought to bind NH_4^+-N, deprotonate it, and use the residual NH_3-N (Raven et al., 1992b). Hence, when considering the NH_3-N/NH_4^+-N source $\delta^{15}N$, it is the source NH_4^+-N signature which is relevant.

In most natural situations the amount of exogenous NH_3-N is minimal, and almost all is in the NH_4^+-N form as it enters the plant (Raven et al., 1992b). However, identification of the source $\delta^{15}N$ is not straightforward. The relative $\delta^{15}N$ values of NH_3-N versus NH_4^+-N are determined by a pH-dependent equilibration, which can shift the two isotopic values by as much as 20‰ (Yoneyama et al., 1993). In general, NH_3-N and NH_4^+-N appear to be more mobile across the cell membrane, in both directions, than NO_3^--N and will quickly establish an equilibrium. However, within the plant cell, the average cytoplasmic pH is 7.3 and the average vacuolar pH can vary between about 3 and 6 (J. Raven, personal communication, 1995). This range of pH environments within the cell implies that potentially the NH_3-N to NH_4^+-N equilibrium ratio can vary by a factor of about 10^4 (i.e., 10^{-6} in the vacuole at pH 3.3 versus 10^{-2} in the cytoplasm at pH 7.3) when a molecule of reduced nitrogen moves across the vacuolar membrane.

Most of the published studies on $\delta^{15}N$ effects of NH_4^+-N/NH_3-N assimilation were done for aquatic single-celled organisms or in vitro assays of enzyme activity (Handley and Raven, 1992). The results are conflicting, with reported whole-organism discriminations relative to source being positive in some reports and negative in others. In addition, because of the equilibration of NH_3-N and NH_4^+-N, there can be a whole-organism ^{15}N-enrichment or ^{15}N-depletion relative to source (Handley and Raven, 1992), depending on whether the discrimination is calculated relative to external NH_3-N or NH_4^+-N.

We are aware of only two research publications dealing with δ^{15}N fractionation of vascular plants assimilating NH_4^+-N. Yoneyama et al. (1993) examined NH_4^+-N assimilation by the chloroplastic form, GS_2, in vitro, for spinach. Relative to source NH_4^+-N, the assimilatory fractionation was $-16.5 \pm 1.5‰$ ($\alpha = ^{14}k/^{15}k = 1.0165$). This fractionation was independent of external NH_4^+-N concentrations and incubation temperatures used in this experiment. For in vivo experiments with rice (Yoneyama et al., 1991a) whole-plant depletions were $-7.5‰$ and $-7.9‰$ relative to 20 mg l^{-1} and 100 mg l^{-1} solutions of NH_4^+-N. In contrast to experiments on NO_3^--N assimilation, there was little δ^{15}N variation among plant parts because little unassimilated free NH_3-N was found in the plant tissues. The little free NH_3-N found in rice was more ^{15}N-enriched than the source NH_4^+-N or the organic N in the plant, indicating assimilatory fractionation. In the same paper they reported the results of giving Synechococcus (a cyanobacterium, one wild-type and two mutant strains) either NO_3^--N or NH_4^+-N. Assimilation of NO_3^--N produced a whole-organism fractionation relative to source of $-2.7‰$ and $-4.8‰$, whereas NH_4^+-N assimilation caused a whole-organism discrimination relative to source of $-17.6‰$. These results are consistent with those reported by Macko et al. (1987) for heterotrophic bacteria fed on different amino acids.

3. Nitrogenase

Despite the many publications estimating N_2-fixation on the basis that its maximum fractionation in plants and among plant parts is about $-2‰$, the actual fractionation factor for nitrogenase, itself, is not known. The value $-2‰$ is also dependent on measurements of a relatively small number of largely rhizobially symbiotic plants. There has been no significant and comparative survey of other modes of N_2-fixation, such as Frankia symbioses, endophytic associations and free-living diazotrophs. We have not been able to locate any published record of fractionation values determined for isolated nitrogenase in vitro or of a satisfactory isotopic mass balance for δ^{15}N in pure culture of N_2-fixing microbes. All of the determinations that we have been able to find (Shearer and Kohl, 1986; Yoneyama et al., 1986; Handley and Raven, 1992) were done on plant tissues, on either tissues of vascular plants which are the host recipients of symbiotically fixed N, or their symbiotically infected nodules or on whole algal cultures where the analysed organism is both the N_2-fixer and the recipient. This means that the commonly published value of $-2‰$ does not represent the fractionation due to nitrogenase alone; it represents, instead, the net fractionation due to any one or combination of several possible steps, e.g. a rate-limiting step prior to nitrogenase (as proposed by Delwiche and Steyn, 1970), to nitrogenase itself, or to assimilatory fractionations and losses downstream from nitrogenase. Wherever the discrimination occurs, there must be a loss of one of the isotopes of N for discrimination to be expressed in the whole organism relative to source.

We have measured $\delta^{15}N$ values (unpublished) for samples of three free-living *Nostoc* colonies from Ireland of $-4.8‰ \pm 0.4$ s.d.; $\delta^{15}N$ for *Nostoc* colonies ($n = 3$) excavated from the host plant, *Gunnera*, were *c.* 0 ‰, and *Gunnera* roots were $-2.64‰ \pm 0.15$; the free-living *Nostoc* values are close to those found by Delwiche and Steyn (1970) for whole cultures of *Azotobacter vinlandii* ($\delta^{15}N = -3.9‰$). Yoneyama *et al.* (1986), however, found that whole plant values for N_2-fixing symbioses ranged from $-0.2‰$ to $-2‰$ and argued that fractionations may occur both during fixation and during N export from nodules. Until better α-values can be obtained *via* mass balance studies, the most conservative course is to assume that nitrogenase has an α-value of 1.000 and that all observed whole-tissue or whole-plant discriminations have other causes. There are also no data on the fractionations of N by the different chemical forms of nitrogenase.

P. The Two-Source Natural Abundance Model for Estimating N_2-Fixation in Natural Systems

It is common in natural systems that $\delta^{15}N$ alone does not distinguish between types of plants (N_2-fixers versus non-fixers), types of mycorrhizal associations (Ecto- or AM) or sources of N (Ledgard *et al.*, 1985; Hansen and Pate, 1987; Pate *et al.*, 1993; Handley *et al.*, 1994b; Högberg and Alexander, 1995; Nadelhoffer *et al.*, 1995; Stock *et al.*, 1995).

The two-source $\delta^{15}N$ method for estimating N_2-fixation has been defended (Shearer and Kohl, 1986) as semi-quantitative. But this avoids the central issue: one cannot know what is being estimated, semi-quantitatively or otherwise. This method is predicated on using $\delta^{15}N$ as a tracer, and it is not a tracer. Where only means or medians of estimates of N_2-fixation have been compared using the two-source $\delta^{15}N$ method versus isotope dilutions (tracer method) in agricultural experiments (Androsoff *et al.*, 1994; Stevenson *et al.*, 1995), there was fairly good agreement between the two methods. But what comprises those means? They are not composed of statistically similar populations of values, because the same two studies showed that there was no correlation between estimates (paired samples) made by these two methods (Figure 16). This suggested some fundamental differences in the details of what was being measured by the two approaches and that a simple comparison of means, ignoring variances, was fallacious. It is noteworthy, also, that in one of these studies (Androsoff *et al.*, 1994), soil NH_4^+-N and NO_3^--N were in approximately equal supply, thereby establishing that even in this agricultural situation there were, at a minimum three significant chemical sources of N in this system, two from the soil (of unknown individual $\delta^{15}N$ values) and one from the atmosphere.

The method has been so extensively published (e.g., Shearer and Kohl, 1986, 1989) that we will not repeat the details here. Briefly, it presumes that the amount of N in a plant which is derived from atmospheric N_2 can be calculated by comparing the foliar $\delta^{15}N$ values of the putative fixer, a non-fixer and the putative fixer grown, hydroponically, without a combined N source. When the system under study approaches having only two N sources the model will work. When it is not applicable to a system under study, what are some of the reasons that it may not be valid? We list below five conditions of the model and a brief discussion of each, most of which resolve into one issue: there is no way to know what is being measured.

(1) *That plants can be treated as having only two major N sources, soil N as a single integrated source and atmospheric N_2.* There has been a large body of evidence, for many decades, that soil is not a single N resource for plants, chemically, spatially or temporally. We proposed the term, N-niche, but the information is not new to us. In Europe, at least, atmospheric N deposition also presents a major extra-soil source of plant N, ranging from about 25 to an extraordinary 300 kg ha^{-1} yr^{-1} (Pearson and Stewart, 1993). This input is also not a single source, having range of values from about +20‰ to −20‰ (Figure 19).

(2) *That a non-fixing reference plant can be chosen to 'integrate' a typical soil N source, which is representative of the soil N used by both the fixer and non-fixer.* Hansen and Pate (1987) reported from a study of pastures in Australia that there was '... little evidence of reference plant N accurately reflecting $\delta^{15}N$ of soil N or of well-nodulated legumes showing lower values than companion reference plants'. Stock *et al.* (1995), Nadelhoffer *et al.* (1995) and Handley *et al.* (1994b) were unable to find suitable reference plants in several different types of vegetation in both Africa and Alaska. If plants fill what we have termed here as N-niches, then there may be no two plants in a mixed stand which fill exactly the same niche and, therefore, no two taxa which use exactly the same soil N at the same time. This principle has been demonstrated with data on plant-available soil N use versus atmospheric N_2-fixation (Ledgard *et al.*, 1985; Hansen and Pate, 1987; Bremer and van Kessel, 1990; Bremer *et al.*, 1993; Pate *et al.*, 1993; Handley *et al.*, 1994b).

If, all but a trivial amount of plant-available soil N occurs in one form (NO_3^--N or NH_4^+-N) so that there are only two isotopic N sources (soil N and atmospheric N_2), and if the soil source remains isotopically steady through the growing season, then the plants will reflect this isotopically. However, these are very stringent conditions to impose or to demonstrate in a natural system, where heterogeneity of N sources and their source signatures is more likely.

(3) *That $\delta^{15}N$ of the fixer can be calibrated by comparing plants totally dependent on fixation with non-fixers of the same or different taxa.* As discussed in the work of Côte *et al.* (1989) the presence/absence of a symbiotic association may alter the N metabolism of the host plant, so that inoculated and uninoculated plants are not equivalent. Ledgard *et al.* (1985) showed that the strength of the N sink presented by a non-fixer growing in the same site with a fixer changed with the proportion of plant N derived from atmospheric fixation by the N_2-fixing symbiosis. Since the effects of this non-fixer sink may be taxon specific, as well as phenotypically and spatially variable, it is difficult to imagine how a suitable 'reference' plant is to be identified nor is it easy to imagine how a 'calibration' of fixing and non-fixing plants can be done in the absence of such competition.

(4) *That soil sources represent reliably a ^{15}N-enriched or -depleted source which can be compared with N_2-fixation having a $\delta^{15}N$ value of 0‰.* Implicit in the two-source, natural abundance method is the unstated, but wholly false, assumption that $^{15/14}N$, at natural abundance level, can be used reliably as tracer isotopes. N_2-fixation does not produce a unique $\delta^{15}N$ value. Many soil mineralisation processes (other than N_2-fixation) lead to *c.* 0‰ $\delta^{15}N$ values for mineral and organic soil N (Feigin *et al.*, 1974a, b; Herman and Runden, 1989; Yoneyama, 1994). Hansen and Pate (1987) also pointed out that N_2-fixation by free-living soil microbes (as also reported by Piccolo *et al.*, 1994, among others) could contribute recently fixed N_2 to both legumes and reference plants, thereby confounding the origin of fixed N in plants. Also, plants commonly use N transferred by fungi from organic (and inorganic) sources, as well as relatively large organic molecules directly (Chapin, 1993; Raven *et al.*, 1993; Kielland, 1994), which are not necessarily more enriched than mineral N.

(5) *That the $\delta^{15}N$ of foliage reflects N_2-fixation.* Without detailed knowledge of the plant's N phenology, $\delta^{15}N$ of foliage cannot be interpreted in terms of N_2-fixation. Phenological studies mentioned earlier (Hansen and Pate, 1987), as well as our own old field data (Figures 8, 10, and 11a, b, c and Wheeler *et al.*, 1979 and 1987), demonstrated the decoupling of near-0‰ foliar signatures with known periods of N_2-fixation.

   The natural abundance two-source-mixing model as widely published by especially Shearer and Kohl (1986, 1989) for estimating N_2-fixation cannot cope with the real diversity of many ecosystems. The model itself is valid, wherever, and only wherever, it can be clearly demonstrated that there are no more than two major isotopic N-sources and that those two sources have significantly different isotopic signatures and where fractionations between source and sink can be quantified.

Q. Conclusions

Much of the present terrestrial δ^{15}N literature is devalued by a blurred perception of what is empirical and what is mechanistic. There has been a tendency in this literature to make the tacit assumption that all ecosystems are alike and that empirical correlations from one study can be generalised to fit all cases. Without substantiating information on possible fractionations between source and sink, δ^{15}N signatures cannot be used as tracers. It is an extremely valuable lateral approach to examining ecosystem function and best used in combination with other methods, such as light ^{15}N-enrichment, traditional ecophysiological and microbiological approaches, to contribute additional insight and to suggest patterns and testable hypotheses.

The techniques of δ^{15}N analyses are ideally suited to research in remote locations; the interpretations are not. Nitrogen isotope interpretations are complicated and depend on a thorough knowledge of the site under study and upon controlled experimentation. The isotopic values, do not, themselves, supply the answers; the answers are supplied through examining patterns of isotopic signatures in the context of what is already known about the system's ecology and through testing hypotheses which arise from analysis of the δ^{15}N patterns. Foliar δ^{15}N values may change during a growing season, and one-trip sampling provides only a snapshot of δ^{15}N patterns.

Over-interpretation is a major problem in the current literature. There will not be any simple, generalisable correlations of δ^{15}N in natural, terrestrial systems; δ^{15}N is a function of the same processes which comprise the N cycle, and this varies both spatially and temporally and within and among ecosystems. Because there are no simple, universal 'laws' governing the site-specific details of the N cycle, there will be no simple, universal 'laws' of δ^{15}N.

It is usually unnecessary to call on extraordinary explanations for observed δ^{15}N. Most observed shifts in plant or soil δ^{15}N can be explained by either a source change and/or kinetic fractionations. A few situations are complicated by equilibrium fractionations. Plants and soil, with or without fungal associates, can have signatures within the range attributed to N_2-fixation, which have no origin in N_2-fixation. In mixed vegetation, plants may occupy N-niches, exploiting N as multiple resources in space and time. This precludes the use of the natural abundance (δ^{15}N) two-source method for estimating N_2-fixation in natural systems unless the N-source attributions are empirically and simultaneously verified by manipulative experimentation which takes into account rigorous statistical analyses.

N_2-fixation is only one N-source among many for plants and soil organisms, and its main study lies historically in agriculture, not ecology. For the future, we think that the more interesting and relevant challenge is to understand the mechanisms of N-partitioning in complex systems. One way to address this need is long-term iterative studies of terrestrial $\delta^{15}N$, focusing on particular ecosystem types, in which: (1) survey patterns are collected and analysed to give rise to (2) hypotheses about ecosystem N partitioning, which are then (3) tested under controlled conditions with *inter alia* ^{15}N labelling techniques and the results (4) verified in the field by manipulative experimentation.

What distinguishes the work which we have discussed as being successful uses of $\delta^{15}N$? For the most part, authors of papers which we deem to have lasting value have: (1) adhered to the possible limits of interpretation; (2) paid careful attention to the existing knowledge in isotope chemistry without resorting to extrapolated correlations from other systems and without attempting to apply univariate isotope theory to multivariate field problems; (3) applied isotope interpretations in the light of detailed background knowledge of the study subject; (4) noted background levels of isotopic signatures versus expected treatment signals; (5) actually measured most of the sources, sinks and fractionations implicit in the system; (6) made use of careful experimental design and combined this with conservative use of statistical analyses; (7) not attempted to quantity processes which are legitimately useful only as qualitative observations; (8) been careful about the level of organisation and logic addressed by their arguments.

ACKNOWLEDGEMENTS

We are especially indebted to Professor Howard Griffiths for his help with the final version of the manuscript. We also especially thank Dr David Robinson, Dr Brian Fry, Dr F.S. Chapin, III, Professor J.A. Raven, Professor P. Högberg, Dr T.H.E. Heaton and Professor D. Schulze for their valuable contributions in reading and criticizing the manuscript. The opinions expressed in this review are solely those of the authors. We also owe a debt of gratitude to Professor H.-L. Schmidt for contributing original data on isotopic enrichment factors associated with nitrate reductase and to Dr Bruce Osborne, University College, Dublin for allowing us to publish $\delta^{15}N$ values on *Nostoc* and *Gunnera* samples collected by him and analysed by us. We thank Dr David Hopkins for providing a soils description at the Scottish old field site, Dr C.T. Wheeler for his help with acetylene reductions and interpretation of broom data, and Ms Sigrun Holdhus, Ms Winnie Stein, and Ms Louise Rennie for help with laboratory analyses and field work. $\delta^{15}N$ research by Dr Handley and Dr Scrimgeour is supported mainly by contracts from the European Commission of the European Union, Mylnefield Research Services and the Scottish Office Agriculture, Environment and Fisheries Department.

REFERENCES

Abbadie, L., Mariotti, A. and Menaut, J.-C. (1992). Independence of savanna grasses from soil organic matter for their nitrogen supply. *Ecology* **73**, 608–613.

Allen, M.F. (1991). *The Ecology of Mycorrhizae.* Cambridge University Press, New York.

Allen, S. and Raven, J.A. (1987). Intracellular pH regulation in *Ricinus communis* grown with ammonium or nitrate as N source: the role of long-distance transport. *J. Exp. Bot.* **38**, 580–596.

Allen, S., Raven, J.A. and Sprent, J.I. (1988). The role of long-distance transport in intracellular pH regulation in *Phaseolus vulgaris* grown with ammonium or nitrate as nitrogen source, or nodulated. *J. Exp. Bot.* **39**, 513–528.

Allison, F.E. (1931). Forms of nitrogen assimilated by plants. *Quart. Rev. Biol.* **6**, 313–321.

Amberger, A. and Schmidt, H.-L. (1987). Natürliche isotopengehalte von Niträt als Indikatoren für dessen Herkunft. *Geochim. Cosmochim. Acta* **51**, 2699–2705.

Anderson, L.S. and Mansfield, T.A. (1979). The effects of nitric oxide pollution on the growth of tomato. *Environ. Poll.* **20**, 113–121.

Andreux, F., Cerri, C., Vose, P.B. and Vitorello, V.A. (1990). Potential of stable isotope, ^{15}N and ^{13}C, methods for determining input and turnover in soils. In: *Nutrient Cycling in Terrestrial Ecosystems: Field Methods, Application and Interpretation* (Ed. by A.F. Harrison and P. Ineson), pp. 259–275. Elsevier Applied Science, London.

Androsoff, G.L., van Kessel, C. and Pennock, D.J. (1994). Landscape-scale estimates of dinitrogen fixation by *Pisum sativum* by nitrogen-15 natural abundance and enriched isotope dilution. *Biol. Fertil. Soils* **20**, 33–40.

Arnebrant, K., Ek, H., Finlay, R.D. and Söderström, B. (1993). Nitrogen translocation between *Alnus glutinosa* (L.) Gaertn. seedlings inoculated with *Frankia* sp. and *Pinus contorta* Doug. ex Loud seedlings connected by a common ectomycorrhizal mycelium. *New Phytol.* **124**, 231–242.

Bååth, E. and Söderström, B. (1979). Fungal biomass and fungal immobilization of plant nutrients in Swedish coniferous forest soils. *Rev. Ecol. Biol. Soil* **16**, 477–489.

Badger, M.R. and Price, G.D. (1994). The role of carbonic anhydrase in photosynthesis. *Annu. Rev. Plant Physiol.* **45**, 369–392.

Barea, J.M., Azcón-Aguilar, C. and Azcón, R. (1991). The role of vesicular-arbuscular mycorrhizae in improving plant N acquisition from soil as assessed with ^{15}N. In: *Stable Isotopes in Plant Nutrition, Soil Fertility and Environmental Studies: Proc. Symp. Vienna, 1–5 October 1990* (Ed. Anonymous), pp. 209–216. IAEA/FAO, Vienna.

Barnes, C.J., Jacobson, G. and Smith, G.D. (1992). The origin of high-nitrate ground waters in Australian arid zone. *J. Hydrol.* **137**, 181–197.

Belsky, A.J., Amundson, R.G., Duxbury, J.M., Riha, S.J., Ali, A.R. and Mwonga, S.M. (1989). The effects of trees on their physical, chemical, and biological environments in a semi-arid savanna in Kenya. *J. Appl. Ecol.* **26**, 1005–1024.

Berg, B. (1988). Dynamics of nitrogen (^{15}N) in decomposing Scots pine (*Pinus sylvestris*) needle litter. Long-term decomposition in a Scots pine forest. VI. *Can. J. Bot.* **66**, 1539–1546.

Berg, B., McClaugherty, C. and Johansson, M.-B. (1993a). Litter mass-loss rates in late stages of decomposition at some climatically and nutritionally different pine sites. Long-term decomposition in a Scots pine forest. VIII. *Can. J. Bot.* **71**, 680–692.

Berg, B., Berg, M.P., Bottner, P., Box, E., Breymeyer, A., De Anta, R.C., Kratz, W., Madeira, M., Mälkönen, E., McClaugherty, C., *et al.* (1993b). Litter mass loss rates in pine forests of Europe and Eastern United States: some relationships with climate and litter quality. *Biogeochem.* **20**, 127–159.

Bergersen, F.J., Turner, G.L., Amarger, N., Mariotti, F. and Mariotti, A. (1986). Strain of *Rhizobium lupini* determines natural abundance of ^{15}N in root nodules of *Lupinus* spp. *Soil Biol. Biochem.* **18**, 97–101.

Bergersen, F.J., Peoples, M.B. and Turner, G.L. (1988). Isotopic discrimination during the accumulation of nitrogen by soybeans. *Aust. J. Plant Physiol.* **15**, 407–420.

Bergersen, F.J., Peoples, M.B., Herridge, D.F. and Turner, G.L. (1990). Measurement of N_2 fixation by ^{15}N natural abundance in the management of legume crops: roles and precautions. In: *Nitrogen Fixation: Achievements and Objectives. Proc. 8th International Congress of Nitrogen Fixation. Knoxville, Tennessee, USA, May 20–26, 1990* (Ed. by P.M. Gresshoff, L.E. Roth, G. Stacey and W.E. Newton), pp. 315–322. Chapman and Hall, New York.

Biggar, J.W. (1978). Spatial variability of nitrogen in soils. In: *Nitrogen in the Environment* (Ed. by D.R. Nielsen and J.G. MacDonald), pp. 201–211. Academic Press, New York.

Binkley, D. and Matson, P. (1983). Ion exchange resin bag method for assessing forest soil nitrogen availability. *Soil Sci. Soc. Am. J.* **47**, 1050–1052.

Binkley, D., Sollins, P. and McGill, W.B. (1985). Natural abundance of nitrogen-15 as a tool for tracing alder-fixed nitrogen. *Soil Sci. Soc. Am. J.* **49**, 444–447.

Black, A.S. and Waring, S.A. (1977). The natural abundance of ^{15}N in the soil-water system of a small catchment area. *Austr. J. Soil Sci.* **15**, 51–57.

Blackmer, A.M. and Bremner, J.M. (1977). Nitrogen isotope discrimination in denitrification of nitrate in soils. *Soil Biol. Biochem.* **9**, 73–77.

Böttcher, J., Strebel, O., Voerkelius, S. and Schmidt, H.-L. (1990). Using isotope fractionation of nitrate-nitrogen and nitrate-oxygen for evaluation of microbial denitrification in a sandy aquifer. *J. Hydrol.* **114**, 413–424.

Bouwman, L.A., Bloem, J., van den Boogert, P.H.J.F., Bremer, F., Hoenderboom, G.H.J. and de Ruiter, P.C. (1994). Short-term and long-term effects of bacterivorous nematodes and nematophagous fungi on carbon and nitrogen mineralization in microcosms. *Biol. Fertil. Soils* **17**, 249–256.

Bremer, E. and van Kessel, C. (1990). Appraisal of the nitrogen-15 natural-abundance method for quantifying dinitrogen fixation. *Soil Sci. Am. J.* **54**, 404–411.

Bremer, E., Gehlen, H., Swerhone, G.D.W. and van Kessel, C. (1993). Assessment of reference crops for the quantification of N_2 fixation using natural and enriched levels of ^{15}N abundance. *Soil Biol. Biochem.* **25**, 1197–1202.

Broadbent, F.E., Rauschkolb, R.S., Lewis, K.A. and Chang, G.Y. (1980). Spatial variability of nitrogen-15 and total nitrogen in some virgin and cultivated soils. *Soil Sci. Soc. Amer. J.* **44**, 524–527.

Bronk, D.A., Glibert, P.M. and Ward, B.B. (1994). Nitrogen uptake, dissolved organic nitrogen release, and new production. *Science.* **265**, 1843–1846.

Caldwell, M.M. and Pearcy, R.W., Eds. (1995). *Exploitation of Environmental Heterogeneity by Plants.* Academic Press, San Diego.

Chapin, F.S.I., Moilanen, L. and Kielland, K. (1993). Preferential use of organic nitrogen for growth by a non-mycorrhizal arctic sedge. *Nature* **361**, 150–153.

Cheng, H.H., Bremner, J.M. and Edwards, A.P. (1964). Variations of nitrogen-15 abundance in soils. *Science.* **146**, 1574–1575.

Cliquet, J.-B. and Stewart, G.R. (1993). Ammonia assimilation in *Zea mays* L. infected with a vesicular-arbuscular mycorrhizal fungus *Glomus fasciculatum.* *Plant Phys.* **10**, 865–871.

Cornell, S., Rendell, A. and Jickells, T. (1995). Atmospheric inputs of dissolved organic nitrogen to the oceans. *Nature* **376**, 243–246.

Côté, B., Vogel, C.S. and Dawson, J.O. (1989). Autumnal changes in tissue nitrogen of autumn olive, black alder and eastern cottonwood. *Plant Soil* **118**, 23–32.

Coûteaux, M.-M., Bottner, P. and Berg, B. (1995). Litter decomposition, climate and litter quality. *Tree* **10**, 63–66.

Crabtree, R.C. and Bernston, G.M. (1994). Root architectural responses of *Betula lenta* to spatially heterogeneous ammonium and nitrate. *Plant Soil* **158**, 129–134.

DeNiro, M.J. and Epstein, S. (1981). Influence of diet on the distribution of nitrogen isotopes in animals. *Geochim. Cosmochim. Acta* **45**, 341–351.

de Ruiter, P.C., Moore, J.C., Zwart, K.B., Bouwman, L.A., Hassink, J., Bloem, J., de Vos, J.A., Marinissen, J.C.Y., Didden, W.A.M., Lebbink, G., *et al.* (1993). Simulation of nitrogen mineralization in the below-ground food webs of two winter wheat fields. *J. Appl. Ecol.* **30**, 95–106.

Delwiche, C.C. and Steyn, P.L. (1970). Nitrogen isotope fractionation in soils and microbial reactions. *Environ. Sci. Tech.* **4**, 929–934.

Devienne, F., Mary, B. and Lamaze, T. (1994a). Nitrate transport in intact wheat roots I. Estimation of cellular fluxes and NO_3^- distribution using compartmental analysis from data of $^{15}NO_3^-$ efflux. *J. Exp. Bot.* **45**, 667–676.

Devienne, F., Mary, B. and Lamaze, T. (1994b). Nitrate transport in intact wheat roots II. Long-term effects of NO_3^- concentration in the nutrient solution on NO_3^- unidirectional fluxes and distribution within the tissues. *J. Exp. Bot.* **45**, 677–684.

Domenach, A.M., Kurdali, F. and Bardin, R. (1989). Estimation of symbiotic dinitrogen fixation in alder forest by the method based on natural ^{15}N abundance. *Plant Soil* **118**, 51–59.

Doube, B.M., Ryder, M.H., Davoren, C.W. and Stephens, P.M. (1994). Enhanced root nodulation of subterranean clover (*Trifolium subterraneum*) by *Rhizobium leguminosarium* biovar *trifolii* in the presence of the earthworm *Aporrectodea trapezoides* (Lumbricidae) *Biol. Fertil. Soils* **18**, 169–174.

Doughton, J.A., Saffigna, P.G. and Vallis, I. (1991). Natural abundance of ^{15}N in barley as influenced by prior cropping or fallow, nitrogen fertilizer and tillage. *Austr. J. Agric. Res.* **42**, 723–733.

Durka, W., Schulze, E.-D., Gebauer, G. and Voerkelius, S. (1994). Effect of forest decline on uptake and leaching of deposited nitrate determined from ^{15}N and ^{18}O measurements. *Nature* **372**, 765–767.

Ehleringer, J.R. (1993). Gas exchange implications of isotopic variation in arid-land plants. In: *Water Deficits: Plant Responses from Cell to Community* (Ed. by J.A.C. Smith and H. Griffiths), pp. 265–284. BIOS Scientific Publishers, Oxford.

Ehleringer, J.R., Hall, A.E., and Farquhar, G.D. Eds. (1993). *Stable Isotopes and Plant Carbon–Water Relations*. Academic Press, San Diego, CA.

Ellenberg, H. (1977). Stickstoff als Standortsfaktor, insbesondere für mitteleuropaische Pflanzengesellschaften. *Oecol. Plant.* **12**, 1–22.

Evans, R.D. and Ehleringer, J.R. (1994). Water and nitrogen dynamics in an arid woodland. *Oecol.* **99**, 233–242.

Farquhar, G.D. and Richards, R.A. (1984). Isotopic composition of plant carbon correlates with water-use efficiency of wheat genotypes. *Aust. J. Plant Physiol.* **11**, 539–552.

Farquhar, G.D., Wetselaar, R. and Weir, B. (1983). Gaseous nitrogen losses from plants. In: *Developments in Plant and Soil Sciences* (Ed. by J.R. Freeney and J.R. Simpson), pp. 159–180. Nijhoff, Den Hague.

Feigin, A., Shearer, G., Kohl, D.H. and Commoner, B. (1974a). Variation in the natural ^{15}N abundance in nitrate mineralized during incubation of several Illinois soils. *Soil Sci Soc. Amer. Proc.* **38**, 90–95.

Feigin, A., Shearer, G., Kohl, D.H. and Commoner, B. (1974b). The amount and nitrogen-15 content of nitrate in soil profiles from two Central Illinois fields in a corn-soybean rotation. *Soil Sci. Soc. Amer. Proc.* **38**, 465–471.

Fitter, A.H. (1987). An architectural approach to the comparative ecology of plant root systems. *New Phytologist (Suppl.)* **106**, 61–77.

Focht, D.D. (1973). Isotope fractionation of 15-N and 14-N in microbiological nitrogen transformations: a theoretical model. *J. Environ. Qual.* **2**, 247–252.

Fogel, M.L. and Cifuentes, L.A. (1993). Isotope fractionation during primary production. In: *Organic Geochemistry: Principles and Applications* (Ed. by M.H. Engel and S.A. Macko), pp. 73–98. Plenum Press, New York.

Fogel, R. (1985) Roots as primary producers in below-ground ecosystems. In: *Ecological Interactions in Soil (Special Publication of the British Ecological Society, No. 4)* (Ed. by A.H. Fitter, D. Atkinson, D.J. Read and M.B. Usher, pp. 23–36. Blackwell, Oxford.

Fogel, R. and Hunt, G. (1979). Fungal and arboreal biomass in a western Oregon Douglas-fir ecosystem. *Can. J. Forest Res.* **13**, 219–232.

Frey, B. and Schüepp, H. (1993). Acquisition of nitrogen by external hyphae of arbuscular mycorrhizal fungi associated with *Zea mays* L. *New Phytol.* **124**, 221–230.

Freyer, H.D. and Aly, A.I.M. (1974). Nitrogen-15 studies on identifying fertilizer excess in environmental studies. In: *Isotope Ratios as Pollutant Source and Behaviour Indicators – Proceedings of a Symposium* (Ed. by IAEA/FAO), pp. 21–33. IAEA/FAO, Vienna.

Fry, B., Jones, D.E., Kling, G.W., McKane, R.B., Nadelhoffer, K.J., and Peterson, B.J. (1995). Adding ^{15}N tracers to ecosystem experiments. In: *Stable Isotopes in the Biosphere* (Ed. by E. Wada, T. Yoneyama, M. Minagawa, T. Ando, and B.D. Fry), pp. 171–192. Kyoto University Press, Kyoto.

Gainey, P.L. (1917). The significance of nitrification as a factor in soil fertility. *Soil Science* **3**, 399–416.

Garcia-Mendez, G., Maass, J., Matson, P.A. and Vitousek, P.M. (1991). Nitrogen transformations and nitrous oxide flux in a tropical deciduous forest in Mexico. *Oecol.* **88**, 362–366.

Garten, C.T., Jr. (1992). Nitrogen isotope composition of ammonium and nitrate in bulk precipitation and forest throughfall. *Intern. J. Environ. Anal. Chem.* **47**, 33–45.

Garten, C.T., Jr. (1993). Variation in foliar ^{15}N abundance and the availability of soil nitrogen on walker branch watershed. *Ecology.* **74**, 2098–2113.

Garten, C.T., Jr. and van Miegroet, H. (1994). Relationships between soil nitrogen dynamics and natural ^{15}N abundance in plant foliage from Great Smoky Mountains National Park. *Can. J. Forest Res.* **24**, 1636–1645.

Gebauer, G. and Dietrich, P. (1993). Nitrogen isotope ratios in different compartments of a mixed stand of spruce, larch and beech trees and of understorey vegetation including fungi. *Isotopes Environ. Health Stud.* **29**, 35–44.

Gebauer, G. and Schulze, E.-D. (1991). Carbon and nitrogen isotope ratios in different compartments of a healthy and a declining *Picea* forest in the Fichtelgebirge, NE Bavaria. *Oecol.* **87**, 198–207.

Gebauer, G., Giesemann, A., Schulze, E.-D. and Jäger, H.-J. (1994). Isotope ratios and concentrations of sulfur and nitrogen in needles and soils of *Picea abies* stands as influenced by atmospheric deposition of sulfur and nitrogen compounds. *Plant Soil* **164**, 267–281.

Gleixner, G., Danier, H.-J., Werner, R.A., and Schmidt, H.-L. (1993). Correlations between the ^{13}C content of primary and secondary plant products in different cell compartments and that in decomposing Basidiomycetes. *Plant Physiol.* **102**, 1287–1290.

Goericke, R., Montoya, J.P. and Fry, B. (1994). Physiology of isotopic fractionation in algae and cyanobacteria. In: *Stable Isotopes in Ecology and Environmental Science* (Ed. by K. Lajtha and R.H. Michener), pp. 187–221. Blackwell Scientific Publications, London.

Goovaerts, P. and Chiang, C.N. (1993). Temporal persistence of spatial patterns for mineralizable nitrogen and selected soil properties. *Soil Sci. Soc. Amer. J.* **57**, 372–381.

Grime, J.P., Hodgson, J.G. and Hunt, R. (1988). *Comparative Plant Ecology: A Functional Approach to Common British Species.* Unwin Hyman, London,

Groffman, P.M., Zak, D.R., Christensen, S., Mosier, A. and Tiedje, J.M. (1993). Early spring nitrogen dynamics in a temperate forest landscape. *Ecology.* **74**, 1579–1585.

Guo-Qing, S., Guang-Xi, X., Ya-Chen, C. and Hua, X. (1991). Study on variation in ^{15}N natural abundance I. Characteristics of variation in ^{15}N natural abundance of forest soils. *Pedosphere* **1**, 277–281.

Handley, L.L. and Raven, J.A. (1992). The use of natural abundance of nitrogen isotopes in plant physiology and ecology. *Plant Cell Environ.* **15**, 965–985.

Handley, L.L., Daft, M.J., Wilson, J., Scrimgeour, C.M., Ingleby, K. and Sattar, M.A. (1993). Effects of the ecto- and VA-mycorrhizal fungi *Hydnagium carneum* and *Glomus clarum* on the δ^{15}N and δ^{13}C values of *Eucalyptus globulus* and *Ricinus communis*. *Plant Cell Environ.* **16**, 375–382.

Handley, L.L., Nevo, E., Raven, J.A., Martínez-Carrasco, R., Scrimgeour, C.M., Pakniyat, H. and Forster, B.P. (1994a). Chromosome 4 controls potential water use efficiency (δ^{13}C) in barley. *J. Exp. Bot.* **45**, 1661–1663.

Handley, L.L., Odee, D. and Scrimgeour, C.M. (1994b). δ^{15}N and δ^{13}C patterns in savanna vegetation: dependence on water availability and disturbance. *Funct. Ecol.* **8**, 306–314.

Handley, L.L., Raven, J.A., Brendel, O., Schmidt, S., Erskine, P., Turnbull, M. and Stewart, G.R. (1996). Natural abundance fractionations of N and C in field-collected fungi. Presented at Stable Isotopes Mass Spectroscopy Users Group, Llandudno, Wales, 13 January 1996. *Rapid Commun. Mass Spectr.* **10**, 974–978.

Handley, L.L., Robinson, D.R., Forster, B.P., Ellis, R.P., Scrimgeour, C., Gordon, D.C. and Nevo, E. (1996) Shoot δ^{15}N correlates with genotype and stress in barley. *Planta* **200** (in press).

Hansen, A.P. and Pate, J.S. (1987). Evaluation of the ^{15}N natural abundance method and xylem sap analysis for assessing N_2 fixation of understorey legumes in jarrah (*Eucalyptus marginata* Donn *ex* Sm.) forest in S.W. Australia. *J. Exp. Bot.* **38**, 1446–1458.

Hatch, D.J. and Murray, P.J. (1994). Transfer of nitrogen from damaged roots of white clover (*Trifolium repens* L.) to closely associated roots of intact perennial ryegrass (*Lolium perenne* L.). *Plant Soil* **166**, 181–185.

Hauck, R.D., Bartholomew, W.V., Bremner, J.M., Broadbent, F.E., Cheng, H.H., Edwards, A.P., Keeney, D.R., Legg, J.O., Olsen, S.R. and Porter, L.K. (1972). Use of variations in natural nitrogen isotope abundance for environmental studies: A questionable approach. *Science* **177**, 453–454.

Heaton, T.H.E. (1985). Isotopic and chemical aspects of nitrate in the groundwater of the Springbok Flats. *Water, S. Africa* **11**, 199–208.

Heaton, T.H.E. (1986). Isotopic studies of nitrogen pollution in the hydrosphere and atmosphere: a review. *Chem. Geol. (Isotope Geoscience Section)* **59**, 87–102.

204   L.L. HANDLEY AND C.M. SCRIMGEOUR

Heaton, T.H.E. and Crossley, A. (1995). Carbon isotope variations in a plantation of Sitka spruce, and the effect of acid mist. *Oecol.* **103**, 109–117.

Heaton, T.H.E., Talma, A.S. and Vogel, J.C. (1983). Origin and history of nitrate in confined groundwater in the western Kalahari. *J. Hydrol.* **62**, 243–262.

Heaton, T.H.E., Spiro, B. and Robertson, S.M.C. (1996). Potential canopy influences on the isotopic composition of nitrogen and sulphur in atmospheric deposition. *Oecologia*, in press.

Hedin, L.O. (1994). Stable isotopes, unstable forest. *Nature* **372**, 725–726.

Hellebust, J.A. (1974). Extracellular products. *Algal Physiology and Biochemistry* (Ed. by W.D.P. Stewart), pp. 838–863. Blackwell Scientific Publications, Oxford.

Herbel, M.J. and Spalding, R.F. (1993). Vadose zone fertilizer-derived nitrate and δ¹⁵N extracts. *Ground Water* **31**: 376–382.

Herman, D.J. and Rundel, P.W. (1989). Nitrogen isotope fractionation in burned and unburned chapparal soils. *Soil Sci. Soc. Amer. J.* **53**, 1229–1236.

Hill, P.W., Handley, L.L., and Raven, J.A. (1996). *Juniperus communis* L. ssp. *communis* at Balnaguard, Scotland: Foliar carbon discrimination (δ¹³C) and 15-N natural abundance (δ¹⁵N) suggests gender-linked differences in water and N use. *Botan. J. Scotland* **48**, 209–224.

Hoch, M.P., Fogel, M.L. and Kirchman, D.L. (1992). Isotope fractionation associated with ammonium uptake by marine bacterium. *Limnol. Oceanogr.* **37**, 1447–1459.

Hoch, M.P., Fogel, M.L. and Kirchman, D.L. (1994). Isotope fractionation during ammonium uptake by marine microbial assemblages. *Geomicrobiol.* **12**, 113–127.

Högberg, P. (1990). ¹⁵N natural abundance as a possible marker of the ectomycorrhizal habit of trees in mixed African woodlands. *New Phytol.* **115**, 483–486.

Högberg, P. and Alexander, I.J. (1995). Roles of root symbioses in African woodland and forest: evidence from ¹⁵N abundance and foliar analysis. *J. Ecol.* **83**, 217–224.

Högberg, P. and Johannisson, C. (1993). ¹⁵N abundance of forests is correlated with losses of nitrogen. *Plant Soil* **157**, 147–150.

Högberg, P., Näsholm, T., Högbom, L. and Ståhl, L. (1994). Use of ¹⁵N labelling and ¹⁵N natural abundance to quantify the role of mycorrhizas in N uptake by plants: importance of seed N and of changes in the ¹⁵N labelling of available N. *New Phytol.* **127**, 515–519.

Högberg, P., Johannisson, C., Högberg, M., Hogbom, L., Nasholm, T. and Hallgren, J.-E. (1995). Measurements of abundances of ¹⁵N and ¹³C as tools in retrospective studies of N balances and water stress in forests: A discussion of preliminary results. *Plant Soil* **168–169**, 125–133.

Høgh-Jensen, H. and Schjoerring, J.K. (1994). Measurement of biological dinitrogen fixation in grassland: comparison of the enriched ¹⁵N dilution and the natural ¹⁵N abundance methods at different nitrogen application rates and defoliation frequencies. *Plant Soil* **166**, 153–163.

Hübner, C., Redl, G. and Wurst, F. (1991). *In situ* methodology for studying N-mineralization in soils using anion exchange resins. *Soil Biol. Biochem.* **23**, 701–702.

Hutchings, M.J. and de Kroon, H. (1994). Foraging in plants: the role of morphological plasticity in resource acquisition. *Adv. Ecol. Res.* **25**, 159–237.

Ingelög, T. and Nohrstedt, H.-O. (1993). Ammonia formation and soil pH increase caused by decomposing fruitbodies of macrofungi. *Oecol.* **93**, 449–451.

Jackson, L.E., Schimel, J.P. and Firestone, M.K. (1989). Short-term partitioning of ammonium and nitrate between plants and microbes in an annual grassland. *Soil Biol. Biochem.* **21**, 409–415.

Johannisson, C. and Högberg, P. (1994). ¹⁵N abundance of soils and plants along an experimentally induced forest nitrogen supply gradient. *Oecol.* **97**, 322–325.

Johnson, H.B. and Mayeux, H.S., Jr. (1990). *Prosopis glandulosa* and the nitrogen balance of rangelands: extent and occurrence of nodulation. *Oecol.* **84**, 176–185.

Karamanos, R.E. and Rennie, D.A. (1978). Nitrogen isotope fractionation during ammonium exchange reactions with soil clay. *Can. J. Soil Sci.* **58**, 53–60.

Karamanos, R.E. and Rennie, D.A. (1980). Changes in natural ^{15}N abundance associated with pedogenic processes in soil. II. changes on different slope positions. *Can. J. Soil Sci.* **60**, 365–372.

Karamanos, R.E., Voroney, R.P., and Rennie, D.A. (1981). Variation in natural N-15 abundance of central Saskatchewan soils. *Soil Sci. Soc. Amer. J.* **45**, 826–828.

Kielland, K. (1994). Amino acid absorption by Arctic plants: implications for plant nutrition and nitrogen cycling. *Ecology.* **75**, 2373–2383.

Kim, K.-R. and Craig, H. (1993). Nitrogen-15 and oxygen-18 characteristics of nitrous oxide: a global perspective. *Science* **262**, 1855–1857.

Knight, J.D., Verhees, F., van Kessel, C. and Slinkard, A.E. (1993). Does carbon isotope discrimination correlate with biological nitrogen fixation? *Plant Soil* **153**, 151–153.

Kohl, D.H. and Shearer, G. (1980). Isotopic fractionation associated with symbiotic N_2 fixation and uptake of NO_3^- by plants. *Plant Physiol.* **66**, 51–56.

Kohls, S.J., van Kessel, C., Baker, D.D., Grigal, D.F. and Lawrence, D.B. (1994). Assessment of N_2 fixation and N cycling by *Dryas* along a chronosequence within the forelands of the Athabasca Glacier, Canada. *Soil Biol. Biochem.* **26**, 623–632.

Komor, S.C. and Anderson, H.W., Jr. (1993). Nitrogen isotopes as indicators of NO_3^- sources in Minnesota sand-plain aquifers. *Ground Water* **31**, 260–269.

Lajtha, K. and Marshall, J.D. (1994). Sources of variation in the stable isotopic composition of plants. In: *Stable Isotopes in Ecology and Environmental Science* (Ed. by K. Lajtha and R.H. Michener), pp. 1–21. Blackwell Scientific Publications, Oxford.

Lajtha, K. and Schlesinger, W.H. (1986). Plant response to variations in nitrogen availability in a desert shrubland community. *Biogeochem.* **2**, 29–37.

Lau, L.S., Young, R.H.F., Konno, S.K., Olnall, R.J., and Lee, H.H. (1975). *Wet-weather Water Quality Monitoring: Kaneohe Bay, Oahu, Hawaii.* Water Resources Research Center, Univ. of Hawaii, Honolulu, Hawaii.

Ledgard, S.F., Freney, J.R. and Simpson, J.R. (1984). Variations in natural enrichment of ^{15}N in the profiles of some Australian pasture soils. *Austr. J. Soil Res.* **22**, 155–164.

Ledgard, S.F., Woo, K.C. and Bergersen, F.J. (1985). Isotopic fractionation during reduction of nitrate and nitrite by extracts of spinach leaves. *Aust. J. Plant Physiol.* **12**, 631–640.

Lee, J.A. and Stewart, G.R. (1978). Ecological aspects of nitrogen assimilation. *Adv. Bot. Res.* **6**, 2–43.

MacKerron, D.K.L. (1994). Meteorological records. In: *1993 Annual Report*, p. 171. Scottish Crop Research Institute. Invergowrie, Dundee.

Macko, S.A. and Estep, M.L.F. (1984). Microbial alteration of stable nitrogen and carbon isotopic compositions of organic matter. *Org. Geochem.* **6**, 787–790.

Macko, S. and Ostrom, N.E (1994). Pollution studies using stable isotopes. In: *Stable Isotopes in Ecology and Environmental Science* (Ed. by K. Lajtha and R.H. Michener), pp. 45–62. Blackwell Scientific Publications, London.

Macko, S.A., Estep, M.L., Engel, M.H. and Hare, P.E. (1986). Kinetic fractionation of stable isotopes during amino acid transamination. *Geochim. Cosmochim. Acta* **50**, 2143–2146.

Macko, S., Fogel, M.L., Hare, P.E., and Hoering, T.C. (1987). Isotopic fractionation of nitrogen and carbon in the synthesis of amino acids by microorganisms. *Chem. Geol.* **65**, 79–92.

Madhavan, S., Markwell, J.P. and O'Leary. (1995). Correlation between leaf chlorophyll content and carbon isotope fractionation in field grown soybean genotypes. *Plant Physiol.* **108** (Suppl.), 89.

Mariotti, A., Mariotti, F., Amarger, N., Pizelle, G., Ngambi, J.-M., Champigny, M.-L. and Moyse, A. (1980a). Fractionnements isotopiques de l'azote lors des processus d'absorption des nitrates et de fixation de l'azote atmosphérique par les plantes. *Physiol. Veg.* **18**, 163–181.

Mariotti, A., Pierre, D., les Bains, T., Vedy, J.C., Bruckert, S. and Guillemot, J. (1980b). The abundance of natural nitrogen-15 in the organic matter of soils along an altitudinal gradient (Chalais, Haute Savoie, France). *Catena* **7**, 293–300.

Mariotti, A., German, J.C., Hubert, P., Kaiser, P., Letolle, R., and Tardieux, A. (1981). Experimental determination of nitrogen kinetic isotopes fractionation: some principles. Illustration for the denitrification and nitrification processes. *Plant Soil* **62**, 413–430.

Mariotti, A., Martiotti, F., Champigny, M.-L., Amarger, N. and Moyse, A. (1982). Nitrogen isotope fractionation associated with nitrate reductase activity and uptake of nitrate by pearl millet *Pennisetum* spp. *Plant Phys. (Bethesda)* **69**, 880–884.

Mariotti, A., Landreau, A. and Simon, B. (1988). ^{15}N isotope bigeochemistry and natural denitrification process in groundwater: application to the chalk aquifer of northern France. *Geochim. Cosmochim. Acta* **52**, 1869–1878.

Martin, F. and Botton, B. (1993). Nitrogen metabolism of ectomycorrhizal fungi and ectomycorrhiza. *Adv. Plant Pathol.* **9**, 83–102.

Martin, F., Cote, R. and Canet, D. (1994). NH_4^+ assimilation in the ectomycorrhizal basidiomycete *Laccaria bicolor* (Maire) Orton, a ^{15}N-NMR study. *New Phytol.* **128**, 479–485.

McLeod, A.R., Holland, M.R., Shaw, P.J.A., Sutherland, P.M., Darrall, N.M., and Skeffington, R.A. (1995). Enhancement of nitrogen deposition to forest trees exposed to SO_2. *Nature* **347**, 277–279.

Medina, R., Olleros, T. and Schmidt, H.-L. (1982). Isotope effects on each, C- and N-atoms, as a tool for the elucidation of enzyme catalyzed amide hydrolyses. In: *Stable Isotopes: Proceedings 4th International Conference, Julich, March 23–26, 1981* (Ed. by H.-L. Schmidt, H. Forstel and K. Heinzinger), pp. 77–82. Elsevier Scientific Publishing Company, Amsterdam.

Melillo, J.M., Aber, J.D., Linkins, A.E., Ricca, A., Fry, B. and Nadelhoffer, K.J. (1989). Carbon and nitrogen dynamics along the decay continuum: plant litter to soil organic matter. *Plant Soil.* **115**, 189–198.

Meints, V.W., Boone, L.V. and Kurtz, L.T. (1975). Natural ^{15}N abundance in soil, leaves, and grain as influenced by long term additions of fertilizer N at several rates. *J. Environ. Qual.* **4**, 486–490.

Michelsen, A. and Sprent, J.I. (1994). The influence of vesicular-arbuscular mycorrhizal fungi on the nitrogen fixation of nursery-grown Ethiopian acacias estimated by the ^{15}N natural abundance method. *Plant Soil* **160**, 249–257.

Michelsen, A., Schmidt, I.K., Jonasson, S., Quarmby, C., and Sleep, D. (1996). Leaf ^{15}N abundance of subarctic plants provides field evidence that ericoid, ectomycorrhizal and non-arbuscular mycorrhizal species access different sources of soil nitrogen. *Oecol.* **105**: 53–63

Millard, P. (1988). The accumulation and storage of nitrogen by herbaceous plants. *Plant, Cell Environ.* **11**, 1–8.

Millard, P. and Proe, M.F. (1992). Storage and internal cycling of nitrogen in relation to seasonal growth of Sitka spruce. *Tree Physiol.* **10**, 33–43.

Minagawa, M. and Wada, E. (1984). Stepwise enrichment of ^{15}N along food chains: further evidence and the relation between δ^{15}N and animal age. *Goechim. Cosmochim. Acta* **48**, 1135–1140.

Murray, P.J. and Hatch, D.J. (1994). *Sitona* weevils (Coleoptera: Curculionidae) as agents for rapid transfer of nitrogen from white clover (*Trifolium repens* L.) to perennial ryegrass (*Lolium perenne* L.). *Ann. Appl. Biol.* **125**, 29–33.

Nadelhoffer, K.J. and Fry, B. (1988). Controls on natural nitrogen-15 and carbon-13 abundances in forest soil organic matter. *Soil Sci. Soc. Amer. J.* **52**, 1633–1640.

Nadelhoffer, K.J. and Fry, B. (1994). Nitrogen isotope studies in forest ecosystems. In: *Stable Isotopes in Ecology and Environmental Science* (Ed. by K. Lajtha and R.H. Michener), pp. 22–44. Blackwell Scientific Publications, London.

Nadelhoffer, K., Shaver, G., Fry, B., Giblin, A., Johnson, L. and McKane, R. (1995). ^{15}N natural abundances and N use by tundra plants. *Oecol.* **107**, 386–394.

Näsholm, T. (1994). Removal of nitrogen during needle senescence in Scots pine (*Pinus sylvestris* L.). *Oecol.* **99**, 290–296.

Nömmik, H., Pluth, D.J., Larsson, K. and Mahendrappa, M.K. (1994). Isotopic fractionation accompanying fertilizer nitrogen transformations in soil and trees of a Scots pine ecosystem. *Plant Soil* **158**, 169–182.

Northup, R.R., Yu, Z., Dahlgren, R.A., and Vogt, K.A. (1995). Polyphenol control of nitrogen release from pine litter. *Nature* **377**, 227–229.

O'Leary, M. (1988). Carbon isotopes in photosynthesis. *BioSci.* **38**, 328–336.

Okano, K., Komaki, S. and Matsuo, K. (1994). Remobilization of nitrogen from vegetative parts to sprouting shoots of young tea (*Camellia sinensis* L.) plants. *Japanese J. Crop Sci.* **63**, 125–130.

Owens, N.J.P. (1987). Natural variations in ^{15}N in the marine environment. *Advances in Marine Biology* **24**, 389–451.

Owens, N.J.P. and Rees, A.P. (1989). Determination of nitrogen-15 at sub-microgram levels of nitrogen using automated continuous-flow isotope ratio mass spectrometry. *Analyst.* **144**, 1655–1657.

Page, H.M. (1995). Variation in the natural abundance of ^{15}N in the halophyte, *Salicornia virginica*, associated with groundwater subsidies of nitrogen in a southern California salt-marsh. *Oecol.* **104**: 181–188.

Palm, C.A. and Sanchez, P.A. (1991). Nitrogen release from the leaves of some tropical legumes as affected by their lignin and polyphenolic contents. *Soil Biol. Biochem.* **23**, 83–88.

Pate, J.S., Stewart, G.R. and Unkovich, M. (1993). ^{15}N natural abundance of plant and soil components of a *Banksia* woodland ecosystem in relation to nitrate utilisation, life form, mycorrhizal status and N_2-fixing abilities of component species. *Plant Cell Environ.* **16**, 365–373.

Pate, J.S., Unkovich, M.J., Armstrong, E.L. and Sanford, P. (1994). Selection of reference plants for ^{15}N natural abundance assessment of N_2 fixation by crop and pasture legumes in south-west Australia. *Austr. J. Agric. Res.* **45**, 133–147.

Pearson, J. and Ji, Y.-M. (1994). Seasonal variation of leaf glutamine synthetase isoforms in temperate deciduous trees strongly suggests different function for the enzymes. *Plant, Cell Environ.* **17**, 1331–1337.

Pearson, J. and Stewart, G.R. (1993). The deposition of atmospheric ammonia and its effects on plants. Tansley Review No. 56. *New Phytol.* **125**, 283–305.

Piccolo, M.C., Neill, C. and Cerri, C.C. (1994). Natural abundance of ^{15}N in soils along forest-to-pasture chronosequences in the western Brazilian Amazon Basin. *Oecol.* **99**, 112–117.

Preston, T. and McMillan, D.C. (1988). Rapid sample throughput for biomedical stable isotope tracer studies. *Biomed. Environ. Mass Spectrom.* **16**, 229–235.

Raven, J.A. (1988). Acquisition of nitrogen by the shoots of land plants: its occurrence and implications for acid-base regulation. *New Phytol.* **109**, 1–20.

Raven, J.A. and Farquhar, G.D. (1990). The influence of N metabolism and organic acid synthesis on the natural abundance of isotopes of carbon in plants. *New Phytol.* **116**: 505–529.

Raven, J.A., Allen, S. and Griffiths, H. (1984). N-source, transpiration rate and stomatal aperture. In: *Ricinus in Membrane Transport in Plants* (Ed. by W.J. Cram, K. Janacek, R. Rybova and K. Sigler), pp. 161–162. Academia, Prague.

Raven, J.A., Wollenweber, B. and Handley, L.L. (1992a). Ammonia and ammonium fluxes between photolithotrophs and the environment in relation to the global nitrogen cycle. *New Phytol.* **121**, 5–18.

Raven, J.A., Wollenweber, B. and Handley, L.L. (1992b). A comparison of ammonium and nitrate as nitrogen sources for photolithotrophs. *New Phytol.* **121**, 19–32.

Raven, J.A., Wollenweber, B. and Handley, L.L. (1993). The quantitative role of ammonia/ammonium transport and metabolism by plants in the global nitrogen cycle. *Physiol. Plantar.* **89**, 512–518.

Read, D.J. (1991). Mycorrhizas in ecosystems – nature's response to 'The Law of the Minimum'. In: *Frontiers in Mycology* (Ed. by D.L. Hawksworth), CAB International, Cambridge.

Rice, E.L. and Pancholy, S.K. (1972). Inhibition of nitrification by climax vegetation. *Am. J. Bot.* **59**, 1033–1040.

Riga, A., van Praag, H.J. and Brigode, N. (1971). Rapport isotopique naturel de l'azote dans quelques sols forestiers et agricoles de belgique soumis a divers traitements culturaux. *Geoderma* **6**, 213–222.

Robertson, G.P. and Gross, K.L. (1994). Assessing the heterogeneity of belowground resources: quantifying pattern and scale. In: *Exploitation of Environmental Heterogeneity by Plants: Ecophysiological Processes Above- and Belowground.* (Ed. by M.M. Caldwell and R.W. Pearcy), pp. 237–253. Academic Press, San Diego.

Robinson, D., Linehan, D.J. and Caul, S. (1991). What limits nitrate uptake from soil? *Plant, Cell Environ.* **14**, 77–85.

Rogers, H.H., Campbell, J.C. and Volk, R.J. (1979a). Nitrogen-15 dioxide uptake and incorporation by *Phaseolus vulgaris* (L.). *Science* **206**, 333–335.

Rogers, H.H., Jeffries, H.E. and Witherspoon, A.M. (1979b). Measuring air pollutant uptake by plants: Nitrogen dioxide. *J. Environ. Qual.* **8**, 551–557.

Roskoski, J.P. (1981). Nodulation and N_2-fixation by *Inga jinicuil*, a woody legume in coffee plantations. I. Measurements of nodule biomass and field C_2H_2 reduction rates. *Plant Soil* **59**, 201–206.

Ryckert, R., Skujins, J., Sorensen, D. and Porcella, D. (1978). Nitrogen fixation by lichens and free-living microorganisms in deserts. In: *Nitrogen in Desert Ecosystems* (Ed. by N.E. West and J. Skujins), pp. 20–30. Dowden, Stroudsburg, PA.

Sandhu, J., Sinha, M. and Ambasht, R.S. (1990). Nitrogen release from decomposing litter of *Leucaena leucocephala* in the dry tropics. *Soil Biol. Biochem.* **22**, 859–863.

Schimel, J.P., Jackson, L.E. and Firestone, M.K. (1989). Spatial and temporal effects on plant-microbial competition for inorganic nitrogen in a California annual grassland. *Soil Biol. Biochem.* **21**, 1059–1066.

Schmidt, H.-L. and Voerkelius, S. (1989). Origin and isotope effects of oxygen in compounds of the nitrogen cycle. *Proceedings 5th Working Meeting. Isotopes in Nature. Leipzig, September 1989*, pp. 613–624.

Schulze, E.-D. (1989). Air pollution and forest decline in a spruce (*Picea abies*) forest. *Science* **244**, 776–783.

Schulze, E.-D., Chapin, F.S., III and Gebauer, G. (1994). Nitrogen nutrition and isotope differences among life forms at the northern treeline of Alaska. *Oecol.* **100**, 406–412.

Schulze, E.-D., Gebauer, G., Schulze, W., and Pate, J.S. (1991a). The utilization of nitrogen from insect capture by different growth forms of *Drosera* from Southwest Australia. *Oecol.* **87**, 240–246.

Schulze, E.-D., Gebauer, G., Ziegler, H. and Lange, O.L. (1991b). Estimates of nitrogen fixation by trees on an aridity gradient in Namibia. *Oecol.* **88**, 451–455.

Scrimgeour, C.M., Gordon, S.C., Handley, L.L. and Woodford, J.A.T. (1995) Trophic levels and anomalous δ^{15}N of insects on raspberry (*Rubus idaeus* L.). *Isotopes Environ. Health. Stud.* **31**, 107–115.

Selles, F., Karamanos, R.E. and Kachanoski, R.G. (1986). The spatial variability of nitrogen-15 and its relation to the variability of other soil properties. *Soil Sci. Soc. Amer. J.* **50**, 105–110.

Shearer, G. and Kohl, D.H. (1986). N_2-fixation in field settings: estimations based on natural ^{15}N abundance. *Aust. J. Plant Physiol.* **13**, 699–756.

Shearer, G. and Kohl, D.H. (1989). Estimates of N_2 fixation in ecosystems: the need for and basis of the ^{15}N natural abundance method. *Stable Isotopes in Ecological Research* (Ed. by P.W. Rundel, J.R. Ehleringer and K.A. Nagy), pp. 343–374. New York, Springer-Verlag.

Shearer, G., Kohl, D.H. and Chien, S.-H. (1978). The nitrogen-15 abundance in a wide variety of soils. *Soil Sci. Soc. Amer. J.* **42**, 899–902.

Shearer, G., Kohl, D.H., Virginia, R.A., Bryan, B.A., Skeeters, J.L., Nilsen, E.T., Sharifi, M.R. and Rundel, P.W. (1983). Estimates of N_2-fixation from variation in the natural abundance of ^{15}N in Sonoran Desert ecosystems. *Oecol.* **56**, 365–373.

Shearer, G., Schneider, J.D. and Kohl, D.H. (1991). Separating the efflux and influx components of net nitrate uptake by *Synechococcus* R2 under steady-state conditions. *J. Gen. Microbiol.* **137**, 1179–1184.

Smirnoff, N. and Stewart, G.R. (1985). Nitrate assimilation and translocation by higher plants: comparative physiology and ecological consequences. *Physiol. Plantar.* **64**, 133–140.

Srivastava, A.K. and Ambasht, R.S. (1995). Biomass, production, decomposition of and N release from root nodules in two *Casuarina equisetifolia* plantations in Sonbhadra, India. *J. Appl. Ecol.* **32**, 121–127.

Stadler, J., Gebauer, G. and Schulze, E.-D. (1993). The influence of ammonium on nitrate uptake and assimilation in 2-year-old ash and oak trees – a tracer study with ^{15}N. *Isotopes Environ. Health Stud.* **29**, 85–92.

Stephens, P.M., Davoren, C.W., Doube, B.M. and Ryder, M.H. (1994). Ability of the earthworms *Aporrectodea rosea and Aporrectodea trapezoides* to increase plant growth and the foliar concentration of elements in wheat (*Triticum aestivum* cv. Spear) in a sandy loam soil. *Biol. Fertil. Soils* **18**, 150–154.

Stevenson, F.C., Knight, J.D. and van Kessel, C. (1995). Dinitrogen fixation in pea: controls at the landscape- and micro-scale. *Soil Sci. Soc. Amer. J.* **59**, 1603–1611.

Stewart, C.R., Mann, A.F. and Fentem, P.A. (1980). Enzymes of glutamate formation: glutamate dehydrogenase, glutamine synthetase, and glutamate synthase. In: *The Biochemistry of Plants* (Ed. by B.J. Miflin), pp. 271–327. Academic Press, New York.

Stewart, G.R., Hegarty, E.E. and Specht, R.L. (1988). Inorganic nitrogen assimilation in plants of Australian rainforest communities. *Physiol. Plantar.* **74**, 26–33.

Stewart, G.R., Pate, J.S. and Unkovich, M.J. (1993). Characteristics of inorganic nitrogen assimilation of plants in fire-prone Mediterranean-type vegetation. *Plant, Cell Environ.* **16**, 351–363.

Stewart, G.R., Schmidt, S., Handley, L.L., Turnbull, M.H., Erskine, P.D. and Joly, C.A. (1994). [15]N natural abundance of vascular rainforest epiphytes: implications for nitrogen source and acquisition. *Plant, Cell Environment* **18**, 85–90.

Stock, W.D., Wienand, K.T. and Baker, A.C. (1995). Impacts of invading N_2-fixing *Acacia* species on patterns of nutrient cycling in two Cape Ecosystems: evidence from soil incubation studies and ^{15}N natural abundance. *Oecol.* **101**, 375–382.

Sutherland, R.A., van Kessel, C., Farrell, R.E. and Pennock, D.J. (1993). Landscape-scale variations in plant and soil nitrogen-15 natural abundance. *Soil Sci. Soc. Amer. J.* **57**, 169–178.

Sutton, M.A., Pitcairn, C.E.R. and Fowler, D. (1993). The exchange of ammonia between the atmosphere and plant-communities. *Adv. Ecol. Res.* **24**, 301–393.

Thornton, B., Millard, P., and Tyler, M.R. (1995). Effects of nitrogen supply on the seasonal remobilization of nitrogen in *Ulex europaeus* L. *New Phytol.* **130**, 557–563.

Tobita, S., Ito, O., Matsunaga, R., Rao, T.P., Rego, T.J., Johansen, C. and Yoneyama, T. (1994). Field evaluation of nitrogen fixation and use of nitrogen fertilizer by sorghum/pigeonpea intercropping on an alfisol in the Indian semi-arid tropics. *Biol. Fertil. Soils* **17**, 241–248.

Trebaçz, K., Simonis, W. and Schonknecht, G. (1994). Cytoplasmic Ca^{2+}, K^+, and NO_3^- activities in the liverwort *Conocephalum conicum* L. at rest and during action potentials. *Plant Physiol.* **106**, 1073–1084.

Treseder, K.K., Davidson, D.W. and Ehleringer, J.R. (1995). Absorption of ant-provided carbon dioxide and nitrogen by a tropical epiphyte. *Nature* **375**, 137–139.

Turner, G.L. and Gibson, A.H. (1980). Measurement of nitrogen fixation by indirect means. In: *Methods for Evaluating Biological Nitrogen Fixation* (Ed. by F.J. Bergersen), pp. 111–138. John Wiley and Sons, Chichester.

Turner, G.L., Gault, R.R., Morthorpe, L. and Chase, D.L. (1987). Differences in natural abundance of ^{15}N in extractable mineral nitrogen of cropped and fallowed surface soils. *Austr. J. Agric. Res.* **38**, 15–25.

Unkovich, M.J., Pate, J.S., Sanford, P. and Armstrong, E.L. (1994). Potential precision of the $\delta^{15}N$ natural abundance method in field estimates of nitrogen fixation by crop and pasture legumes in south-west Australia. *Austr. J. Agric. Res.* **45**, 119–132.

van Kessel, C., Farrell, R.E. and Pennock, D.J. (1994a). Carbon-13 and nitrogen-15 natural abundance in crop residues and soil organic matter. *Soil Sci. Soc. Am. J.* **58**, 382–389.

van Kessel, C., Farrell, R.E., Roskoski, J.P. and Keane, K.M. (1994b). Recycling of the naturally-occurring ^{15}N in an established stand of *Leucaena leucocephala*. *Soil Biol. Biochem.* **26**, 757–762.

van Kessel, C., Singleton, P.W. and Hoben, H.J. (1985). Enhanced N-transfer from a soybean to maize by vesicular arbuscular mycorrhizal (VAM) fungi. *Plant Physiol.* **79**, 562–563.

Velinsky, D.J., Fogel, M.L., Todd, J.F. and Bradley, M.T. (1991). Isotopic fractionation of dissolved ammonium at the oxygen-hydrogen sulfide interface in anoxic waters. *Geophys. Res. Lett.* **18**, 649–652.

Virginia, R.A. and Jarrell, W.M. (1983). Soil properties in a mesquite-dominated Sonoran Desert ecosystem. *Soil Sci. Soc. Amer. J.* **47**, 138–144.

Vitousek, P.M. and Matson, P.A. (1984). Mechanisms of nitrogen retention in forest ecosystems: a field experiment. *Science.* **225**, 51–52.

Vitousek, P.M., Shearer, G. and Kohl, D.H. (1989). Foliar ^{15}N natural abundances in Hawaiian rainforest: patterns and possible mechanisms. *Oecol.* **78**, 383–388.

Vogel, J.C., Talma, A.S. and Heaton, T.H.E. (1981). Gaseous nitrogen as evidence for denitrification in groundwater. *J. Hydrol.* **50**, 191–200.

Wada, E., Kadonaga, T. and Matsuo, S. (1975). ^{15}N abundance in nitrogen of naturally occurring substances and global assessment of denitrification from isotopic viewpoint. *Geochem. J.* **9**, 139–148.

Wada, E., Mizutani, H. and Minagawa, M. (1991). The use of stable isotopes for food web analysis. *Crit. Rev. Food Sci. Nutr.* **30**, 361–371.

Wada, E., Ando, T. and Kumazawa, K. (1995). Biodiversity of stable isotope ratios. In: *Stable Isotopes in the Biosphere* (Ed. by E. Wada, T. Yoneyama, M. Minagawa, T. Ando and B. Fry). pp. 7–16.

Wedin, D.A., Tieszen, L.L., Dewey, B. and Pastor, J. (1995). Carbon isotope dynamics during grass decomosition and soil organic matter formation. *Ecol.* **76**, 1383–1392.

Wendler, R., Carvalho, P.O., Pereira, J.S. and Millard, P. (1995). The role of nitrogen remobilisation from old leaves for the growth of *Eucalyptus globulus*. *Tree Physiol.* **15**, 679–683.

Wheeler, C.T. and Dickson, J.H. (1990). Symbiotic nitrogen fixation and distribution of *Spartocytisus supranubius* on Las Canadas, Tenerife. *Vieraea* **19**, 309–314.

Wheeler, C.T., Perry, D.A., Helgerson, O. and Gordon, J.C. (1979). Winter fixation of nitrogen in Scotch broom (*Cytisus scoparius* L.). *New Phytol.* **82**, 697–701.

Wheeler, C.T., Helgerson, O.T., Perry, D.A. and Gordon, J.C. (1987). Nitrogen fixation and biomass accumulation in plant communities dominated by *Cytisus scoparius* in Oregon and Scotland. *J. Appl. Ecol.* **24**, 231–237.

Wheeler, P.A. (1983). Phytoplankton nitrogen metabolism. In: *Nitrogen in the Marine Environment* (Ed. by E.J. Carpenter and D.G. Capone), pp. 309–346. Academic Press, New York.

Yoneyama, T. (1995). Nitrogen metabolism and fractionation of nitrogen isotopes in plants. In: *Stable Isotopes in the Biosphere* (Ed. by E. Wada, T. Yoneyama, M. Minagawa, T. Ando and B.D. Fry), pp. 92–102. Kyoto University Press, Kyoto.

Yoneyama, T. and Kaneko, A. (1989). Variations in the natural abundance of ^{15}N in nitrogenous fractions of komatsuna plants supplied with nitrate. *Plant Cell Physiol.* **30**, 957–962.

Yoneyama, T. and Sasakawa, H. (1991). Enrichment of natural ^{15}N abundance in *Frankia*-infected nodules. *Soil Sci. Plant Nutr.* **37**, 741–743.

Yoneyama, T., Fujita, K., Yoshida, T., Matsumoto, T., Kambayashi, I., and Yazaki, J. (1986). Variation in natural abundance of ^{15}N among plant parts and in ^{15}N/^{14}N fractionation during N$_2$ fixation in legume–rhizobia symbiotic system. *Plant Cell Physiology* **27**, 791–799.

Yoneyama, T., Kouno, K. and Yazaki, J. (1990a). Variation of natural ^{15}N abundance of crops and soils in Japan with special reference to the effect of soil conditions and fertilizer application. *Soil Sci. Plant Nutr.* **36**, 667–675.

Yoneyama, T., Murakami, T., Boonkerd, N., Wadisirisuk, P., Siripin, S. and Kouno, K. (1990b). Natural ^{15}N abundance in shrub and tree legumes, *Casuarina* and non-N$_2$-fixing plants in Thailand. *Plant and Soil* **128**, 287–292.

Yoneyama, T., Omata, T., Nakata, S. and Yazaki, J. (1991a). Fractionation of nitrogen isotopes during the uptake and assimilation of ammonia by plants. *Plant Cell Physiol.* **32**, 1211–1217.

Yoneyama, T., Uchiyama, T., Sasakawa, H., Gamo, T., Ladha, J. and Watanabe, I. (1991b). Nitrogen accumulation and changes in natural ^{15}N abundance in the tissues of legumes with emphasis on N$_2$ fixation by stem-nodulating plants in upland and paddy fields. *Soil Sci. Plant Nutrition* **37**, 75–82.

Yoneyama, T., Kamachi, K., Yamaya, T. and Mae, T. (1993). Fractionation of nitrogen isotopes by glutamine synthetase isolated from spinach leaves. *Plant Cell Physiol.* **34**, 489–491.

Yoshida, N. (1988). [15]N depleted N_2O as a product of nitrification. *Nature* **335**, 528–529.

Young, R.H.F., Lau, L.S., Konno, S.K., and Lee, H.H. (1976). *Water quality monitoring: Kaneohe Bay and Selected Watershed, July to December 1975.* Water Resources Research Center, Univ. of Hawaii, Honolulu, Hawaii.

Zak, D.R., Groffman, P.M., Pregitzer, K.S., Christensen, S. and Tiedje, J.M. (1990). The vernal dam: plant–microbe competition for nitrogen in northern hardwood forests. *Ecology.* **71**, 651–656.

Age-Related Decline in Forest Productivity: Pattern and Process

M.G. RYAN, D. BINKLEY AND J.H. FOWNES

ADVANCES IN ECOLOGICAL RESEARCH VOL. 27
ISBN 0–12–013927–8

Academic Press Limited, 1997

I. SUMMARY

An understanding of the pattern of forest growth is fundamental for determining the role of forests as sources and sinks in global carbon budgets. Such knowledge is also important for constructing mechanistic models of forest growth, for predicting the response of forests to changing climate, and for understanding the production and sustainability of commercial forests.

In even-aged forests, growth and biomass accumulation decline after reaching a peak (generally coinciding with the peak in stand leaf area) relatively early in a stand's life. The literature shows that this decline nearly always occurs, but the amount and timing of the decline vary. In general, more productive stands show an earlier peak growth and a steeper growth decline.

Initial work suggested that the fraction of assimilates available for wood production declined as woody biomass and respiration of the living cells increased. However, recent direct and indirect tests do not support this respiration hypothesis. Other potential causes include: (1) reduced photosynthesis because of the increasing hydraulic resistance of taller trees; (2) decreasing nutrient supply as a result of nutrient immobilization in living and decaying biomass, leading to lower leaf area, a shift in carbon allocation to root production, or reduced photosynthetic capacity; (3) reduced leaf area from abrasion in the crowns of taller trees with longer branches; (4) increased mortality of older trees; (5) physiological changes associated with changes in genetic expression (i.e, maturation of tissues); and (6) increased reproductive effort.

Reductions in wood production coincide with reductions in stand leaf area, and reduced leaf area is likely to be responsible for a portion of the age-related growth decline. Reduced growth efficiency (wood production per unit of leaf area) also coincides with growth decline and points to the existence of other mechanisms.

Foliage on older trees shows lower photosynthetic capacity and lower diurnal assimilation than foliage of younger trees. Either increased hydraulic resistance or reduced nutrient availability may decrease photosynthesis; current evidence supports the hydraulic mechanism. Allocation to fine roots may increase in older stands, but information is too limited for any general insights. While reduced leaf area, reduced photosynthesis, and increased fine-root allocation implicate reduced nutrient availability in older stands, the few available studies of nutrient availability show no clear trend of increasing or decreasing nutrient supply with stand age.

Juvenile and mature shoots differ in form and physiological behavior, and these differences generally cause grafted shoots from mature trees to grow less than those from juvenile trees. The bulk of maturational changes occur early in a tree's life, while declines in growth and growth efficiency occur in mature trees. Therefore, maturation may play only a limited role in age-related growth decline. Reproductive effort may increase with increasing tree age, but reproduction is sporadic, and the carbon costs of reproduction appear to be only a small portion of annual assimilation.

A syndrome of changes in structure and function are associated with age-related growth decline, but reduced leaf area and reduced photosynthetic capacity appear to be the most consistent features of the pattern. The causes of these reductions need direct examination and experimentation.

II. INTRODUCTION

The growth of forests typically changes with forest age, reaching a peak relatively soon in stand development followed by a substantial decline. Foresters have long been interested in these changes because they affect the economies of wood production. More recently, issues of global carbon cycling have focused the attention of scientists from many fields on the carbon fluxes of forests and how these fluxes change with forest age. These age-related changes in forest growth (and C flux) are fundamental to discussions of whether tropical forests are net sources or sinks for C (Lugo and Brown, 1992), whether conversion of old-growth forests to young plantations will increase rates of C storage (Harmon et al., 1990), and whether deposition of N will increase C storage in temperate forests (Kauppi et al., 1992).

Biomass accumulation and growth of even-aged forests follows a universal pattern as the trees increase in size. Growth is slow initially, increases as leaf area develops, peaks as leaf area reaches its maximum, and then declines for the majority of the stand's life span (Figure 1). The pattern was first identified in studies of forest growth and yield (Assmann, 1970), but also from ecological studies (Forrest and Ovington, 1970; Turner and Long, 1975; Tadaki et al., 1977; Gholz and Fisher, 1982; Long and Smith, 1992; Ryan and Waring, 1992). Mechanisms responsible for early growth and the growth peak may be well understood, but there is a lack of agreement about the cause of growth decline.

While much of the research in forestry and forest ecology has focused on aggrading forests, forests change in structure (Spies and Franklin, 1991) and function (Grier et al., 1981; Yoder et al., 1994) after dry matter production and leaf area reach their peak. Understanding why growth declines after reaching a peak is important for several reasons. First, mechanistic models of forest production require knowledge of changes in leaf area, nutrition, carbon allocation, and physiology to predict forest growth and function in varied environments (Ryan et al., 1996b). These models are an important tool for assessing the role of forests as terrestrial carbon sources and sinks in the global carbon budget and for understanding the response of forests (and terrestrial carbon storage) to global changes in climate and atmospheric CO_2. Second, regional or global estimates of forest productivity may be possible

M.G. RYAN *ET AL.*

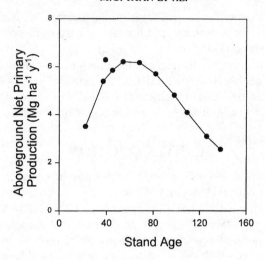

Fig. 1. Growth of *Picea abies* with stand age. Stands were in the vicinity of Karelia, USSR (62°N 34°E). Growing season length was 150 days, mean temperature during the growing season was 11.9°C, and growing season precipitation was 380 mm. From data in DeAngelis *et al.* (1980).

using remotely-sensed estimates of leaf area and climate coupled to models that estimate the efficiency of the conversion of absorbed energy to biomass (Landsberg *et al.*, 1996). Conversion efficiency varies with forest age, and understanding this variation will be vital for the success of these models. Third, because of their slower growth, older forests may be more susceptible to insects and pathogens (Waring, 1987), and may respond differently to changes in climate (Ryan *et al.*, 1996c). Finally, growth decline may be linked with the sustainability of commercial forest production (Murty *et al.*, 1996).

In this review, we re-examine the evidence for the pattern of growth decline with age and summarize the evidence for the mechanisms that may be responsible. We start with an overview of the proposed mechanisms. We then present a framework for understanding the changes in stand productivity with age, because many of the proposed mechanisms are linked and affect carbon allocation. We present the available information on the importance of various mechanisms behind growth decline, in the context of the stand carbon cycle. We conclude with suggestions for experiments that can distinguish among alternative hypotheses.

III. CAUSES OF DECLINING PRODUCTIVITY AS FORESTS AGE

The changes in productivity over time in a forest must derive from changes in components of:

$$NPP = \sum_{i=1}^{n} \varepsilon_i P_i S_i - R \qquad (1)$$

where NPP is net primary production, S_i is the supply rate of a required resource (light, nutrients, water), P_i is the proportion of S_i obtained, ε_i is the conversion efficiency for the resource, and R is respiration (Scurlock et al. 1985). Changes in NPP and stem growth may derive from a variety of processes that underlie the production equation, including changes in nutrient supplies, allocation among tree tissues, and stand structure; the hydraulic resistance to water flow through stems; and perhaps some genetic limitation of the physiology of older individuals.

The decline in growth rates as forests age has been commonly attributed to a changing balance between photosynthesis and respiration (Yoda et al., 1965; Whittaker and Woodwell, 1967). This hypothesis argues that leaf area will reach a plateau (limited by light, water, and nutrient supply), stand photosynthesis will remain constant (because of constant leaf area), but respiration will continue to increase as respiring biomass accumulates. Although the photosynthesis/respiration explanation of this pattern has received almost no verification for forest ecosystems, it is accepted almost as universally as the pattern itself (Kramer and Kozlowski, 1979; Waring and Schlesinger, 1985; Aber and Melillo, 1991; Oliver and Larson, 1996).

The focus on respiration as the key to growth declines assumed that the first three components of the production equation (supply of resources, acquisition of resources, and efficiency of resource use) remained constant as trees grew larger, leaving only respiration to balance observed declines in net primary production or stem growth. This argument can be traced back to early ecosystem research when woody biomass was found to be many times larger than leaf biomass. Respiration was thought to be a major fraction of gross primary production (60–65% according to Möller et al., 1954; Yoda et al., 1965), and later dimension analysis clearly showed that woody biomass and wood surface area increased much faster than leaf area per tree or stand (Whittaker and Woodwell, 1967).

Recent evidence does not support increasing respiration as the cause of growth decline. Careful estimates of woody-tissue respiration for a chronosequence of lodgepole pine stands suggested that increased maintenance respiration in older stands was too small to account for decreased growth rate. In fact, decreases in growth respiration resulted in lower total stem respiration in the oldest stand (Ryan and Waring, 1992). In addition, S.T. Gower et al. (unpublished data) found that woody-tissue maintenance respiration was similar across a chronosequence in a temperate deciduous forest. Indirect evidence against the respiration hypothesis has come from growth analysis of spacing studies, where trees at wider spacings continued to accelerate in growth rate while supporting higher wood:leaf ratios than closer spacings already in growth decline (Fownes and Harrington, 1990; Harrington and Fownes, 1996).

A variety of other explanations are possible, and several may be more consistent with available information:

(1) *Increased hydraulic resistance.* The total hydraulic resistance between the soil and leaves increases as trees grow taller (Mattson-Djos, 1981; Yoder *et al.*, 1994; Mencuccini and Graces, 1996) and branches lengthen (Waring and Silvester, 1994; Walcroft *et al.*, 1996). Increased resistance may force closure of stomata earlier in the day or earlier in a drought cycle (to prevent cavitation in xylem) and lower time-integrated photosynthesis (Yoder *et al.*, 1994; Ryan and Yoder, 1996).

(2) *Decreased nutrient supply.* The supply of nutrients from the soil may decline as available nutrients accumulate in biomass (Binkley *et al.*, 1995; Schulze *et al.*, 1995) or become immobilized during the decomposition of wood (Murty *et al.*, 1996; Gower *et al.* 1996a). Reduced nutrient supply can lower photosynthetic capacity (Waring and Schlesinger, 1985) and leaf area (Gower *et al.*, 1992, 1995, 1996b) and increase carbon allocation to fine roots (Gower *et al.*, 1995), all of which reduce carbon available for wood growth.

(3) *Reduced leaf area caused by crown abrasion.* Crown interaction increases with tree height, and such interaction may increase gaps between crowns and stand lower leaf area (Putz *et al.*, 1984; Long and Smith, 1992).

(4) *Increased mortality of older trees.* If the mortality of individuals increased after stand leaf area reached peak, net biomass production of the stand would decrease even if the growth of surviving trees was unaffected.

(5) *Increased reproductive effort.* The carbon costs of reproduction can be up to 15% of that of annual wood growth (Linder and Troeng, 1981). If seed production increases with tree age, the increased carbon sink might lower the carbon available for wood production.

(6) *Genetic changes with meristem age.* Genetic expression may change as tissue ages and the number of cell divisions increases (Greenwood and Hutchinson, 1993). These changes in 'maturation' may slow growth in both height and diameter.

The current paradigm that respiration accounts for declining production in older forests has rarely been tested, but the available evidence indicates that respiration changes probably account for only a small portion of the decline in growth (Ryan and Waring, 1992). Management of forests for wood production is often based on expected rates of productivity, and expectations of changes in those rates with stand age. If the decline in growth in older forests results from respiration increases or maturation, then the decline may be unavoidable unless tree genetics are changed. If the decline depends in part on changes in soil nutrient supply, then management activities that increase or decrease soil nutrient supply could substantially alter forest production, optimal rotation ages, and other aspects of stand management. Further, the response of existing forests to changing climate or increased nitrogen deposition may depend strongly on which components dominate the pattern of declining productivity with age. Rates of photosynthesis, respiration and soil nutrient mineralization are interdependent, but respond differently to changes in temperature, and soil moisture. Accurate predictions of forest responses to changing climate will require a functional understanding of the causes of declining forest growth in older stands.

IV. EMPIRICAL EVIDENCE FOR CHANGES IN STAND GROWTH

That growth declines in the latter stages of the life of a forest is a general principle in forestry. However, a systematic examination of the data to support growth decline and to identify mechanisms is lacking. In this section, we examine evidence for growth decline in studies of forest growth and yield, in ecological studies of forest production, and in ecological studies of individual trees.

A. Growth and Yield Studies

The classic pattern of stand growth decline was first identified in early studies of forest growth (see Assmann, 1970, for background). These 'growth and yield' studies used measurements of biomass in chronosequences of forest stands, together with the past growth of individual trees, to estimate standing biomass and growth of ideal stands under a variety of growing conditions. In our small survey of growth and yield studies, net growth rate measured between 30 and 100 years after the peak was, on average, 57% of the peak rate (Table 1; range = 19–100%). In another study, Grier et al. (1989) summarized biomass and growth information for 75 second-growth forests of Douglas-fir in the Pacific Northwest U.S. They concluded that biomass increment increases sharply during the first 20 years of stand development, then drops by 20–50% in the next 20 years, reaching a sustained level for the next 60 years.

Because tree height growth peaks earlier and declines more rapidly on higher-quality sites (Beck, 1971), we expected that biomass growth would also decline more rapidly on sites with higher quality. This pattern was found in four of the eight studies we examined (*Betula papyrifera, Pinus banksiana, Pinus ponderosa, Populus tremuloides*). In other studies, the rate of growth decline was unaffected by site quality (*Picea mariana*), increased with site quality (*Pinus contorta* var *latifolia*), or varied with site quality, but showed no consistent pattern (*Quercus alba*).

Table 1

Growth decline and site quality in forestry growth-and-yield studies

Species	Age at maximum Growth (year)	Reference age[1] (year)	Growth at reference age (% of peak growth)		Reference
			High site quality	Low site quality	
Betula papyrifera	55	90	19	62	Gregory and Haack (1965)
Picea mariana	70	110	24	24	Perala (1971)
Pinus banksiana	30	60	85	69	Rudolph and Laidly (1990)
Pinus contorta var. *latifolia*	30	110	71	28	Dahms (1964)
Pinus ponderosa	25	125	24	33	Meyer (1938)
Populus tremuloides (Lake States)	30	60	86	92	Perala (1977)
Populus tremuloides (Alaska)	40	90	100	100	Gregory and Haack (1965)
Quercus alba	30	60	48	50	Gingrich (1971)

[1]Age when a medium-quality site would be harvested under normal management. Further declines in growth occurred in all studies except *Populus tremuloides* (Alaska).

Yield tables are typically constructed using data from relatively few stands and are very subjective (Meyer, 1938). In these studies, growth is estimated from differences in accumulated biomass in different stands and not measured directly. Nonetheless, yield tables based on forest stand data overwhelmingly show that woody biomass accumulation slows as stands age.

Silvicultural studies of spacing and thinning effects are other sources of information on the pattern and controls of growth decline. Generally, the peak and decline in annual volume increment follows canopy closure (Evans, 1982) and occurs sooner at higher tree densities (Harrington and Fownes, 1996). Coppice stands, having many shoots per stump, may peak in biomass accumulation in one year or less under favorable year-round growing conditions (Harrington and Fownes, 1993). These and other plantation studies (Dudley 1990) support the universality of the pattern of peak and decline in even-aged stand growth.

B. Ecological Studies

A survey of chronosequence studies (Table 2, Figure 2) shows that wood production for all stands declines after reaching its peak coinciding with the peak in stand leaf area. Wood production at the end of the studied chronosequence was lower than 50% of peak for eight of the 12 studies (Table 2).

In section VI.A, we show that the declines in wood production often coincide with reductions in stand leaf area, and also that declining leaf area is normal as stands age. Because reduced leaf area will lower carbon assimilation, we wondered if the declines in wood production were solely caused by reduced leaf area. The influence of leaf area on production can be assessed by estimating the growth efficiency (Waring, 1983), or wood production per unit leaf area. If growth efficiency declines with stand age, then the reduction in wood growth exceeds the reduction in leaf area. Also, because light absorption varies exponentially with leaf area, marginal reductions in leaf area may actually improve growth efficiency. For 10 of the 12 chronosequences surveyed (the exceptions were *Pinus elliottii* and *Pinus radiata*), growth efficiency was lower at the end of the chronosequence than when the stand had maximum leaf area (Figure 3). For *Larix gmelinii* and *Eucalyptus grandis* (good site), the peak in growth efficiency did not coincide with the peak in stand leaf area. In high-density stands of fast-growing tropical trees, growth efficiency declined continuously with stand leaf area at leaf area indices above 1.0, both before and after peak biomass growth (Fownes and Harrington, 1990; Harrington and Fownes, 1996). These declines in growth efficiency convincingly show that reduced leaf area is not solely responsible for declines in wood production.

Table 2

Chronosequence studies: length of chronosequence, age at maximum leaf area, maximum leaf area or leaf biomass, and maximum annual above-ground wood production

Species	Range of chronosequence (years)	Age at maximum leaf area (years)	Peak leaf area ($m^2\ m^{-2}$)	Peak leaf biomass ($Mg\ ha^{-1}$)	Peak annual above-ground wood production	Reference
Abies balsamea	5-60	20		15.5	4.9 Mg ha^{-1}	Sprugel (1984)
Abies veitchii	3-127	37		6.8	6.2 Mg ha^{-1}	Tadaki et al. (1977)
Eucalyptus grandis (good site)	1-6	3.1		8.5	23.5 Mg ha^{-1}	Reis et al. (1985)
Eucalyptus grandis (poor site)	2-6	2.7		4.2	5.5 Mg ha^{-1}	Reis et al. (1987)
Eucalyptus regnans	50-230	50	3.4			Dunn and Connor (1993)
Larix gmelinii	49-380	49		4.3	3.3 mol m^{-2}	Schulze et al. (1995)
Picea abies	22-138	68	3.8		6.3[1] Mg ha^{-1}	DeAngelis et al. (1980)
Pinus contorta var. *latifolia* (Colorado)	40-245	40	3.8		168 g C m^{-2}	Ryan and Waring (1992)
Pinus contorta var. *latifolia* (Wyoming)	10-120	35	4.0		9.5 m^3 ha^{-1}	Long and Smith (1992)
Pinus elliottii	3-34	14		5.7	6.5 Mg ha^{-1}	Gholz and Fisher (1982)
Pinus radiata	3-12	7		11.6	17.6 Mg ha^{-1}	Forrest and Ovington (1970)
Pinus taeda	10-60	30		4.6	7.3 Mg ha^{-1}	Switzer et al. (1966)
Prunus pennsylvanica	1-14	6	3.1		16.6[1] Mg ha^{-1}	Marks (1974)

[1]Above-ground net primary production (includes foliage production of <25% of total).

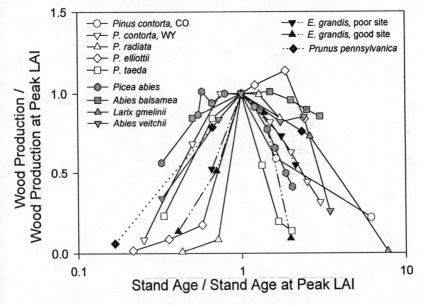

Fig. 2. Wood production relative to wood production when the stand had maximum leaf area plotted against stand age divided by stand age at stand maximum leaf area for 12 chronosequence studies. Actual values of wood production, length of the chronosequence, tree age at peak stand leaf area, and references for the chronosequence studies are given in Table 2.

C. Individual Trees

Comparing the growth of trees of different ages and sizes is difficult because larger trees have more leaf area, larger root mass, and generally produce more wood than smaller trees. Comparing growth efficiencies removes the effect of tree size. Kaufmann and Ryan (1986) found that growth efficiency of individual subalpine conifer trees declined with age in *Pinus contorta* var. *latifolia*, *Picea engelmannii*, and *Abies lasiocarpa*. Of the three species, *Pinus contorta* had the highest growth efficiency when young, but showed the greatest decline with age. Yoder *et al.* (1994) also found substantially lower growth efficiency in older *Pinus ponderosa* compared with adjacent younger trees.

D. Conclusions

We conclude that strong evidence exists (from both the forestry and ecological literature) for a decline in wood production relatively early in the development of stands. The decline in wood production coincides with a decline in leaf area, but concurrent declines in growth efficiency show that leaf area is

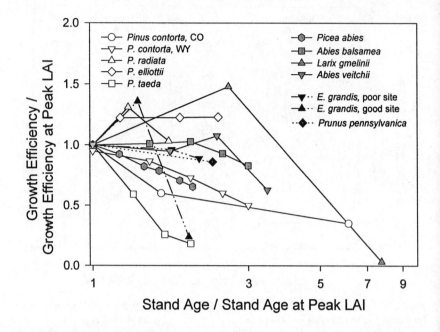

Fig. 3. Growth efficiency relative to growth efficiency when the stand had maximum leaf area plotted against stand age divided by stand age at stand maximum leaf area for 12 chronosequence studies. Only values after the stand reached maximum leaf area are plotted. Actual values of leaf area or biomass, wood production, length of the chronosequence, tree age at peak stand leaf area, and references for the chronosequence studies are given in Table 2.

not solely responsible for declining wood production. The timing and magnitude of declines in growth and growth efficiency vary widely, as may the processes underlying these changes.

V. CONTROLS OF BIOMASS PRODUCTION

In sections VI–X, we discuss the changes in function and structure that occur with forest stand development. Although we focus on specific processes, such as photosynthesis, nutrient supply, and respiration, we recognize that the processes are linked and can change in concert. To put these linkages in perspective, we briefly review the carbon cycle and its controls (Figure 4; see also Cannell, 1989; Cannell and Dewar, 1994; Waring and Ryan, 1995).

The ability to capture and use solar energy determines the amount of photosynthesis and plant carbon gain. Leaf area supplies the surface to absorb radiation, and leaf protein content, plant water status, and climate

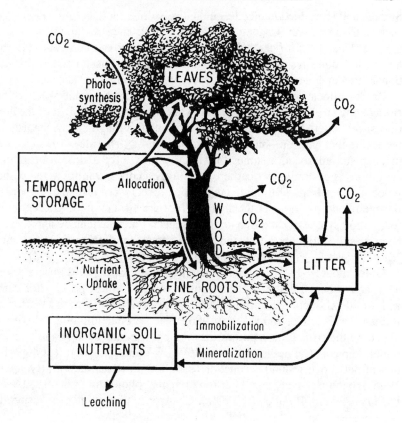

Fig. 4. Carbon balance of a forest ecosystem.

(temperature and humidity) largely determine the rate of photosynthesis. Maintaining existing cells and supporting growth of new cells requires energy, and a large portion of the energy fixed in photosynthesis is lost to respiration. Biomass production, therefore, results from the balance between photosynthesis and respiration. Only a small portion of the energy incident upon leaves is converted into sugars and then to biomass. For forests, 1 MJ of absorbed photosynthetically active radiation yields 1 g biomass, an efficiency of about 1.8% (Harrington and Fownes, 1995; Landsberg *et al.*, 1996). Of the carbon fixed in photosynthesis in forests, roughly 50–70% is lost to respiration of foliage, woody tissue, and fine roots and associated symbionts (Ryan, 1991).

The carbon remaining after respiration can be allocated to support (coarse roots, stemwood, branches), light capture (leaves), nutrient uptake (fine roots, root exudates, mycorrhizas), reproduction (seeds and fruiting structures), and storage (carbohydrates in ray parenchyma cells, fine roots and foliage).

Storage and reproduction may be qualitatively different than 'utilization sinks', such as tissue growth (Cannell and Dewar, 1994). The structural parts of a tree are interdependent; for example, leaves require water and nutrients from roots and a conducting system to supply them. Allocation is constrained by this functional interdependence so that plant parts are maintained in balance.

Carbon allocation is poorly understood, particularly in woody plants where storage and structure are large (Cannell and Dewar, 1994). Most of the attention of modellers has focused on root:shoot partitioning, and mechanistically based models can reproduce observed responses (Thornley, 1972a, b). The mechanistic approach assumes that relative growth of a sink (roots, shoots, cambium) is co-limited by carbon and nitrogen concentrations at the point of phloem unloading. Differential allocation occurs because the carbon and nitrogen concentrations at the shoot or root are a function of the supply rate (e.g., photosynthesis) and transport resistance. The mechanistic approach has been extended to trees with wood growth treated similarly to roots and shoots (Rastetter *et al.*, 1991; Thornley, 1991; Thornley and Cannell, 1992).

Other models (FOREST-BGC; Running and Gower, 1991) use a priority approach, where foliage and fine roots have the highest priority for carbon and wood production has the lowest priority for fixed carbon (Waring and Pitman, 1985; Oliver and Larson, 1996). Wood production might also be regulated by the structural requirements of supporting the canopy and transpirational demand (Cannell and Dewar, 1994). Wood production is a large fraction of net above-ground production (e.g., 80% in *Pinus radiata*; Ryan *et al.*, 1996a), but a relatively small portion of annual photosynthesis (e.g., 15–30% in *Pinus radiata*; Ryan *et al.*, 1996a). Changes in photosynthesis, respiration, or allocation could all potentially alter wood production.

The carbon cycle is strongly linked to water and nutrient cycles and any change in soil moisture or nutrition with forest development will likely alter both carbon uptake and allocation. Water and nutrient availability largely control leaf area and leaf photosynthetic capacity, and climate controls the soil moisture, temperature, and atmospheric humidity whic constrain photosynthesis. In addition to influencing carbon acquisition, nutrition can also alter carbon allocation by promoting changes in fine root biomass or turnover. Given the small fraction of the carbon budget used for wood production, many processes could interact to yield lower wood production with stand development.

VI. CHANGES IN PHOTOSYNTHESIS WITH STAND DEVELOPMENT

If photosynthesis declines as trees and stands develop, and the carbon costs of respiration and root and foliage production remain constant, wood production will also decline. Three mechanisms could lower photosynthesis in older trees and stands: (1) reduced leaf area, (2) reduced photosynthetic capacity, and (3) reduced photosynthetic performance. With lower leaf area, less surface area is available to absorb energy and fix carbon. Photosynthetic capacity controls the amount of carbon fixed per unit of light absorbed, so a lowered capacity will yield less carbon fixed under the same light regime. Photosynthetic 'performance' is the diurnal integration of photosynthesis for a leaf. Leaf area and photosynthetic capacity may be similar for leaves on young and old trees, but the leaves might show different patterns of assimilation if the soil to leaf hydraulic pathways differ. Photosynthesis is strongly linked with nutrition (because nutrition may control leaf area and photosynthetic capacity) and with respiration (because foliage respiration and photosynthetic capacity are coupled, and because the balance between photosynthesis and respiration determines dry matter production).

A. Reduced Leaf Area

Kira and Shidei (1967) defined the model of growth of leaf area after forest establishment for even-aged stands; a similar pattern was found in a model analysis of hardwood succession (Bormann and Likens, 1979). Leaf area develops slowly as stems build supporting structure and plant canopies occupy available space. Leaf area peaks relatively early in the life of a stand at a level greater than the final equilibrium value. In hardwood succession, the initial peak may result from co-existance of early-successional, short-lived species before full establishment of shade-tolerant, longer-lived species (Covington and Aber, 1980). In even-aged stands, a lower leaf area at equilibrium than at peak may result from crown abrasion in taller trees or intense competition. At equilibrium leaf area, mortality is matched by growth of new trees or crown expansion of existing trees.

Leaf-area data from chronosequence and monitored plots often shows a pattern that differs from the concept presented above. For example, most developing forests show a gradual, persistent decline after reaching a peak relatively early in the life of a stand, rather than a rapid post-peak decline in leaf area followed by equilibrium (Figure 5). Even though forests differ greatly in growth rates and peak leaf biomass, nearly all of the chronosequences in Figure 5 show a remarkably consistent *relative* pattern in leaf area development and decline. *Eucalyptus* and *Pinus radiata* (selected for commercial use) build leaf area very rapidly compared with other species, but leaf area also declines relatively more rapidly.

Decline in leaf area with stand development appears to be the rule, and equilibrium leaf area the exception. Of the 13 chronosequences plotted, only one (*Abies balsamea:* Sprugel, 1984) maintained relatively constant leaf area after reaching the peak. Another exception appears to be high-density stands of tropical trees, which decline in stand biomass growth without a decrease in

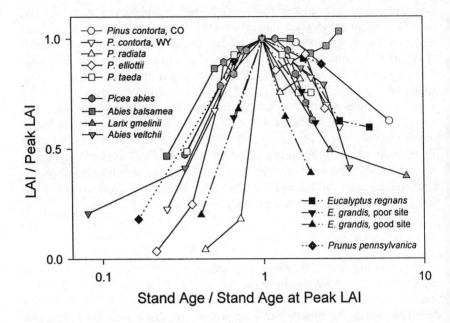

Fig. 5. Stand leaf area or biomass relative to maximum stand leaf area or biomass plotted against stand age divided by stand age at stand maximum leaf area for 13 chronosequence studies. Actual values of leaf area or biomass, wood production, length of the chronosequence, tree age at peak stand leaf area, and references for the chronosequence studies are given in Table 2.

stand leaf area (Harrington and Fownes, 1993, 1996). Extensive analysis of woody biomass and leaf area allometry rule out the possibility that these findings resulted from inappropriate extrapolation of leaf area equations (Fownes and Harrington, 1992; Harrington and Fownes, 1993), which could explain the lack of leaf area decline in some other studies (Fownes and Harrington, 1990).

A decline in leaf area through time will lower photosynthesis, decrease litterfall, increase light in the understory, and perhaps alter the understory microclimate. Because the relationship between leaf area and energy absorption is nonlinear, carbon fixation would not be expected to decline at the same rate as the decline in leaf area. For example, if a forest has a peak leaf area index of 4 (e.g., 40-yr-old lodgepole pine), a 25% decrease in leaf area may yield only a 9% decline in absorbed radiation (Ryan and Waring, 1992) if the light extinction coefficient remains constant. Increased radiation in the understory could increase the growth of understory plants that may compete with the trees for water and nutrients. Altered litter inputs and microclimate could affect decomposition and nutrient availability.

Three mechanisms could lower leaf area in a stand after the leaf area peak: altered nutrition, increased mechanical abrasion of crowns, and mortality – either as increased death of individuals or lack of replacement of dead individuals by new seedlings.

Several studies have shown that nutrient availability can fall in the late stages of the life of an even-aged stand (Grier *et al.*, 1981; Binkley *et al.*, 1995; Schulze *et al.*, 1995). Nutrient availability may decline if nutrient capital becomes bound in vegetation (primarily in wood) and unavailable for new growth (Schulze *et al.*, 1995). Because ratios of carbon to nutrients in woody tissues are much higher than in soil organic matter (> 200 versus 20–30), decomposition of woody litter (more common in the latter stages of a stand's life) will immobilize nutrients (Murty *et al.*, 1996).

Leaf area and nutrient availability are tightly linked (Nadelhoffer *et al.*, 1985; Gower *et al.*, 1992, 1995; Herbert and Fownes, 1995), but the precise mechanism is less clear. Reduced nitrogen availability generally increases allocation below-ground which increases the tree's ability to acquire nutrients (Gower *et al.*, 1995). Poor nutrient availability may also lower concentrations of photosynthetic enzymes in foliage which would lower photosynthetic capacity and annual carbon fixation. As carbon fixation declines or the carbon that is fixed is allocated to root production, less carbon may be available to build new foliage.

As tree height increases, crown interaction increases because the arc described by the top of a crown moving in the wind is larger. The resultant mechanical abrasion will prune crowns so that their shape becomes narrower and the gaps in between crowns widens (Putz *et al.*, 1984; Long and Smith, 1992). Foliage depth may be greater as gaps form between crowns, but increasing crown length may not compensate for the foliage lost in gap creation (Long and Smith, 1992). We speculate that crown abrasion may be a larger factor in sparser canopies because denser canopies should lessen the force of wind on individual trees.

A third possibility for explaining reductions in leaf area for an ageing stand is mortality coupled with a lack of replacement. In the earlier stages of stand development, survivors can occupy the space vacated by the death of a neighbor by increasing their crown radius. As trees grow larger, death may result in gaps so large that mechanical constraints on branch length prevent trees from fully exploiting the newly available resources. Seedlings may take many years to develop leaf area and reach the canopy in gaps. These unfilled gaps will lower stand leaf area.

B. Reduced Photosynthetic Capacity

Few studies have compared gas exchange in young and old trees, but those studies suggest that photosynthetic capacity or foliar nitrogen content (closely related to photosynthetic capacity; Field and Mooney, 1986) are often lower in foliage from old trees. Schoettle (1994) found that old bristle-cone pine (*Pinus aristala*) trees had lower photosynthetic capacity, stomatal conductance, and foliar nitrogen in 1-year-old foliage than in young trees. For sequoia (*Sequoiadendron giganteum*), Grulke and Miller (1994) found declines in maximum assimilation, dark respiration, stomatal conductance, and the ratio of intercellular to ambient CO_2 concentration on same-aged foliage from tree ages of < 1 to 2000 years. Kull and Koppel (1987) also found lower photosynthesis in Norway spruce (*Picea abies*) foliage from older trees, accompanied by lower foliar nitrogen content.

With a reduced photosynthetic capacity, older forests will fix less carbon per unit of light absorbed. Litter quality may also be poorer, and litter with a high C:N ratio may immobilize more nitrogen as it decomposes. Lower nitrogen availability may promote lower foliar nitrogen and further decrease photosynthetic capacity. Lower photosynthetic capacity could therefore initiate positive feedback that will continue to force itself downward. This hypothesis has not been tested experimentally, and the mechanism for initiating reduced photosynthetic capacity remains unknown. We note, however, that differences in litter quality within species have not been demonstrated to influence nutrient supply (Prescott, 1995).

C. Reduced Photosynthetic Performance

Foliage from young and old trees can have similar maximum photosynthesis rates but a different diurnal pattern of photosynthesis. For example, Yoder *et al.* (1994) observed that the diurnal pattern of photosynthesis differed between same-aged foliage from young and old trees in two species of conifers (Figure 6). The differences in photosynthesis did not appear to be caused by differences in photosynthetic capacity because foliar nitrogen contents and peak photosynthesis rates were similar. However, stomata on the needles from old trees closed earlier in the day, leading to a lower diurnal total carbon assimilation. Fredericksen *et al.* (1996) found that the diurnal pattern of photosynthesis also differed between foliage from young and old cherry (*Prunus serotina*) trees.

Differences in water use or carbon fixation between young and old strands have also been measured. In a study of 40- and 140-year-old Norway spruce, whole-tree sapflow measurements showed that transpiration was about 25% lower for the older trees in dry air (Tenhunen *et al.*, 1996). These differences between individual young and old trees scale to substantial differences at the stand level and are even more striking because the leaf area was about 50% greater in the older stand. The lower transpiration and higher leaf area in the old forest indicate a lower stomatal conductance. If stomatal conductance is lower, photosynthesis will also be lower. Another recent study (Sellers *et al.*,

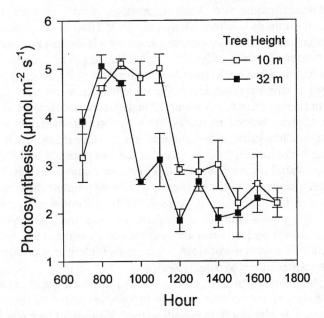

Fig. 6. Diurnal pattern of photosynthesis of 1-year-old foliage from 55-year-old (10 m tall) and 229-year-old (32 m tall) *Pinus ponderosa* foliage from the upper one-third of the canopy. Redrawn from Yoder *et al.* (1994).

1995) compared eddy-correlation measurements of CO_2 and energy fluxes between young and old *Pinus banksiana* forests in northern Canada. Preliminary results from this study indicate that midday net CO_2 uptake and latent energy loss (i.e., transpiration for a dry canopy) were consistently higher for the young stand. Again, the older stand had higher leaf area, suggesting that the flux differences resulted from differences in photosynthetic performance or capacity.

Yoder *et al.* (1994) speculated that differences in photosynthetic performance between young and old trees might be caused by increased hydraulic resistance in taller trees with longer branches. The link between stomatal conductance and the hydraulic resistance of a tree's vascular system has been demonstrated (Sperry *et al.*, 1993; Whitehead *et al.*, 1996) and there is evidence that hydraulic resistance increases with height (Yoder *et al.*, 1994) and branch length (Waring and Silvester, 1994). Sperry and colleagues (Sperry and Pockman, 1993; Sperry *et al.*, 1993; Sperry, 1995) showed that stomata will close to maintain leaf water potential above a level that would cause irreversible cavitation in xylem. Experimental increases in hydraulic resistance have induced stomatal closure without affecting leaf water potential (Sperry and Pockman, 1993). These lines of evidence indicate that stomata may close to restrict transpiration, not simply in response to dry air (see Monteith, 1995). Photosynthesis declines as stomata close (Ball *et al.*, 1987), so any diurnal reduction in stomatal opening caused by a hydraulic limitation in the vascular system decreases daily carbon assimilation by the canopy.

Hydraulic limitation could contribute to the slowing of the growth of individual trees as size increases. Hydraulic resistance increases with the overall path length (height of tree and length of branches) and decreases as the permeability of the sapwood increases. As a tree grows, hydraulic resistance increases as the tree gains height and branches lengthen (Mattson-Djos, 1981; Mencuccini and Grace, 1996). Because of the link between hydraulic resistance and stomatal opening, and the link between stomatal opening and photosynthesis, a larger hydraulic resistance will likely promote different diurnal and perhaps different seasonal patterns of carbon assimilation – patterns that result in lower total assimilation. If foliage on older trees photosynthesizes less, wood growth will be lower because the other carbon costs (respiration, foliage production, root growth) are similar to or slightly greater than those in younger trees (Ryan and Waring, 1992).

Hydraulic limitation also suggests a positive feedback that could ultimately limit tree height: if trees lose a portion of sapwood each year (to heartwood formation) and less new xylem is built as trees gain height (because less carbon is fixed), sapwood permeability and the effective conducting area of sapwood will be reduced. Therefore, the hydraulic resistance might progressively deteriorate to slow and ultimately stop vertical growth. We can speculate that reduced diurnal photosynthesis near the tops and branch tips of older trees promotes reduced allocation of nitrogen to that foliage and lower photosynthetic capacity.

D. Conclusions

In general, increases in the size and age of trees and stands lead to a syndrome of changes that tend to reduce carbon assimilation. At the stand level, leaf area decreases after reaching a peak rather early in a stand's life, leading to less light absorption and lower photosynthesis. At the tree level, leaf area may still increase as trees grow in size, but photosynthetic capacity or performance may decrease. These changes in photosynthesis result in less carbon being available for wood production.

VII. CHANGES IN NUTRIENT SUPPLY

Larger forests tend to have more nutrients accumulated in biomass and many case studies have documented that the rate of increase in nutrient storage in biomass decreases in older stands. For example, Pearson *et al.* (1987) showed that accumulation rates for nitrogen and other nutrients peaked near the time of peak stemwood increment and then declined. The declining rate of accumulation of nutrients in older forests could result from a declining ability to use the nutrients, or from a declining nutrient supply.

Nutrients are most usually available at the initiation of an even-aged stand because of reduced plant uptake, soil disturbance, or fertilizer application (Binkley, 1986). Soils may also be warmer (from increased insolation) and wetter (from decreased interception and transpiration) early in stand development, favoring microbial activity. In addition, litter inputs are decreased, therefore decreasing the microbial immobilization of N and resulting in less competition between trees and microbes for available soil N. Nutrient availability may decline with stand age as soils become cooler and drier, heterotrophic microbes compete more effectively with trees for nutrients, and nutrients become increasingly stored in the heartwood of live trees, in dead trees, and in woody litter.

Nutrient supply and NPP are linked (Miller, 1984; Nadelhoffer *et al.*, 1985; Waring and Schlesinger, 1985) and changes in nutrient supply often result in changes in NPP (Gower *et al.*, 1992). Any decrease in nutrient supply with stand age may reduce wood accumulation rates both by decreasing assimilation (through reductions in leaf area and leaf area efficiency) and by reducing allocation to wood growth as root production increases to scavenge for nutrients.

Does nutrient supply decline with stand age? In a review of biogeochemical budgets and ecosystem succession, Gorham *et al.* (1979) assumed that declining net ecosystem production in older stands would lead to increased outputs of nutrients as the ability of the stand to use nutrients declined. Vitousek and Reiners (1975) documented higher nitrate concentrations in some streams draining old stands of red spruce (*Picea rubens*) relative to younger stands, suggesting that either the nutrient supply was higher in the old stands, or that the forests were indeed less capable of using nutrients. Chapin *et al.* (1986) noted that old individuals may have a lower ability to respond to increasing nutrient supply than younger individuals of the same species; old aspen (*Populus tremuloides*) stands did not respond as well to fertilization as younger stands. These data contradict the expectation of increasing nutrient limitation in older stands.

Many studies have examined changes in soil nutrient supplies shortly after major disturbances (such as clearcutting), and a few chronosequence studies have documented patterns that might occur with stand development. In this section, we discuss a wide variety of case studies and then discuss the extent to which changes in nutrient supply might explain declines in stemwood production.

A. Comparisons of old and regenerating forests

Several studies have found that nutrient mineralization rates in regenerating forests can be equal to or greater than that in adjacent old forests (Vitousek and Denslow, 1986; Matson *et al.*, 1987; Zou *et al.*, 1992). Vitousek and Denslow (1986) examined net N mineralization and acid-fluoride-extractable phosphate in intact lowland rainforests and forest gaps at La Selva Biological Station in Costa Rica. They found no evidence of differences in N supply and only slight, nonsignificant increases in phosphorus in the gaps. Matson *et al.* (1987) also examined net N mineralization following logging and burning at another site in Costa Rica. In the first year after treatment, net N mineralization was elevated, but a 5-year post-cutting site showed little difference from the old-forest site. Zou *et al.* (1992) examined N mineralization in two adjacent forests in the La Selva and found no differences in net or gross mineralization or nitrification between the old-growth forest and a 4-year-old successional forest.

Several case studies have documented greater nutrient mineralization in regenerating forests than in old forests. For example, Frazer *et al.* (1990) compared net N mineralization in an old-growth mixed-conifer forest in the Sierra Nevada of California with adjacent stands of 5- and 17-year-old trees. The youngest site had the highest net N mineralization rates, followed closely by the 17-year-old site. The old-growth forest had less than 20% of the net N mineralization rate of the younger forests. In wind-created gaps, net N mineralization 3 years after gap formation declined by about 15% in sugar maple (*Acer saccharum*) forests but increased 20% in hemlock (*Tsuga canadensis*) forests (Mladenoff, 1987).

Clearcutting and site preparation (e.g. burning, plowing) tend to elevate nutrient availability compared with adjacent, older forests. Studies in loblolly pine (*Pinus taeda*) found that *in situ* rates of net N mineralization (Vitousek and Matson, 1985; Fox *et al.*, 1986) and nitrification (Vitousek and Matson, 1985) increased substantially with the intensity of logging disturbance. Krause and Ramlal (1986) used ion exchange resin bags to examine the effects of clearcutting and site preparation on nutrient supply in a balsam fir and black spruce (*Picea mariana*) stand. In the first summer after treatment, the resins collected about twice as much nitrate and phosphate in the cut site than in the uncut forest. In British Columbia, Binkley (1984) compared adjacent cut and uncut stands at three elevations on Vancouver Island. Laboratory incubations indicated no differences in net N mineralization between cut and uncut sites, but on-site resin bags indicated much greater N availability in cut sites. Burger and Pritchett (1984) examined the effects of clearcut logging of slash pine (*Pinus elliottii*) and longleaf pine (*Pinus palustris*) in Florida with a range of subsequent site preparation activities. Aerobic laboratory incubations showed that potential net N mineralization decreased in the sequence from uncut forest to cut-and-burned treatment, to the cut, burned-and-plowed treatment. However, simulations based on differences in soil temperature and moisture suggested that *in-situ* rates of net N mineralization may not differ among treatments.

B. Comparisons of Age Sequences of Forests

Measurements of nutrient mineralization in forest chronosequences show varied patterns of nutrient availability with stand age. Nutrient availability can decrease (Matson and Boone, 1984; Davidson *et al.*, 1992; Binkley *et al.*, 1995), increase (Sasser and Binkley, 1989), or be lowest in middle-aged stands, compared with younger or older stands (Ryan and Waring, 1992; Olsson, 1996).

Matson and Boone (1984) examined net N mineralization in a wave-pattern regeneration sequence in a mountain hemlock (*Tsuga mertensiana*) ecosystem in Oregon, an ecosystem with very low soil N content. Net mineralization doubled from the old-growth stage into the dead-forest zone, and then declined back to the old-growth stage over the next several decades. Ion exchange resin bags also showed declining soil N supply with age, while K and P supplies appeared to increase (Waring *et al.*, 1987). In this case, the decline in soil N may lead not only to declining growth, but to increasing susceptibility to root pathogens because of lack of carbon with which to fight infection.

Binkley *et al.* (1995) examined relationships between nutrient availability, nutrient limitation, and stem growth in a replicated age sequence of lodgepole pine (*Pinus contorta*) in Wyoming. Net nitrification in in-field incubations was more than four times greater in stands younger than 43 years than in older stands. The oldest stands (112–117 years) had significantly lower net N mineralization than younger stands. No trends were apparent in measures of soil P, Ca, Mg, or K. Single-tree fertilization trials indicated that nutrient supply did not limit leaf area or stem growth of forests less than 43 years old, whereas N supply limited leaf and stem growth in older forests. Olsson (1996) used ion exchange resin bags to examine soil N supply over a longer age sequence in lodgepole pine in the same area, and found that N supply was high in forests younger than 50 years, low in forests between 50 and 100 years, and increased in forests older than 200 years. Ryan and Waring (1992) found a pattern similar to that of Olsson (1996) for lodgepole pine in Colorado. While no statistical difference in nitrogen availability existed among adjacent 40-, 60-, and 245-year-old stands, mean N availability was greater in the 245-year and 40-year stands than the 60-years stand. Foliar N reflected N availability, with the 60-year stand having the lowest foliar N.

Sasser and Binkley (1989) examined net N mineralization in two age sequences of fir (*Abies fraseri* and *A. balsamea*) generated by a wave-pattern regeneration sequence. In an environment with high rates of N deposition from the atmosphere (> 10 kg N ha^{-1} yr^{-1}), net N mineralization was high in the dead zone, low to moderate in the regeneration and juvenile stages, and high in the mature (*c.* 80 year old) stands. Sprugel (1984) found that aboveground NPP remained high in balsam fir forests for at least 60 years; bole respiration remained constant, and N supply remained high.

Davidson *et al.* (1992) provided one of the most thorough examinations of the processes behind changes in N turnover in forests of different ages. Net mineralization rates were somewhat higher in a 10-year-old mixed conifer plantation than in a nearby old-growth mixed conifer forest. This higher rate of net mineralization resulted from lower microbial immobilization of the mineralized N (based on ^{15}N pool dilution) in the younger stand, rather than from a higher rate of N release from organic pools. The rate of gross mineralization (total N released from organic pools) was about two to three times higher in the old-growth forest. The younger forest also had four to ten times as much net nitrate production as the old-forest, but again this resulted from lower microbial immobilization of nitrate rather than from greater production of nitrate.

Herbert (1984) followed the productivity of black wattle (*Acacia mearnsii*) plantations through three rotations with and without supplemental fertilization. In all rotations, the rate of basal area (and volume) growth declined after about the first five years. Fertilization increased growth rates but had no effect on the overall pattern of growth with stand age.

C. Old-field Succession Sequences

No clear pattern of nutrient availability and forest age emerges from the literature on old-field succession. Montes and Christensen (1979) examined an age sequence in North Carolina of an abandoned agricultural site (old-field), a 40-year-old loblolly pine stand, and an old-growth oak-hickory stand. Rates of net mineralization and nitrification were higher in the old-field and in the old-growth stand than in the intermediate-aged pine stand. A more thorough, replicated study by Christensen and MacAller (1985) found no consistent trend in soil N supply with successional stages from old-fields through hardwood stages. Pastor *et al.* (1987) examined four old-field sites in Minnesota ranging in age from 16 to > 100 years after agricultural abandonment. Net N mineralization increased through the age sequence. The proportion of total soil N mineralized declined with time, however, suggesting a declining quality of organic matter over time. Robertson and Vitousek (1981) examined net mineralization and nitrification rates in an old-field succession sequence in New Jersey. Both rates tended to increase with time, reaching a maximum in the old-growth forest stage. Thorne and Hamburg (1985) examined an old-field sequence in New Hampshire. They found no relationship between several chemical measures of labile N pools with stand age, although net nitrification did decline with stand age.

D. Conclusions

Nutrient supply (particularly N) does increase after disturbances in most cases, but this is not a universal pattern. Beyond the initial disturbance responses, no clear trend of increasing or decreasing nutrient supply was apparent in the case studies we examined. Two case studies with fir forests in the eastern United States showed increasing N supply with stand age (Sasser and Binkley, 1989), and case studies with lodgepole pine in the Rocky Mountains showed higher N supply in young and old forests than in interme-diate-aged forests. The ^{15}N tracer work of Davidson *et al.* (1992) demonstrat-ed that commonly measured indexes of soil N supply may miss important dynamics in the soil N cycle. We conclude that, in some cases, changing nutrient supply plays a role in the declining growth of older forests, but this explanation is not universal.

VIII. CHANGES IN RESPIRATION

Respiration by the living cells in foliage, fine roots, and wood (ray parenchyma cells in sapwood) can use 50–70% of the carbon fixed in photosynthesis (Ryan, 1991; Ryan *et al.*, 1994b). Because of the large respiration costs reported in forests, early hypotheses about the decline of wood production with forest age focused on respiration. The reasoning was: (1) leaf area, and presumably the capacity of a forest canopy to assimilate carbon, peaks early in a stand's life, but (2) wood production declines as woody biomass increas-es, and (3) respiration increases with biomass in crop plants (Thornley, 1970). Therefore, respiration of the woody biomass was thought to use an increasing fraction of assimilation, and wood production declined (Yoda *et al.*, 1965; Kira and Shidei, 1967). A similar reasoning was used by Whittaker and Woodwell (1967), except they used surface area rather than woody biomass. The 'respiration hypothesis' is the textbook explanation for productivity decline (Kramer and Kozlowski, 1979; Waring and Schlesinger, 1985; Cannell, 1989), although only indirect evidence exists (Whittaker and Woodwell, 1967; Waring and Schlesinger, 1985) to support the explanation.

Direct measurements and estimation from published respiration rates argue against the role of woody-tissue respiration in causing age-related pro-ductivity decline. Respiration costs were similar (Ryan and Waring, 1992) for an old lodgepole pine forest and an adjacent younger stand (88 g C m^{-2} year^{-1} in a 245-year-old stand versus 103 g C m^{-2} year^{-1} in a 40-year-old stand), even though wood production was substantially lower in the old forest (37 g C m^{-2} y^{-1} in a 245-yr-old stand versus 168 g C m^{-2} yr^{-1} in a 40-year-old stand). Similarly, S. T. Gower *et al.* (unpublished data) found little difference in woody-tissue respiration costs in a chronosequence in a temperate deciduous forest. Another study (Sprugel, 1984) used published respiration rates and allometric equations to estimate branch and bole respiration for a balsam-fir chronosequence and found nearly constant woody-tissue respiration after age 25 years, while bole wood production declined from 325 to 210 g m^{-2} yr^{-1}. In the balsam-fir chronosequence, bole surface area actually declined with stand

238   M.G. RYAN *ET AL.*

age as fewer (but larger) trees and branches replaced the more numerous smaller trees and branches. Finally, maintenance respiration for woody tissue in old and young stands of *Tsuga heterophylla* and *Pseudotsuga menziesii* on highly productive sites was similar, and respiration was not implicated in the low wood production in the old stand (Runyon *et al.*, 1994).

Other indirect evidence argues against woody-tissue causing productivity decline. If respiration is partitioned into the functional components of growth and maintenance (see Amthor, 1986), growth respiration will decline along with wood growth. Maintenance respiration rates of woody tissues generally appear to be low (Ryan *et al.*, 1994b), and maintenance respiration of wood uses a small fraction (5–12%) of annual carbon fixation (Ryan *et al.*, 1994a, 1995). Growth analysis of spacing studies showed that trees at wider spacings (with higher wood:leaf ratios) continued to grow and have higher growth efficiency, while trees with narrower spacings declined in growth and growth efficiency (Fownes and Harrington, 1990; Harrington and Fownes, 1996). Finally, extrapolating rates based on surface area may be biased because of the variability caused by the underlying sapwood (Ryan, 1990).

Maintenance respiration rates are low for woody-tissues, because the stems contain few living cells and these have low activity (Ryan, 1990); also, in many trees the fraction of sapwood (which contains the living cells) can be only a small portion of the total woody biomass. The sapwood fraction appears to be high in small, drought-tolerant trees and low in large, drought-intolerant trees. Because of this trend, little of the woody biomass in forests with high biomass is in sapwood. For example, the above-ground sapwood fraction for a tropical wet forest was estimated as 23% of total above-ground biomass (Ryan *et al.*, 1994a).

Although respiration rates for foliage and fine roots are much larger than those of woody tissue (Ryan *et al.*, 1994b; Sprugel *et al.*, 1995), the fraction of assimilation used by foliage and fine root respiration is unlikely to change with forest development. If leaf area and photosynthetic capacity remain constant after canopy closure, foliage respiration is unlikely to change. If leaf area declines (see section VI.A), foliage respiration as a fraction of assimilation is likely to decline, not increase (Ryan *et al.*, 1994b). Fine root biomass and nutrient content largely determine fine root respiration (Ryan *et al.*, 1996a), but potential changes in fine-root biomass with stand development remain largely unknown. At a coarse (global) scale, fine root biomass may correlate with foliage (Raich and Nadelhoffer, 1989), but at the ecosystem level, relationships between foliage and fine roots are strongly dependent on nutrition and water (Gower *et al.*, 1996b). In forest development, changes in root turnover or carbon allocation to root exudates and mycorrhizas may be more important than changes in respiration of fine roots (see section IX.B).

IX. CHANGES IN ALLOCATION

A peak and decline in stem growth with stand age could also result from a constant rate of net primary production but reduced allocation to stem production. For example, a comparison of young and old *Abies amabilis* stands showed allocation to fine roots increased as a fraction of NPP (Grier *et al.*, 1981). A similar trend was observed for an age sequence of slash pine (Gholz *et al.*, 1986), where both the relative and absolute allocation to roots increased with stand age.

Another possibility is that other turnover processes increase with stand age. When biomass accumulation rates are estimated by successive stand inventories, stem mortality is often not included, resulting in an underestimation of actual stem production. As tree crowns converge during canopy closure, lateral branches may become suppressed, resulting in a rearrangement of the leaf area distribution within crowns (Ford, 1984). In a model of *Eucalyptus globulus* growth, the fractional leaf turnover rate increased from 0 to 0.5 year^{-1} at 6 years (unfertilized) and 4 years (fertilized) to simulate this effect (Linder *et al.*, 1985). Patterns in NPP allocation also may interact strongly with nutrient availability.

A. Per Tree

The factors that directly control the allocation of photosynthate in individual trees remain largely unknown; important factors include genetic, biochemical, and physiological regulation of transport from sources and sinks (Cannell and Dewar, 1994; Gower *et al.*, 1995), and competition among trees (Nilsson and Albrektson, 1993). At a coarser scale, a great deal is known about biomass production of stems, leaves, and in some cases roots. Growth of stemwood raises the tree's canopy above competitors; stern growth also responds to wind stress (Coutts, 1983). Allocation to leaf production depends in part on availability of site resources, such as water and nutrients (Cannell and Dewar, 1994; Gower *et al.*, 1995), and the acquisition of these resources depends on allocation to fine roots and mycorrhizas. Young trees allocate more photosynthate to foliage production than wood production, and this ratio declines as canopy closure develops. Allocation to large support roots often follows a simple allometric relationship, such as 20% of total tree biomass (Jackson and Chittenden, 1981; Gower *et al.*, 1995). Trees that experience strong competitive stress may allocate larger fractions of assimilate to stem wood production. For example, Nilsson and Albrektson (1993) found that suppressed Scots pine (*Pinus sylvestris*) trees showed greater stem growth per gram of needle mass than dominant trees, whereas allocation to needles and branches did not differ between competitive classes. Higher wood growth for suppressed conifers was also found by Kaufmann and Ryan (1986).

B. Changes in Fine Root Production and Longevity

Fine root production is difficult to measure, and the strengths and weaknesses of the major approaches have been debated extensively (Fairley and Alexander, 1985; Nadelhoffer *et al.*, 1985; Kurz and Kimmins, 1987; Santantonio and Grace, 1987; Raich and Nadelhoffer, 1989). Most studies that focused on root production omitted the photosynthate requirement for root construction and maintenance respiration (Ewel and Gholz, 1991), which may be 50% of below-ground allocation (Ryan, 1991). Some generalizations about root production appear to be supported from research over the past two decades, and these patterns depend strongly on scale. At a global scale, the allocation of photosynthate to fine root production and respiration generally increases with increasing ecosystem productivity (Raich and Nadelhoffer, 1989). At a regional scale, fine root production (and probably respiration) may (or may not) increase with increasing N availability. For example, Nadelhoffer *et al.* (1985) used estimates of net mineralization and N uptake and allocation to above-ground tissues to estimate what fine root production must have been to balance the N budget. Across nine stands (and nine species) in south-central Wisconsin, they estimated that root production must increase with increasing rates of net N mineralization in the soils. This calculation, coupled with decreases in standing biomass of fine roots on high N sites indicated that root longevity must be much shorter on richer sites (a finding confirmed by Pregitzer *et al.*, 1993).

Within single species, the pattern of root production appears to be the opposite of the regional and global patterns. Studies of fine root production that compared the same species growing in high and low productivity sites generally found equal or greater fine root production and increased relative below-ground allocation on the poorer sites (Keyes and Grier, 1981; Santantonio and Hermann, 1985; Beets and Pollock, 1987; Comeau and Kimmins, 1989; Gower *et al.*, 1996a; but see Nadelhoffer *et al.*, 1985). Most of these studies attribute differences in site productivity to water supply, but nutrient supply may also be important. Greater nitrogen availability appears to reduce allocation to fine roots even if plants become deficient in other nutrients (Ericsson 1995).

Several intensive experiments have also demonstrated lower absolute and relative below-ground production following fertilization of Scots pine (*Pinus sylvestris*; Linder and Axelsson, 1982), Norway spruce (*Picea abies*, Persson *et al.*, 1995; Clemensson-Lindell and Persson, 1995), Douglas-fir (*Pseudotsuga menziesii*; Vogt *et al.*, 1990; Gower *et al.*, 1992), and loblolly pine (*Pinus taeda*; H.L. Allen and P. Dougherty, personal communication). None of these fertilization experiments examined patterns with stand age; however, if soil fertility changes with stand development (and the evidence is mixed, see section VII), below ground allocation would also likely change.

Only a few studies examined root biomass and production with stand age. Berish (1982) found that fine root biomass (production was not examined) increased from 220 to 1300 to 1550 kg ha^{-1} in an age sequence of 1-, 8- and 70-year-old successional forests in Costa Rica. Vogt *et al.* (1983) looked at fine root biomass (but not production) on two age sequences of Douglas fir. Stands younger than 15 years had low fine root biomass on both types of sites, but no further trends were evident with stand age. Grier *et al.* (1981) compared 23-year-old and 180-year-old stands of Pacific silver fir (*Abies amabilis*) and found that ecosystem NPP was similar between the stands (about 17 to 18 Mg ha^{-1} yr^{-1}), as was total below-ground NPP (about 12 Mg ha^{-1} yr^{-1}). The major difference between the stands was the greater contribution of shrubs in the production budget of the younger stand. Comparing only tree production, the younger stand showed much greater above-ground production and lower below-ground production (especially of fine roots) than the older stand. Fine root production by trees in a 27-year-old slash pine (*Pinus elliottii*) stand was substantially higher than in younger stands (Gholz *et al.*, 1986). However, the total below-ground production (trees + understory) of the younger stand matched that of the older stand.

Some current work in lodgepole pine forests in Wyoming has found that below-ground allocation (as measured by the difference between soil respiration and above-ground litter inputs) declines substantially in old stands, but the timing of the decline does not match the timing of the decline in above-ground NPP (S. Resh and F.W. Smith, personal communication, 1996). Recent work with loblolly pine in South Carolina is demonstrating a strong response of proportional allocation below-ground in response to fertilization and irrigation, but very little response in actual below-ground NPP (NCSFNC, 1995; H.L. Allen and P. Dougherty, personal communication, 1995). Irrigation increased NPP of a young plantation by about 11%; allocation to roots declined from 41% in control plots to 34%. Fertilization increased NPP by 65%, with just 26% going below-ground. Fertilization plus irrigation more than doubled NPP, with just 21% allocated below-ground. The actual below-ground NPP (Mg ha^{-1} yr^{-1}) varied little (±10%) among treatments.

The paucity of studies on root production with stand age allows no generalization of trends. Any decrease in soil N supply with stand age would probably increase relative allocation to fine root production, but the evidence for a decline in N supply with stand age is mixed. Studies with Pacific silver fir (Grier *et al.*, 1981) and slash pine (Gholz *et al.*, 1986) indicated a major role for root dynamics of understory species; changes in root production by competing species can alter resource supplies to trees and presumably alter allocation patterns of the overstory.

C. Allocation to Symbionts

Most tree species develop obligatory or facultative symbioses with mycorrhizal fungi (Allen, 1991), and a few genera are capable of symbiotic N fixation in association with prokaryotic bacteria (Werner, 1992). The mycorrhizal symbioses are usually viewed as mutualistic associations where fungi receive carbohydrates from the trees and the trees receive resources such as N, P, and water. The carbohydrate cost of sustaining the mycorrhizal network can be substantial and probably varies among species and across resource gradients (such as soil P supply). While complete carbon budgets for mycorrhizas have not been developed, allocation to symbionts will change with stand age only if mycorrhizal growth is sensitive to soil nutrition and if soil nutrition changes with stand age.

Mycorrhizas can use up to 20% of assimilation (Soderstrom, 1991; Ryan *et al.*, 1996a), but the carbon cost can be offset by increased assimilation. For example, Reid *et al.* (1983) provided $^{14}CO_2$ to Scots pine seedlings with and without mycorrhizas. After 10 weeks, the mycorrhizal seedlings had fixed 2.5 times as much C as the non-mycorrhizal seedlings and accumulated 2.1 times as much biomass. Respiration as a fraction of total assimilation was much greater in the mycorrhizal seedlings (30%) than in the non-mycorrhizal seedlings (18%).

Early work on mycorrhizas suggested that mycorrhizal associations declined as the external supply of nutrients increased (Marx *et al.*, 1977), but the solution concentrations of nutrients were often one or more orders of magnitude higher than those encountered in forest soils. Mycorrhizal associations probably remain important across normal ranges of soil fertility, but low supplies of soil P may increase proportional allocation of C to mycorrhizas (cf. Ekblad, 1995). Termorshuizen (1993) examined the effects of N fertilization (at rates up to 60 kg N ha^{-1} yr^{-1}, for 3 years) on mycorrhizas, and concluded that high supplies of N reduced production of fungal fruiting bodies, but did not change the frequency of mycorrhizal infection of root tips or the volume of mycorrhizal hyphae per unit of soil volume. Similarly, Soderstrom (1991) found that fertilization with up to 600 kg N ha^{-1} did not alter the proportion of root tips colonized by mycorrhizas.

We know of only one study that attempted to characterize the C allocation to mycorrhizas across stand age. Vogt *et al.* (1982) estimated the biomass production of mycorrhizal fungi in a 23-year-old and a 180-year-old stand of Pacific silver fir (*Abies amabilis*). For both stands, the production of the mycorrhizal mantle and sporocarps accounted for about 15% of NPP; inclusion of the extramatrical hyphae that permeate the soil would raise the estimate of both total NPP and proportional allocation to mycorrhizas. In addition, the turnover rates of mycorrhizal tissues are commonly more rapid than those of plant tissues (Allen, 1991), so inclusion of respiration losses may further increase the proportion of GPP allocated to developing and sustaining the mycorrhizal association. Termorshuizen (1993) suggested that mycorrhizal diversity should change with ecosystem development because diversity declined with stand age in Scots pine forests. However, reviews by Brundrett (1991) and Molina *et al.* (1992) do not support generalizations about changes in diversity with ecosystem development.

Over 600 tree species have been documented to be capable of symbiotic N-fixation (MacDicken, 1994), including dozens of genera of legumes and several notable non-legume genera such as *Alnus*. Rates of N fixation are typically about 75–200 kg N ha^{-1} yr^{-1} in monoculture stands, representing a C cost of about 400–1000 kg C ha^{-1} yr^{-1} (Binkley *et al.*, 1994; Binkley and Giardina, 1996). However, the investment of C to aid symbionts in fixing N substantially increases assimilation. Some greenhouse studies have shown that high concentrations of ammonium or nitrate can impede N fixation by symbionts. However, little if any feedback inhibition of N fixation is found where the supply rate of N is high but soil concentrations are realistically low (Binkley *et al.*, 1994). All studies that examined N fixation in old (> 50 year) stands of red alder (*Alnus rubra*) have found rates that match those of younger stands (Binkley *et al.*, 1994). The situation is less clear for leguminous trees; declining N fixation with stand age has been suggested for several field studies (J. Sprent, personal communication, 1995).

Considerably more work is needed to quantify the role of mycorrhizas in the C budgets of forests. However, current evidence does not suggest that changes in mycorrhizal associations account for the decline in above-ground forest production with age. Similarly, there seems to be no reason to speculate that changes in the N-fixing symbiosis could account for the age-related decline in productivity.

D. Allocation to Foliage and Branches

As described earlier (section VI.A), the allocation of photosynthate to leaf production is high in young stands (before canopy closure) and typically declines later. This shift in allocation to foliage can partially explain the early increase in stem production, but is generally in the wrong direction to account for any decline in stem production in later stages. Although total canopy leaf area declines in older stands, reductions in the number of trees typically leads to increased leaf area per tree, and this concentration of more leaves on each stem requires more substantial branch development. This increase in the branch biomass:leaf area ratio has not been examined widely as a possible mechanism for reduced stem production, but at least one study indicated an important role for this change in canopy structure (Long and Smith, 1992).

E. Allocation to Reproduction

The carbon cost of reproduction (seeds, supporting fruiting structures, and associated respiration) is typically a low fraction of annual assimilation. For example, Linder and Troeng (1981) estimated that cone production and respiration in a 120-year-old Scots pine (*Pinus sylvestris*) stand required 157 kg C ha^{-1} y^{-1}, roughly 10–15% of the carbon allocated to annual above-ground wood production and woody-tissue respiration. Respiration of cones is low

(< 3% of net photosynthesis in *Pinus contorta*) because cone biomass is relatively low and cone photosynthesis lowers carbon loss (Dick *et al.*, 1990). Reproduction can lower the growth of foliage and wood (Eis *et al.*, 1965; Tappeiner, 1969). In *Pseudotsuga menziesii*, heavy cone crops reduced diameter increment up to 25% (Eis *et al.*, 1965; Tappeiner, 1969) and also reduced shoot growth (*c.*50%; Tappeiner, 1969) and needle length (40%; Tappeiner, 1969).

Little information exists about whether reproductive effort changes with stand development. In a literature review, Greene and Johnson (1994) found that seed production generally increased with tree size up to a stem diameter (at 1.4 m height) of 0.36 m. Seed production in trees larger than 0.36 m diameter showed a variety of responses to increasing tree size (modal, asymptotic, and increasing linear). Seed production is sporadic (Eis *et al.*, 1965) while growth decline is not (Figures 2 and 3), so reproductive costs alone could not account for declines in stand growth with age. The lack of data on changes in reproductive effort with stand development and the impact of reproduction on diameter growth suggest that further work is needed in this area.

X. OTHER CHANGES IN STRUCTURE AND FUNCTION

A. Maturation

Woody plants undergo changes in developmental behavior (termed maturation or phase change) as they age (Haffner *et al.*, 1991; Greenwood and Hutchinson, 1993; Greenwood, 1995), and these changes can alter physiology and growth. The effects of maturation can be observed on intact trees of different ages and on rooted cuttings or grafted scions taken from different-aged trees. Evidence is sketchy, but observations such as the persistence of mature characteristics in scions grafted onto juvenile rootstock (Greenwood, 1984) suggest that maturation involves genetic changes in the meristematic tissue of plants (Greenwood and Hutchinson, 1993). Changes in gene expression in grafted scions did vary with parent age in eastern larch (*Larix laricina*; Hutchinson *et al.*, 1990b), but no causal relationship between genetic expression and maturation has been established.

The developmental changes associated with maturation are quite varied. Maturation affects branching, foliage morphology, physiology, and biochemistry, and these changes generally slow shoot height and diameter growth (Greenwood and Hutchinson, 1993). For example, in loblolly pine (Greenwood, 1984), Douglas fir (Ritchie and Keeley, 1994), and larch (Greenwood *et al.*, 1989), diameter growth, shoot length, and the number of branches decreased in grafted scions as the age of the parent increased (Figure 7). Photosynthesis may increase (Hutchinson *et al.*, 1990b) or decrease (Hutchinson *et al.*, 1990a) in scions grafted to juvenile rootstock, but differences in leaf area were more important than changes in photosynthesis in determining growth. Mature branches are also more likely to flower, and the proportion of male cones increases as branches age (Greenwood and Hutchinson, 1993). Finally, the ability of scions to root decreases with parent age (Greenwood *et al.*, 1989; Steele *et al.*, 1989) and with branch height in individual trees (Foster and Adams, 1984).

Because of the reductions in growth with parent age in grafted scions, maturation may explain the decrease in vertical growth with age (Greenwood, 1989), and the abrupt decline in vertical growth for fast-growing trees (Greenwood and Hutchinson, 1993). If so, differences in individual tree growth could accumulate to affect stand growth, and maturation may be involved in age-related productivity decline. As judged by the performance of rooted scions, however, transition from juvenility to maturity appears to occur early in the life of a tree, before the age at which peak stand leaf area would occur. For example, the greatest maturational changes occur between 1 and 4 years or loblolly pine (Greenwood, 1984) and Douglas fir (Ritchie and Keeley, 1994), and between 1 and 20 years in larch (Greenwood *et al.*, 1989). Slowing of growth in stands occurs after these ages in these species, suggesting that maturation is not a factor — trees that are slowing in growth rate are already 'mature'.

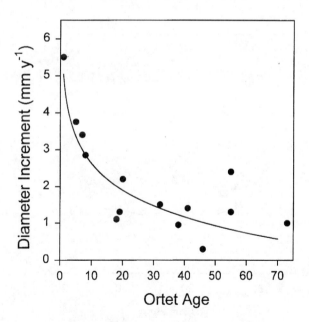

Fig. 7. One-year diameter growth of grafted scions from different-aged ortets (parent trees). Redrawn from Greenwood and Hutchinson (1993).

The living portion of a tree is quite young because the canopy and fine roots are replaced frequently, even in conifers. The oldest cells are ray parenchyma found in the sapwood, which rarely exceed 80 years even in very old trees (Connor and Lanner, 1990). Because of the young age of living tissue, it is unlikely that the individual's age is a factor in growth (Noodén, 1988). Animal cells have a mechanism that slows cell division as cells age and limits the total number of cell divisions that can occur. If plant cells have a similar mechanism, meristems would continuously decrease in activity with age (Greenwood, 1984). However, no direct evidence exists linking maturation and mitotic activity in the apical meristem.

B. Tree Mortality

Most of the data on age-related productivity decline focuses on stands, even though growth per unit of leaf area also declines in individual trees (see section IV). Mortality is a stand-level phenomenon that could alter net wood growth. An increase in mortality with stand age, if it occurred, would lower net wood production. Data on mortality are rare because large plots and long time spans are needed to obtain reliable estimates. However, the sparse existing data do not appear to support an increased mortality with stand age as a factor in age-related productivity decline.

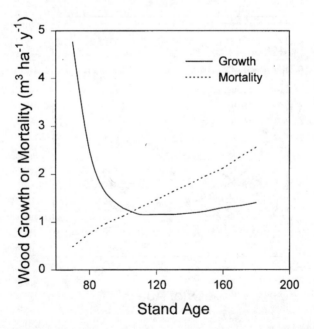

Fig. 8. Annual gross stem growth and mortality for black spruce, drawn from yield tables in Perala (1971).

Forestry yield tables rarely calculate mortality or the number of individuals, but an example from black spruce (Perala, 1971) shows a sharp decline in productivity accompanied by only a modest increase in mortality (Figure 8). An examination of inventory data from the U.S. Rocky Mountain region showed no trend of mortality with stand age (W. K. Olsen, personal communication, 1995): 79% of the stands had no mortality in the last 10 years, and mortality in the other 21% of the stands was variable.

Low net growth rates in old-growth stands apparently result from poor growth, not high mortality. Mortality probabilities for Douglas fir trees in even-aged stands are highest during the first century of stand development (Franklin and Hemstron, 1981), when growth is at its highest (Grier et al., 1989). Long-term studies of tree mortality in old-growth stands (DeBell and Franklin, 1987; Franklin and DeBell, 1988; Hofgaard, 1993) have shown that mortality is roughly equal to production on the surviving trees, so biomass accumulation is near zero. However, growth rates of surviving trees were still low compared with those of younger stands. For example, annual growth of live trees in a mixed western hemlock and Douglas fir old-growth forest was 6.5 m^3 ha^{-1} (DeBell and Franklin, 1987), while peak productivity of a younger Douglas fir stand on a less productive site, was 15 m^3 ha^{-1} (Turner and Long, 1975). In lodgepole pine, growth of survivors in an old-growth forest was 0.74–0.94 Mg ha^{-1} yr^{-1}, while mortality ranged from 0.32 to 1.03 Mg ha^{-1} yr^{-1} (W. Moir, personal communication, 1996). The growth rate was comparable to that found for an old-growth stand in another study on the same site; however, an adjacent younger stand had a growth rate of 4.2 Mg ha^{+1} yr^{-1} (Ryan and Waring, 1992). Binkley and Greene (1983) examined patterns in above-ground NPP for 50 years in stands of red alder, conifers (mostly Douglas fir), and a mixture of red alder and conifers. Above-ground NPP declined in all three stands, whereas mortality was relatively constant after age 25. In 8- to 50-year-old stands of loblolly pine, mortality rate was a constant proportion of total tree number across a 20-fold range of initial stand density (Christensen and Peet, 1981).

In dense monocultures, total stand biomass and the number of individuals are strongly related, with population size decreasing as stand biomass and the size of the average individual increases. One formulation of this relationship (Westoby, 1984), in terms of stand biomass (B) and the number of individuals, (N) is:

$$B = CN^{0.5} \tag{2}$$

where C is a constant. This *self-thinning* rule defines an upper limit for the relationship between density and biomass that many even-aged stands follow (Westoby, 1984). While time is not explicit in the equation, we do know that older stands have fewer, larger individuals. Therefore, it may be instructive to examine the pattern of mortality that would occur if a stand were moving along a self-thinning trajectory. What we find is that with a unit increase in biomass, the percentage of stems dying decreases as stand biomass (or average size of individual) increases, but the biomass of the trees that die is constant as stand biomass increases. Because stands accumulate biomass faster earlier in their life, stands following a self-thinning trajectory should show lower mortality when trees are older and growing slowly. We note that the death of a single large tree leaves a bigger gap than the death of a small tree, and the time required to 'fill' the gap could be longer in older forests. Therefore, a constant rate of mortality of individuals could lower greater productivity in older forests.

C. Stand Structure

Stand structure changes dramatically with stand development (Bingham and Sawyer, 1991; Spies and Franklin, 1991). The height, diameter, and size of the average individual increase through time, together with stand biomass (Bingham and Sawyer, 1991; Spies and Franklin, 1991). Stand density declines (see section X.B) as the crowns of larger trees shade smaller competitors and the smaller trees die. Understory and herbaceous cover increase (Bingham and Sawyer, 1991; Spies and Franklin, 1991) as the stand leaf area declines (see section VI.A) and gaps form from dead trees (Spies *et al.*, 1990). Crown damage from windstorms can increase (Spies and Franklin, 1991), and the amount of woody debris may increase (Spies and Franklin, 1991) or decrease (Bingham and Sawyer, 1991), most likely reflecting the conditions under which stands were formed. Decreases in density and increases in tree size, woody debris, and gap fraction continue to occur even as old-growth stands age (Tyrrell and Crow, 1994).

Reduction in leaf area is the structural change most likely to affect wood production (see section VI.A) because of the loss in carbon assimilation. Tree height may also be a significant factor if photosynthesis is limited by hydraulic conductivity of the stem (section VI.C). However, if woody litter increases, nitrogen will be immobilized by the decaying wood (Murty *et al.*, 1996), lowering nitrogen availability for the overstory (see section VII). Low nutrient availability may lower the photosynthetic capacity of leaves and perhaps also the total leaf area. However, lower leaf area and a higher gap fraction will allow more solar radiation to the forest floor, perhaps warming it and increasing decomposition.

XI. CONCLUDING REMARKS

Declining growth after an even-aged stand reaches some limit appears to be universal, but the timing and magnitude of the decline vary. In our literature search, we found no support for the textbook explanation that growth decline results from increased respiration. Maintenance costs of woody-tissue respiration are low, and the amount of living cells in wood is small and changes little with biomass accumulation. There is also evidence that the balance of assimilation and respiration is conservative, so respiratory load declines as growth declines. The ultimate test of this hypothesis (which yet to be done) would be in the wet tropics, where high biomass and warm temperatures should yield the highest respiration rates.

In our search for mechanistic explanations, we identified two promising candidates:

(1) A decline in stand leaf area usually accompanies a decline in above-ground wood growth, and some of the growth decline is undoubtedly caused by decreased assimilation of a canopy with lower leaf area. Leaf area decline may be caused (a) by the loss of branches through abrasion or the formation of gaps too large for existing trees to exploit, (b) as a response to photosynthesis lowered by hydraulic limitation, (c) through a complex interaction with nutrition, or (d) through maturational changes. Branch abrasion and gap formation occur in even-aged forests, but it is not known whether these processes are common. Lower hydraulic conductance of older trees may reduce photosynthesis and ultimately leaf area. If nutrients become bound in living biomass or are immobilized in decaying wood, nutrient availability will be lower. Fertilization experiments show that improved nutrition will encourage more leaf area, so it may be reasonable that lowered nutrition will encourage shedding of foliage. However, the evidence for declines in nutrient availability that accompany stand development is mixed. The bulk of maturational changes appear to occur very early in the life of a tree, while growth decline generally occurs in mature trees.

(2) Limited evidence suggests that older trees have lower photosynthetic capacity and total diurnal assimilation than younger trees, which will lower canopy assimilation and provide less carbohydrate for wood production. Reductions in capacity may be caused by a feedback with the tree's hydraulic system, lower nutrient availability, or maturation changes. If stomatal behavior limits maximum transpiration and maximum transpiration declines as tree height and branch length increase, stomatal conductance and photosynthesis will be lower in taller trees. Consistently lower assimilation may lower photosynthetic capacity via a mechanism similar to that promoting lower capacity in older conifer needles.

The common patterns of a decline in stand leaf area and leaf photosynthetic capacity suggest a new model of carbon balance with stand development (Figure 9). In this model, photosynthesis and above-ground dry-matter production increase with canopy development. After the forest reaches a maximum leaf area, photosynthesis and dry-matter production decline as leaf

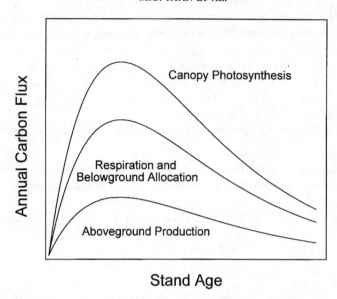

Fig. 9. Proposed model of carbon flux during forest stand development. Canopy photosynthesis increases as leaf area develops and decreases after maximum as leaf area and photosynthetic rates decline.

area, photosynthetic capacity, and photosynthesis decline. The model assumes that allocation to respiration and below ground to roots and symbionts is a constant fraction of assimilation over the life of a forest stand.

Other potential mechanisms appeared less promising as general explanations for growth decline. These mechanisms can affect forest carbon balance and may reduce the productivity of older forests in some cases. Experimental tests of these hypotheses are rare:

(1) Below-ground carbon allocation (fine root production or allocation to symbionts) may increase with stand age, but the evidence is mixed. Increased fine-root production or allocation to symbionts could result from lower nutrient availability and decrease the carbohydrate available for wood production. Because our literature review showed that there was no consistent pattern of nutrition with stand age, we suspect that below-ground allocation will not universally increase with stand age, although it may feature in the development of some forests.

(2) Mortality of individual trees (as per cent of the current population) decreases as stands accumulate biomass. If stands follow a self-thinning curve, the biomass lost to mortality is constant as biomass accumulates. Therefore, it is unlikely that increased mortality contributes to growth decline unless larger gaps formed in older stand require substantially more time for in-filling.

(3) Reproductive costs can reduce stem growth. However, annual reproductive costs vary dramatically while growth declines continuously with stand age.

As forests age, then, they develop a syndrome of changes (lower leaf area, lower photosynthetic capacity, perhaps changes in allocation) that acts to lower primary production. If hydraulic resistance does constrain stomatal conductance and assimilation (a very active area of research), then all trees will suffer this limitation as they grow. However, nutrient limitations and shifts in allocation may also be important.

Experiments are needed to better understand how and why forests function differently as they age. We believe that the most critical questions are:

(1) Why does leaf area decline with stand age?
(2) Does nutrition change with stand development? What is the role of nutrition in forcing changes in below-ground allocation, leaf area, and photosynthesis?
(3) Do hydraulic limitations always lead to reduced photosynthesis as trees grow in height and branches lengthen?
(4) Does below-ground allocation (fine-roots and symbionts) change with stand development? If so, why? Are any changes in below-ground allocation substantial enough to account for changes in above-ground wood growth?

Guying trees (to prevent crown interaction) and fertilization of old forests can discriminate between some of the causes of leaf area decline. Fertilization studies in old, slow-growing forests can determine the role of nutrition in growth decline, allocation to foliage and roots, and photosynthesis; studies in aggrading forests can determine whether nutrition will alter the timing or presence of a growth decline. Leaf and stand measurements of CO_2 and water, together with experimental manipulations of leaf-specific hydraulic conductivity, can determine the role of hydraulic limitation. To determine whether below-ground allocation changes with stand development, we suggest that it be examined (perhaps using the carbon-balance approach) for a wide variety of chronosequences, and over time for rapidly growing stands. While not a critical issue, the respiration hypothesis should be examined in a tropical ecosystem, where respiration costs are expected to be high.

Given the evidence for differences in structure and function between young, mature, and old forests, it is surprising how little information is available. We hope this review will stimulate further efforts to understand growth decline and the changes that occur as a forest ages. Information on nutrition, carbon allocation, and physiological differences is sparse and much needed.

ACKNOWLEDGEMENTS

We acknowledge the support of NSF-DEB93-06356 to J.H. Fownes, D. Binkley, and M.G. Ryan, and USDA-NRICGP-9401021 to M.G. Ryan and B.J. Yoder.

REFERENCES

Aber, J.D. and Melillo, J.M. (1991). *Terrestrial Ecosystems.* Saunders College Publishing, Philadelphia, PA.

Allen, M.F. (1991). *The Ecology of Mycorrhizae.* Cambridge University Press, Cambridge.

Amthor, J.S. (1986). Evolution and applicability of a whole plant respiration model. *J. Theor. Biol* **122**, 473–490.

Assmann, E. (1970). *The Principles of Forest Yield Study.* Pergamon Press, Oxford.

Ball, J.T., Woodrow, I.E. and Berry, J.A. (1987). A model predicting stomatal conductance and its contribution to the control of photosynthesis under different environmental conditions. In: *Progress in Photosynthesis Research, vol. 4* (Ed. by J. Biggens), pp. 221–224. Martinus Nijhoff Publishers, Dordrecht, The Netherlands.

Beck, D.E. (1971). Height-growth patterns and site index of white pine in the southern Appalachians. *For. Sci.* **17**, 252–260.

Beets, P. and Pollock, D. (1987). Accumulation and partitioning of dry matter in *Pinus radiata* as related to stand-age and thinning. *NZ J. For. Sci.* **17**, 246–271.

Berish, C.W. (1982). Root biomass and surface area in three successional tropical forests. *Can. J. For. Res.* **12**, 699–704.

Bingham, B.B. and Sawyer, J.O., Jr. (1991). Distinctive features and definitions of young, mature, and old-growth Douglas-fir/hardwood forests. In: *Wildlife and Vegetation of Unmanaged Douglas-fir Forests* (Ed. by L.F. Ruggiero, K.B. Aubry, A.B. Carey and M.H. Huff), pp. 363–373. US For. Serv. Gen. Tech. Rep. PNW-285, Portland, OR.

Binkley, D. (1984). Does forest removal increase rates of decomposition and nitrogen release? *For. Ecol. Manage.* **8**, 229–233.

Binkley, D. (1986). *Forest Nutrition Management.* John Wiley & Sons Inc., New York, NY.

Binkley, D. and Giardina, C. (1996). Biological nitrogen fixation in plantations. In: *Management of Soil; Water, and Nutrients in Tropical Plantation Forests* (Ed. by E.K.S. Nambiar and A. Brown), Chapter 9, CSIRO/CIFOR, in press.

Binkley, D. and Greene, S. (1983). Production in mixtures of conifers and red alder: the importance of site fertility and stand age. In: *IUFRO Symposium on Forest Site and Continuous Productivity* (Ed. by R. Ballard and S.P. Gessel), pp. 112–117. US For. Serv. Gen. Tech. Rep. PNW-163, Portland, OR.

Binkley, D., Cromack, K., Jr. and Baker, D.D. (1994). Nitrogen fixation by red alder: biology, rates and controls. In: *The Biology and Management of Red Alder* (Ed. by D. Hibbs, D. DeBell and R. Tarrant), pp. 57–72. Oregon State University Press, Corvallis, OR.

Binkley, D., Smith, F.W. and Son, Y. (1995). Nutrient supply and declines in leaf area and production in lodgepole pine. *Can. J. For. Res.* **25**, 621–628.

Bormann, F.H. and Likens, G.E. (1979). *Pattern and Process in a Forested Ecosystem.* Springer-Verlag, New York, NY.

Brundrett, M. (1991). Mycorrhizas in natural ecosystems. *Adv. Ecol. Res.* **21**, 171–313.

Burger, J.A. and Pritchett, W.L. (1984). Effects of clearfelling and site preparation on nitrogen mineralization in a southern pine stand. *Soil Sci. Soc. Am. J.* **48**, 1432–1437.

Cannell, M.G.R. (1989). Physiological basis of wood production: a review. *Scand. J. For. Res.* **4**, 459–490.

Cannell, M.G.R. and Dewar, R.C. (1994). Carbon allocation in trees: a review of concepts for modelling. *Adv. Ecol. Res.* **25**, 59–104.

Chapin, F.S. III, Vitousek, P.M. and Van Cleve, K. (1986). The nature of nutrient limitation in plant communities. *Am. Nat.* **127**, 48–58.

Christensen, N.L. and MacAller, T. (1985). Soil mineral nitrogen transformations during succession in the piedmont of North Carolina. *Soil Biol. Biochem.* **17**, 675–681.

Christensen, N.L. and Peet, R.K. (1981). Secondary forest succession on the North Carolina Piedmont. In: *Forest Succession: Concepts and Applications* (Ed. by D.C. West, H.H. Shugart, and D.B. Botkin), pp. 230–245. Springer-Verlag, New York, NY.

Clemensson-Lindell, A. and Persson, H. (1995). Fine-root vitality in a Norway spruce stand subjected to various nutrient supplies. *Plant Soil* **168–169**, 167–172.

Comeau, P.G. and Kimmins, J.P. (1989). Above- and below-ground biomass and production of lodgepole pine on sites with differing soil moisture regimes. *Can. J. For. Res.* **19**, 447–454.

Connor, K.F. and Lanner, R.M. (1990). Effects of tree age on secondary xylem and phloem anatomy in stems of Great Basin bristlecone pine (*Pinus longaeva*). *Am. J. Bot.* **77**, 1070–1077.

Coutts, M.P. (1983). Root architecture and tree stability. *Plant Soil* **71**, 171–188.

Covington, W.W. and Aber, J.D. (1980). Leaf production during secondary succession in northern hardwoods. *Ecology* **61**, 200–204.

Dahms, W.G. (1964). *Gross and Net Yield Tables for Lodgepole Pine.* US For. Serv. Gen. Tech. Rep. PNW-8, Portland, OR.

Davidson, E.A., Hart, S.C. and Firestone, M.K. (1992). Internal cycling of nitrate in soils of a mature conifer forest. *Ecology* **73**, 1148–1156.

DeAngelis, D.L., Gardner, R.H. and Shugart, H.H. (1980). Productivity of forest ecosystems studies during the IBP: the woodlands data set. In: *Dynamic Properties of Forest Ecosystems* (Ed. by D.E. Reichle), pp. 567–672, International Biological Programme 23. Cambridge University Press, Cambridge.

DeBell, D.S. and Franklin, J.F. (1987). Old-growth Douglas-fir and western hemlock: a 36-year record of growth and mortality. *West. J. Appl. For.* **2**, 111–114.

Dick, J.McP., Smith, R. and Jarvis, P.G. (1990). Respiration rate of male and female cones of *Pinus contorta. Trees* **4**, 142–149.

Dudley, N.S. (1990). Performance and Management of Fast-growing Tropical Trees in Diverse Hawaiian Environments. MS Thesis, University of Hawaii at Manoa, Honolulu, HI.

Dunn, G.M. and Connor, D.J. (1993). An analysis of sap flow in mountain ash (*Eucalyptus regnans*) forests of different age. *Tree Physiol.* **13**, 321–336.

Eis, S., Garman, H. and Ebell, L.F. (1965). Relation between cone production and diameter increment of Douglas fir (*Pseudotsuga menziesii* (Mirb.) Franco), grand fir (*Abies grandis* (Dougl.) Lindl.), and Western white pine (*Pinus monticola* Dougl.). *Can. J. Bot.* **43**, 1553–1559.

Ekblad, A. (1995). Actinorhizal and ectomycorrhizal activities in symbioses of *Alnus incana* and *Pinus sylvestris*. Ph.D. Dissertation, Department of Plant Physiology, Umea University, Sweden.

Ericsson, T. (1995). Growth and shoot:root ratio of seedlings in relation to nutrient availability. *Plant Soil* **168–169**, 205–214.

Evans J. (1982). *Plantation Forestry in the Tropics.* Clarendon Press, Oxford.

Ewel, K.C. and Gholz, H.L. (1991). A simulation model of the role of belowground dynamics in a Florida pine plantation. *For. Sci.* **37**, 397–438.

Fairley, R.I. and Alexander, I.J. (1985). Methods of calculating fine root production in forests. In: *Ecological Interactions in Soils: Plants, Microbes and Animals* (Ed. by A.H. Fitter), pp. 37–42. Blackwell Scientific, Oxford.

Field, C. and Mooney, H.A. (1986). The photosynthesis-nitrogen relationship in wild plants. In: *On the Economy of Plant Form and Function* (Ed. by T.J. Givnish), pp. 25–55. Cambridge University Press, Cambridge.

Ford, E.D. (1984). The dynamics of plantation growth. In: *Nutrition of Plantatation Forests* (Ed. by G.B. Bowen and E.K.S. Nambiar), pp. 17–52. Academic Press, New York, NY.

Forrest, W.G. and Ovington, J.D. (1970). Organic matter changes in an age series of *Pinus radiata* plantations. *J. Appl. Ecol.* **7**, 177–186.

Foster, G.S. and Adams, W.T. (1984). Heritability, gain and C effects in rooting western hemlock cuttings. *Can. J. For. Res.* **14**, 628–638.

Fownes, J.H. and Harrington, R.A. (1990). Modelling growth and optimal rotations of tropical multipurpose trees using unit leaf rate and leaf area index. *J. Appl. Ecol.* **27**, 886–896.

Fownes, J.H. and Harrington, R.A. (1992). Allometry of woody biomass and leaf area in five tropical multipurpose trees. *J. Trop. For. Sci.* **4**, 317–330.

Fox, T.R., Burger, J.A. and Kreh, R.E. (1986). Effects of site preparation on nitrogen dynamics in the southern piedmont. *For. Ecol. Manage.* **15**, 241–256.

Franklin, J.F. and DeBell, D.S. (1988). Thirty-six years of tree population change in an old-growth *Pseudotsuga–Tsuga* forest. *Can. J. For. Res.* **18**, 633–639.

Franklin, J.F. and Hemstrom, M.A. (1981). Aspects of succession in the coniferous forests of the Pacific Northwest. In: *Forest succession: Concepts and Applications* (Ed. by D.C. West, H.H. Shugart, and D.B. Botkin), pp. 212–229. Springer-Verlag, New York, NY.

Frazer, D.W., McColl, J.G. and Powers, R.F. (1990). Soil nitrogen mineralization in a clearcutting chronosequence in a northern California conifer forest. *Soil Sci. Soc. Am. J.* **54**, 1145–1152.

Fredericksen, T.S., Steiner, K.C., Skelly, J.M., Joyce, B.J., Kolb, T.E., Kouterick, K.B. and Ferdinand, J.A. (1996). Diel and seasonal patterns of leaf gas exchange and xylem water potentials of different-sized *Prunus serotina* Ehrh. trees. *For. Sci.* **42**, 350–365.

Gholz, H.L. and Fisher, R.F. (1982). Organic matter production and distribution in slash pine *Pinus elliottii* plantations. *Ecology* **63**, 1827–1839.

Gholz, H.L., Hendry, L. and Cropper, W.P., Jr. (1986). Organic matter dynamics of fine roots in plantations of slash pine (*Pinus elliottii*) in north Florida. *Can. J. For. Res.* **16**, 529–538.

Gingrich, S.F. (1971). *Management of Young and Intermediate Stands of Upland Hardwoods.* US For. Serv. Res. Pap. NE-195, Upper Darby, PA.

Gorham, E., Vitousek, P.M. and Reiners, W.A. (1979). The regulation of chemical budgets over the course of terrestrial ecosystem succession. *Ann. Rev. Ecol. Syst.* **10**, 53–84.

Gower, S.T., Isebrands, J.G. and Sheriff, D.W. (1995). Carbon allocation and accumulation in conifers. In: *Resource Physiology of Conifers: Acquisition, Allocation, and Utilization* (Ed. by W.K. Smith and T.M. Hinckley), pp. 217–254. Academic Press, New York, NY.

Gower, S.T., McMurtrie, R.E., and Murty, D. (1996a). Aboveground net primary production decline with stand age: potential causes. *Trend Ecol. Evol*, **11**, 378–382.

Gower, S.T., Running, S.W., Gholz, H.L., Haynes, B.E., Hunt, E.R., Jr., Ryan, M.G., Waring, R.H. and Cropper, W.P., Jr. (1996b). Influence of climate and nutrition on carbon allocation and net primary production of four conifer forests. *Tree Physiol*, in press.

Gower, S.T., Vogt, K.A. and Grier, C.C. (1992). Carbon dynamics of Rocky Mountain Douglas-fir: influence of water and nutrient availability. *Ecol. Monogr.* **62**, 43–65.

Greene, D.F. and Johnson, E.A. (1994). Estimating the mean annual seed production of trees. *Ecology* **75**, 642–647.

Greenwood, M.S. (1984). Phase change in loblolly pine: Shoot development as a function of age. *Physiol. Plant.* **61**, 518–522.

Greenwood, M. (1989). The effect of phase change on annual growth increment in eastern larch (*Larix laricina* (Du Roi) K. Koch). *Ann. Sc. For.* **46** (Suppl.), 171s–177s.

Greenwood, M.S. (1995). Juvenility and maturation in conifers. *Tree Physiol.* **15**, 433–438.

Greenwood, M.S. and Hutchinson, K.W. (1993). Maturation as a developmental process. In: *Clonal forestry I: Genetics and Biotechnology* (Ed. by M.R. Ahuja and W.J. Libby), pp. 14–33. Springer-Verlag, Berlin.

Greenwood, M.S., Hopper, C.A. and Hutchinson, K.W. (1989). Maturation in larch. I. Effects of age on shoot growth, foliar characteristics, and DNA methylation. *Plant Physiol.* **90**, 406–412.

Gregory, R.A. and Haack, P.M. (1965). *Growth and Yield of Well-stocked Aspen and Birch Stands in Alaska*. US For. Serv. Res. Pap. NOR-2, Juneau, AK.

Grier, C.C., Vogt, K.A., Keyes, M.R. and Edmonds, R.L. (1981). Biomass distribution and above-and below-ground production in young and mature *Abies amabilis* zone ecosystems of the Washington Cascades. *Can. J. For. Res.* **11**, 155–167.

Grier, C.C., Lee, K.M., Nadkarni, N.M., Klock, G.O. and Edgerton, P.J. (1989). *Productivity of Forests of the United States and its Relation to Soil and Site Factors and Management Practices: A Review*. US For. Serv. Gen. Tech. Rep. PNW-222, Portland, OR.

Grulke, N.E. and Miller, P.R. (1994). Changes in gas exchange characteristics during the life span of giant sequoia: implications for response to current and future concentrations of atmospheric ozone. *Tree Physiol.* **14**, 659–668.

Haffner, V., Enjalric, F., Lardet, L. and Carron, M.P. (1991). Maturation of woody plants: a review of metabolic and genomic aspects. *Ann. For. Sci.* **48**, 615–630.

Harmon, M.E., Ferrell, W.K. and Franklin, J.F. (1990). Effects on carbon storage of conversion of old-growth forests to young forests. *Science* **247**, 699–702.

Harrington, R.A. and Fownes, J.H. (1993). Allometry and growth of planted versus coppice stands of four fast-growing tropical tree species. *For. Ecol. Manage.* **56**, 315–327.

Harrington, R.A. and Fownes, J.H. (1995). Radiation interception and growth of planted and coppice stands of four fast-growing tropical trees. *J. Appl. Ecol.* **32**, 1–8.

Harrington, R.A. and Fownes, J.H. (1996). Predicting spacing effects on growth and optimal rotations of tropical multipurpose trees. *Agric. Sys.* **50**, 377–390.

Herbert, D.A. and Fownes, J.H. (1995). Phosphorus limitation of forest leaf area and net primary production on a highly weathered soil. *Biogeochemistry* **29**, 223–235.

Herbert, M.A. (1984). Variation in the growth of and responses to fertilizing black wattle with nitrogen, phosphorus, potassium and lime over three rotations. In:

Symposium on Site and Productivity of Fast Growing Plantations (Ed. by D.G Grey, A.P.G., Schönau, C.J., Schutz, and A. van Laar), pp. 907–920. South African Forest Research Institute, Pretoria, South Africa.

Hofgaard, A. (1993). 50 years of change in a Swedish boreal old-growth *Picea abies* forest. *J. Vegetation Sci.* **4**, 773–782.

Hutchinson, K.W., Greenwood, M.S., Sherman, C., Rebbeck, J. and Singer, P. (1990a). The molecular genetics of maturation in eastern larch [*Larix laricina* (Du Roi) K. Koch]. In: *Plant Aging, Basic and Applied Approaches* (Ed. by R. Rodrigues), pp. 141–145. Plenum Press, New York, NY.

Hutchinson, K.W., Sherman, C.D., Weber, J., Smith, S.S., Singer, P.B. and Greenwood, M.S. (1990b). Maturation in larch. II. Effects of age on photosynthesis and gene expression in developing foliage. *Plant Physiol.* **94**, 1308–1315.

Jackson, D.S. and Chittenden, J. (1981). Estimation of dry matter in *Pinus radiata* root systems. 1. Individual trees. *NZ J. For. Sci.* **11**, 164–182.

Kaufmann, M.R. and Ryan, M.G. (1986). Physiographic, stand, and environmental effects on individual tree growth and growth efficiency in subalpine forests. *Tree Physiol.* **2**, 47–59.

Kauppi, P.E., Mielikainen, K. and Kuusela, K. (1992). Biomass and carbon budget of European forests, 1971 to 1990. *Science* **256**, 70–74.

Keyes, M.R. and Grier, C.C. (1981). Above- and below-ground production in 40-year-old Douglas-fir stands on low and high productivity sites. *Can. J. For. Res.* **11**, 599–605.

Kira, T. and Shidei, T. 1967. Primary production and turnover of organic matter in different forest ecosystems of the Western Pacific. *Jpn. J. Ecol.* **17**, 70–87.

Kramer, P.J. and Kozlowski, T.T. (1979). *Physiology of Woody Plants.* Academic Press, New York, NY.

Krause, H.H. and Ramlal, D. (1986). *In situ* nutrient extraction by resin from forested, clear-cut and site-prepared soil. *Can. J. Soil Sci.* **67**, 943–952.

Kull, O. and Koppel, A. (1987). Net photosynthetic response to light intensity of shoots from different crown positions and age in *Picea abies* (L.) Karst. Scand. *J. For. Res.* **2**, 157–166.

Kurz, W.A. and Kimmins, J.P. (1987). Analysis of some sources of error in methods used to determine fine root production in forest ecosystems: a simulation approach. *Can. J. For. Res.* **17**, 909–912.

Landsberg, J.J., Prince, S.D., Jarvis, P.G., McMurtrie, R.E., Luxmoore, R. and Medlyn, B.E. (1995). Energy conversion and use in forests: an analysis of forest production in terms of radiation utilization efficiency (ϵ). In: *The Use of Remote Sensing in the Modeling of Forest Productivity at Scales from the Stand to the Globe* (Ed. by H.L. Gholz, K. Nakane, H. Shimoda), pp. 273–298. Kluwer Academic Publ., Dordrecht, The Netherlands.

Linder, S. and Axelsson B. (1982). Changes in carbon uptake and allocation pattens as a result of irrigation and fertilization in a young *Pinus sylvestris* stand. In: *Carbon Uptake and Allocation in Subalpine Ecosystems as a Key to Management: Proceedings of an IUFRO Workshop* (Ed. by R.H. Waring), pp. 38–44. Forest Research Laboratory, Oregon State University, Corvallis, OR.

Linder, S. and Troeng, E. (1981). The seasonal course of respiration and photosynthesis in strobili of Scots pine. *For. Sci.* **27**, 267–276.

Linder, S., McMurtrie, R.E. and Landsberg, J.J. (1985). Growth of Eucalyptus: A mathematical model applied to *Eucalyptus globulus*. In: *Crop Physiology of Forest Trees* (Ed. by P.M.A. Tigerstedt, P. Puttonen and V. Koski), pp. 117–126. University of Helsinki, Helsinki, Finland.

Long, J.N. and Smith, F.W. (1992). Volume increment in *Pinus contorta var. latifolia*: the influence of stand development and crown dynamics. *For. Ecol. Manage.* **53**, 53–64.

Lugo, A.E. and Brown, S. (1992). Tropical forests as sinks of atmospheric carbon. *For. Ecol. Manag.* **54**, 239–255.

MacDicken, K.G. (1994). Selection and management of nitrogen-fixing trees. Winrock International Institute for Agricultural Development, Morrilton, Arkansas, USA and UNFAO, Bangkok, Thailand.

Marks, P.L. (1974). The role of pin cherry (*Prunus pennsylvanica* L.) in the maintenance of stability in northern hardwood ecosystems. *Ecology* **44**, 73–88.

Marx, D.H., Hatch, A.B. and Mendicino, J.F. (1977). High soil fertility decreases sucrose content and susceptibility of loblolly pine roots to ectomycorrhizal infection by *Pisolithus tinctorius. Can. J. Bot.* **55**, 1569–1574.

Matson, P.A. (1987). Nitrogen transformations following tropical forest felling and burning on a volcanic soil. *Ecology* **68**, 491–502.

Matson, P.A. and Boone, R.D. (1984). Natural disturbance and nitrogen mineralization: wave-form dieback of mountain hemlock in the Oregon Cascades. *Ecology* **65**, 1511–1516.

Mattson-Djos, E. (1981). The use of pressure-bomb and porometer for describing plant water status in tree seedlings. In: *Proceedings of a Nordic Symposium on Vitality and Quality of Nursery Stock* (Ed. by P. Puttonen), pp. 45–57. Department of Silviculture, University of Helsinki, Helsinki, Finland.

Mencuccini, M. and Grace, J. (1996). Hydraulic conductance, light interception and needle nutrient concentration in Scots pine stands and their relation with net primary productivity. *Tree Physiol* **16**, 459–468.

Meyer, W.H. (1938). *Yield of Even-aged Stands of Ponderosa Pine.* US Dep. Agric. Tech. Bull. 630, Washington, D.C.

Miller, H.G. (1984). Dynamics of nutrient cycling in plantation ecosystems. In: *Nutrition of Plantation Forests* (Ed. by G.B. Bowen and E.K.S. Nambiar), pp. 53–78. Academic Press, New York, NY.

Mladenoff, D.J. (1987). Dynamics of nitrogen mineralization and nitrification in hemlock and hardwood treefall gaps. *Ecology* **68**, 1171–1180.

Molina, R., Massicotte, H. and Trappe, J.M. (1992). Specificity phenomena in mycorrhizal symbioses: community-ecological consequences and practical implications. In: *Mycorrhizal Functioning: an Integrative Plant-Fungal Process* (Ed. by M. Allen), pp. 357–423. Chapman & Hall, New York.

Möller, C.M., Müller, D. and Nielsen, J. (1954). Graphic presentation of dry matter production in European beech. *Forstl. Forsoegsvaes. Dan.* **21**, 327–335.

Monteith, J.L. (1995). A reinterpretation of stomatal responses to humidity. *Plant Cell Environ.* **18**, 357–364.

Montes, R.A. and Christensen, N.L. (1979). Nitrification and succession in the piedmont of North Carolina. *For. Sci.* **25**, 287–297.

Murty, D., McMurtrie, R.E. and Ryan, M.G. (1996). Declining forest productivity in ageing forest stands – a modeling analysis of alternative hypotheses. *Tree Physiol.* **16**: 187–200.

Nadelhoffer, K.J., Aber, J.D. and Melillo, J.M. (1985). Fine roots, net primary production, and soil nitrogen availability: A new hypothesis. *Ecology* **66**, 1377–1390.

NFSFNC (1995). *Twenty-fourth Annual Report*, North Carolina State Forest Nutrition Cooperative, Raleigh, NC.

Nilsson, U. and Albrekstson, A. (1993). Productivity of needles and allocation of growth in young Scots pine trees of different competitive status. *For. Ecol. Manag.* **62**, 173–187.

258 M.G. RYAN *ET AL.*

Noodén, L.D. (1988). Whole plant senescence. In: *Senescence and Aging in Plants* (Ed. by L.D. Noodén and A.D. Leopold), pp. 391–439. Academic Press, San Diego, CA.

Oliver, C.D. and Larson, B.C. (1996). *Forest Stand Dynamics*. John Wiley & Sons Inc., New York, NY.

Olsson, U.R. (1996). *Nitrogen Cost of Production along a Lodgepole Pine Chronosequence*. MS Thesis, Colorado State University, Ft Collins, CO.

Pastor, J., Stillwell, M.A. and Tilman, D. (1987). Nitrogen mineralization and nitrification in four Minnesota old fields. *Oecologia* **71**, 481–485.

Pearson, J.A., Knight, D.H. and Fahey, T.J. (1987). Biomass and nutrient accumulation during stand development in Wyoming lodgepole pine forests. *Ecology* **68**, 1966–1973.

Perala, D.A. (1971). *Growth and Yield of Black Spruce on Organic Soils in Minnesota*. US For. Ser. Res. Pap. NC-56, Minneapolis, MN.

Perala, D.A. (1977). *Manager's Handbook for Aspen in the North Central States*. US For. Ser. Gen. Tech. Rep. NC-36, Saint Paul, MN.

Persson, H., Hooshang, M. and Clemensson-Lindell, A. (1995). Effects of acid deposition on tree roots. *Ecol. Bull.* **44**, 158–167.

Pregitzer, K., Hendrick, R. and Fogel, R. (1993). The demography of fine roots in response to patches of water and nitrogen. *New Phytol.* **125**, 575–580.

Prescott, C.E. (1995). Does nitrogen availability control rates of litter decomposition in forests? *Plant Soil* **168–169**, 83–88.

Putz, F., Parker, G.G. and Archibald, R.M. (1984). Mechanical abrasion and intercrown spacing. *Am. Midl. Nat.* **112**, 24–28.

Raich, J.W. and Nadelhoffer, K.J. (1989). Belowground carbon allocation in forest ecosystems: global trends. *Ecology* **70**, 1346–1354.

Rastetter, E.B., Ryan, M.G., Shaver, G.R., Melillo, J.M., Nadelhoffer, K.J., Hobbie, J.E. and Aber, J.D. (1991). A general biogeochemical model describing the responses of the C and N cycles in terrestrial ecosystems to changes in CO_2, climate, and N deposition. *Tree Physiol.* **9**, 101–126.

Reid, C.P.P., Kidd, F.A. and Ekwebelam, S.A. (1983). Nitrogen nutrition, photosynthesis, and carbon allocation in ectomycorrhizal pine. *Plant Soil* **71**, 415–432.

Reis, M.G.F., Kimmins, J.P., Rezende, G.C. and Barros, N.F. (1985). Acúmulo de biomass em uma sequéncia de idade de *Eucalyptus grandis* plantado no cerrado em duas áreas com diferentes produtividades. *Rev. Árvore* **9**, 149–162.

Reis, M.G.F., Barros, N.F. and Kimmins, J.P. (1987). Acúmulo de nutrientes em uma equencia de idade de *Eucalyptus grandis* W. Hill (ex-Maiden) plantado no cerrado, em duas áreas com diferentes produtividades, em Minas Gerais. *Rev. Arvore* **11**, 1–15.

Ritchie, G.A. and Keeley, J.W. (1994). Maturation in Douglas-fir: I. Changes in stem, branch and foliage characteristics associated with ontogenetic ageing. *Tree Physiol.* **14**, 1245–1259.

Robertson, G.P. and Vitousek, P.M. (1981). Nitrification potentials in primary and secondary succession. *Ecology* **62**, 376–386.

Rudolph, T.D. and Laidly, P.R. (1990). *Pinus banksiana* Lamb. In: *Silvics of North America, Volume 1. Conifers* (Ed. by R.M. Burns and B.H. Honkala), pp. 280–293. US Dep. Agric. For. Ser. Agric. Hdbk. 654, Vol. 1. Washington, D.C.

Running, S.W. and Gower, S.T. (1991). Forest-BGC, a general model of forest ecosystem processes for regional applications II. Dynamic carbon allocation and nitrogen budgets. *Tree Physiol.* **9**, 147–160.

Runyon, J., Waring, R.H., Goward, S.N. and Welles, J.M. (1994). Environmental limits on net primary production and light-use efficiency across the Oregon Transect. *Ecol. Appl.* **4**, 226–237.

Ryan, M.G. (1990). Growth and maintenance respiration in stems of *Pinus contorta* and *Picea engelmannii*. *Can. J. For. Res.* **20**, 48–57.

Ryan, M.G. (1991). The effect of climate change on plant respiration. *Ecol. Appl.* **1**, 157–167.

Ryan, M.G. and Waring, R.H. (1992). Maintenance respiration and stand development in a subalpine lodgepole pine forest. *Ecology* **73**, 2100–2108.

Ryan, M.G. and Yoder, B.J. (1996). Hydraulic limits to tree height and tree growth. *Bioscience*. In Press.

Ryan, M.G., Hubbard, R.M., Clark, D.A. and Sanford, R.L., Jr. (1994a). Woody tissue respiration for *Simarouba amara* and *Minquartia guianensis*, two tropical wet forest trees with different growth habits. *Oecologia* **100**, 213–220.

Ryan, M.G., Linder, S., Vose, J.M. and Hubbard, R.M. (1994b). Dark respiration in pines. *Ecol. Bull.* **43**, 50–63.

Ryan, M.G., Gower, S.T., Hubbard, R.M., Waring, R.H., Gholz, H.L., Cropper, W.P. and Running, S.W. (1995). Woody tissue maintenance respiration of four conifers in contrasting climates. *Oecologia* **101**, 133–140.

Ryan, M.G., Hubbard, R.M., Pongracic, S., Raison, R.J. and McMurtrie, R.E. (1996a). Foliage, fine-root, woody-tissue and stand respiration in *Pinus radiata* in relation to nitrogen status. *Tree Physiol.* **16**, 333–343.

Ryan, M.G., Hunt, E.R., Jr., McMurtrie, R.E., Ågren, G.I., Aber, J.D., Friend, A.D., Rastetter, E.B., Pulliam, W.M., Raison, R.J. and Linder, S. (1996b). Comparing models of ecosystem function for temperate conifer forests. I. Model description and validation. In: *Effects of Climate Change on Production and Decomposition in Coniferous Forests and Grasslands (SCOPE)* (Ed. by J.M. Melillo, G.I. Ågren and A. Breymeyer), pp.313–362. John Wiley and Sons Inc., London.

Ryan, M.G., McMurtrie, R.M., Ågren, G.I., Hunt, E.R., Jr., Aber, J.D., Friend, A.D., Rastetter, E.B. and Pulliam, W.M. (1996c). Comparing models of ecosystem function for temperate conifer forests. II. Simulations of the effect of climate change. In: *Effects of Climate Change on Production and Decomposition in Coniferous Forests and Grasslands (SCOPE)* (Ed. by J.M. Melillo, G.I. Ågren and A. Breymeyer), pp. 363–387. John Wiley and Sons Inc., London.

Santantonio, D. and Grace, J.C. (1987). Estimating fine-root production and turnover from biomass and decomposition data: a compartment-flow model. *Can. J. For. Res.* **17**, 900–908.

Santantonio, D. and Hermann, R.K. (1985). Standing crop, production, and turnover of fine roots on dry, moderate, and wet sites of mature Douglas-fir in western Oregon. *Ann. Sci. For.* **42**, 113–142.

Sasser, C.L. and Binkley, D. (1989). Nitrogen mineralization in high-elevation forests of the Appalachians. II. Patterns with stand development. *Biogeochemistry* **7**, 147–156.

Schoettle, A.W. (1994). Influence of tree size on shoot structure and physiology of *Pinus contorta* and *Pinus aristata*. *Tree Physiol.* **14**, 1055–1068.

Schulze, E.D., Schulze, W., Kelliher, F.M., Vyodskaya, N.N., Ziegler, W., Kobak, K.I., Koch, H., Arneth, A., Kusnetsova, W.A., Sogatchev, A., Issajev, A., Bauer, G. and Hollinger, D.Y. (1995). Aboveground biomass and nitrogen nutrition in a chronosequence of pristine Dahurian *Larix* stands in eastern Siberia. *Can. J. For. Res.* **25**, 943–960.

Scurlock, J.M.O., Long, S.P., Hall, D.O. and Coombs, J. (1985). In: *Techniques in Bioproductivity and Photosynthesis* (Ed. by. J. Coombs, D.O. Hall, S.P. Long, and J.M.O. Scurlock), Introduction, pp. xxi–xxiv. Pergamon Press, Oxford.

Sellers, P., Hall, F.G., Margolis, H.A., Kelly, R., Baldocchi, D., den Hartog, J., Cilar, J., Ryan, M.G., Goodison, B.G., Crill, P., Ranson, J., Lettenmaier, D. and Wickland, D. (1995). The Boreal Ecosystem-Atmosphere Study (BOREAS): an overview and early results from the 1994 field year. *Bull. Am. Meteorol. Soc.* **76**, 1549–1577.

Soderstrom, B. (1991). The fungal partner in the mycorrhizal symbiosis. In: *Ecophysiology of Ectomycorrhizae of Forest Trees,* Marcus Wallenberg Foundation Symp. Proc. Falun, Sweden. No. 7. pp. 5–26.

Sperry, J.S. (1995). Limitations on stem water transport and their consequences. In: *Plant Stems: Physiology and Functional Morphology* (Ed. by B.L. Gartner), pp. 105–124. Academic Press, San Diego, CA.

Sperry, J.S. and Pockman, W.T. (1993). Limitation of transpiration by hydraulic conductance and xylem cavitation in *Betula occidentalis. Plant, Cell Environ.* **16**, 279–287.

Sperry, J.S., Alder, N.N. and Eastlack, S.E. (1993). The effect of reduced hydraulic conductance on stomatal conductance and xylem cavitation. *J. Exp. Bot.* **44**, 1075–1082.

Spies, T.A. and Franklin, J.F. (1991). The structure of natural young, mature, and old-growth Douglas-fir forests in Oregon and Washington. In: *Wildlife and Vegetation of Unmanaged Douglas-fir Forests* (Ed. by L.F. Ruggiero, K.B. Aubry, A.B. Carey and M.H. Huff), pp. 91–110. US For. Serv. Gen. Tech. Rep. PNW–285, Portland, OR.

Spies, T.A., Franklin, J.F. and Klopsch, M. (1990). Canopy gaps in Douglas-fir forests of the Cascade Mountains. *Can. J. For. Res.* **20**, 649–658.

Sprugel, D.G. (1984). Density, biomass, productivity, and nutrient-cycling changes during stand development in wave-regenerated balsam fir forests. *Ecol. Monogr.* **54**, 165–186.

Sprugel, D.G., Ryan, M.G., Brooks, J.R., Vogt, K.A. and Martin, T.A. (1995). Respiration from the organ level to the stand. In: *Resource Physiology of Conifers* (Ed. by W.K. Smith and T.M. Hinckley), pp. 255–299. Academic Press, San Diego, CA.

Steele, M.J., Coutts, M.P. and Yeoman, M.M. (1989). Developmental changes in Sitka spruce as indices of physiological age. II. Rooting of cuttings and callusing of needle explants. *New Phytol.* **114**, 111–120.

Switzer, G.L., Nelson, L.E. and Smith, W.H. (1966). The characterization of dry matter and nitrogen accumulation by loblolly pine (*Pinus taeda* L.). *Soil Sci. Soc. Am. Proc.* **30**, 114–119.

Tadaki, Y., Sato, A., Sakurai, S., Takeuchi, I. and Kawahara, T. (1977). Studies on the primary structure of forest. XVIII. Structure and primary production in subalpine 'dead trees strips' Abies forest near Mt. Asahi. *Jpn. J. Ecol.* **27**, 83–90.

Tappeiner, J.C. II, (1969). Effect of cone production on branch, needle, and xylem ring growth of Sierra Nevada Douglas-fir. *For. Sci.* **15**, 171–174.

Tenhunen, J., Alsheimer, M., Falge, E., Heindl, B., Joss, U., Kostner, B., Lischeid, G., Mandersheid, B., Ostendorf, B., Peters, K., Ryel, R. and Wedler, M. (1996). Water fluxes in a spruce forest ecosystem: a framework for process study integration. In: *Processes in Managed Ecosystems* (Ed. by R. Hantschel, F. Beese and R. Lenz). Springer-Verlag, Berlin, in press.

Termorshuizen, A.J. (1993). The influence of nitrogen fertilisers on ectomycorrhizas and their fungal carpophores in young stands of *Pinus sylvestris. For. Ecol. Manage.* **57**, 179–189.

Thorne, J.F. and Hamburg, S.P. (1985). Nitrification potentials of an old-field chronosequence in Campton, New Hampshire. *Ecology* **66**, 1333–1338.

Thornley, J.H.M. (1970). Respiration, growth and maintenance in plants. *Nature* **227**, 304–305.

Thornley, J.H.M. (1972a). A balanced quantitative model for root:shoot ratios in vegetative plants. *Ann. Bot.* **36**, 431–441.

Thornley, J.H.M. (1972b). A model to describe the partitioning of photosynthate during vegetative growth. *Ann. Bot.* **36**, 419–430.

Thornley, J.H.M. (1991). A transport-resistance model of forest growth and partitioning. *Ann. Bot.* **68**, 211–226.

Thornley, J.H.M. and Cannell, M.G.R. (1992). Nitrogen relations in a forest plantation-soil organic matter ecosystem model. *Ann. Bot.* **70**, 137–151.

Turner, J. and Long, J.N. (1975). Accumulation of organic matter in a series of Douglas-fir stands. *Can. J. For. Res.* **5**, 681–690.

Tyrrell, L.E. and Crow, T.R. (1994). Structural characteristics of old-growth hemlock-hardwood forests in relation to age. *Ecology* **75**, 370–386.

Vitousek, P.M. and Denslow, J.S. (1986). Nitrogen and phosphorus availability in treefall gaps of a lowland tropical rainforest. *J. Ecol.* **74**, 1167–1178.

Vitousek, P.M. and Matson, P.A. (1985). Disturbance, nitrogen availability, and nitrogen losses in an intensively managed loblolly pine plantation. *Ecology* **66**, 1360–1376.

Vitousek, P.M. and Reiners, W.A. (1975). Ecosystem succession and nutrient retention: a hypothesis. *Bioscience* **25**, 376–381.

Vogt, K.A., Grier, C.C., Meier, C.E. and Edmonds, R.L. (1982). Mycorrhizal role in net production and nutrient cycling in *Abies amabilis* ecosystems in western Washington. *Ecology* **63**, 370–380.

Vogt, K.A., Moore, E.E., Vogt, D.J., Redlin, M.J. and Edmonds, R.L. (1983). Conifer fine root and mycorrhizal root biomass within the forest floors of Douglas-fir stands of different ages and site productivities. *Can. J. For. Res.* **13**, 429–437.

Vogt, K.A., Vogt, D.J., Gower, S.T. and Grier, C.C. (1990). Carbon and nitrogen interactions for forest ecosystems. In: *Above and Belowground Interactions in Forest Trees in Acidified Soils* (Ed. by H. Persson), pp. 203–235. Pollution Report Series of the Environmental Research Programme, Commission of the European Communities, Belgium.

Walcroft, A.S., Silvester, W.B., Grace, J.C., Carson, S.D. and Waring, R.H. (1996). Effects of branch length on carbon isotope discrimination in *Pinus radiata. Tree Physiol.* **16**, 281–286.

Waring, R.H. (1983). Estimating forest growth and efficiency in relation to canopy leaf area. *Adv. Ecol. Res.* **13**, 327–354.

Waring, R.H. (1987). Characteristics of trees predisposed to die. *Bioscience* **37**, 569–574.

Waring, R.H. and Pitman, G.B. (1985). Modifying lodgepole pine stands to change susceptibility to mountain pine beetle attack. *Ecology* **66**, 889–897.

Waring, R.H. and Ryan, M.G. (1995). Carbon balance modeling. In: *Encyclopedia of Energy Technology and the Environment*, pp. 480–491. John Wiley & Sons Inc., New York, NY.

Waring, R.H. and Schlesinger, W.H. (1985). Forest Ecosystems: Concepts and Management. Academic Press, Orlando, FL.

Waring, R.H. and Silvester, W.B. (1994). Variation in foliar ^{13}C values within the crowns of *Pinus radiata* trees. *Tree Physiol.* **14**, 1203–1213.

Waring, R.H., Cromack, K., Jr., Matson, P.A., Boone, R.D. and Stafford, S.G. (1987). Responses to pathogen-induced disturbance: decomposition, nutrient availability, and tree vigour. *Forestry* **60**, 219–227.

Werner, D. (1992). *Symbiosis of Plants and Microbes*. Chapman & Hall, London, England.

Westoby, J. (1984). The self-thinning rule. *Adv. Ecol. Res.* **14**, 167–225.
Whitehead, D., Livingston, N.J., Kelliher, F.M., Hogan, K.P., Pepin, S., McSeveny, T.M. and Byers, J.M. (1996). Response of transpiration and photosynthesis to a transcient change in illuminated foliage area for a *Pinus radiata* D. Don tree. *Plant Cell Environ.* **19**, 949–957.
Whittaker, R.H. and Woodwell, G.M. (1967). Surface area relations of woody plants and forest communities. *Am. J. Bot.* **54**, 931–939.
Yoda, K., Shinozaki, K., Ogawa, H., Hozumi, K. and Kira, T. (1965). Estimation of the total amount of respiration in woody organs of trees and forest communities. *J. Biology, Osaka City University* **16**, 15–26.
Yoder, B.J., Ryan, M.G., Waring, R.H., Schoettle, A.W. and Kaufmann, M.R. (1994). Evidence of reduced photosynthetic rates in old trees. *For. Sci.* **40**, 513–527.
Zou, X., Valentine, D.W., Sanford, R.L., Jr. and Binkley, D. (1992). Resin-core and buried-bag estimates of nitrogen transformations in Costa Rican lowland rainforests. *Plant Soil* **139**, 275–283.

Advances in Ecological Research
Volumes 1–27

Cumulative List of Titles

Aerial heavy metal pollution and terrestrial ecosystems, **11**, 218
Age-related decline in forest productivity: pattern and process, **27**, 213
Analysis of processes involved in the natural control of insects, **2**, 1
Ant–plant–homopteran interactions, **16**, 53
Biological strategies of nutrient cycling in soil systems, **13**, 1
Bray-Curtis ordination: an effective strategy for analysis of multivariate ecological data, **14**, 1
Can a general hypothesis explain population cycles of forest lepidoptera? **18**, 179
Carbon allocation in trees: a review of concepts for modelling, **25**, 60
A century of evolution in *Spartina anglica*, **21**, 1
The climatic response to greenhouse gases, **22**, 1
Communities of parasitoids associated with leafhoppers and planthoppers in Europe, **17**, 282
Community structure and interaction webs in shallow marine hard-bottom communities: tests of an environmental stress model, **19**, 189
The decomposition of emergent macrophytes in fresh water, **14**, 115
Dendroecology: a tool for evaluating variations in past and present forest environments, **19**, 111
The development of regional climate scenarios and the ecological impact of greenhouse gas warming, **22**, 33
Developments in ecophysiological research on soil invertebrates, **16**, 175
The direct effects of increase in the global atmospheric CO_2 concentration on natural and commercial temperate trees and forests, **19**, 2
The distribution and abundance of lake-dwelling Triclads—towards a hypothesis, **3**, 1
The dynamics of aquatic ecosystems, **6**, 1
The dynamics of field population of the pine looper, *Bupalis piniarius* L. (Lep., Geom.), **3**, 207
Earthworm biotechnology and global biogeochemistry, **15**, 379
Ecological aspects of fishery research, **7**, 114
Ecological conditions affecting the production of wild herbivorous mammals on grasslands, **6**, 137
Ecological implications of dividing plants into groups with distinct photosynthetic production capabilities, **7**, 87

Index

Biomass production, forest ecosystem *see*
    Forest productivity
Bioturbation, 110–111
Bixa orellana, 119
Bradysia confinis, 106
Bromus erectus, 30
Burnt chapparal soil, 154, 155, 156, 162

Cacti, root lifespan, 15
Carbon allocation, 226
    fine root production, 240–241, 250, 251
    foliage/branches, 239, 243
    forest age-related changes, 239–244,
        249–250
    regulatory factors, 239
    reproduction, 243–244, 251
    stem wood, 239, 243
    support roots, 239
    to symbionts, 241–243, 250, 251
        mycorrhiza, 241–243
        N-fixing symbionts, 243
Carbon budget
    mycorrhizas, 47
    role of forests, 215, 224–226
    root exudates, 43, 44
    root lifespan, 13–15
Carbon cycle
    forest ecosystem, 224–226
    root system growth/maintenance, 3
    soil microfoodwebs function, 104
Casuarina equisetifolia, 151
Cereals, root lifespan, 13, 14
Chlorella vulgaris, 191
Choristoneura fumiferana (spruce bud-
    worm)
    atmospheric vertical layering, 63
    gust front transport, 72
Choristoneura horistoneura fumiferana,
    66
Chortoicetes terminifera (Australian
    plague locust), 64
Chuniodrilus zielae, 112
Circulifer tenellus (beet leafhopper), 82
Citrus
    carbon cost/nutrient uptake simulation,
        20, 22, 23
    mycorrhizas, 46
    root hairs, 43
    root lifespan, 6–7, 8, 12, 13, 14
        adult versus juvenile roots, 38–39
        dry soil response, 38–39
        local soil fertility effect, 34

root diameter relationship, 30, 31
root maintenance/construction costs, 17,
    18, 20
see also Volkamer lemon seedlings
Coastal winds, 66–68
Collembola
    mycorrhizal grazing, 47
    soil mesofauna, 98, 99
    soil micropredator foodwebs, 103
Continous-flow isotope mass spectromtry
    (CF-IRMS), 139
Corn, root lifespan in dry soil, 37–38
Cost–benefit (root efficiency) analysis,
    19–20, 22, 24, 48
C units, 17
    carbon cost parameters, 20–22
    dry soil, 39–41
    mycorrhizas, 47
    nutient uptake, 22–25
    root diameter, 29–31
    root efficiency modeling, 19–20, 22, 24
    root exudates, 43, 44
    root lifespan, 1, 4
    root maintenance/construction trade-
        offs, 17
    soil fertility influence, 34–35
    specific root length (λ), 29
    tissue density, 31
Cotton, root lifespan, 14, 36
Crataegus monogyna (hawthorn), 138
    field sampling, 139
    $\delta^{15}N$
        age-related patterns, 147, 148
        seasonal means, 141, 142–144
        soil N levels, 141
Crown abrasion, 218, 229, 249
Cucumber, mycorrhizal nutrient uptake,
    45
Cytisus scoparius, 138, *see also* N_2-fixing
    broom

Dactylella, soil microfoodwebs, 102
Dactylis glomerata, 30
Defoliation, 14
Dipteran litter transformers, 106
Direct observation methods, 4
Down-slope winds *see* Drainage winds
Drainage winds, 73–75, 83, 84
    down valley insect concentration, 75
    valley bottom, 74–75
Dry soil
    root hair function, 42